Lecture Notes in Control and Information Sciences

Edited by A. V. Balakrishnan and M. Thoma

For further listing of published volumes please turn over to inside of back cover.

Lecture Notes in Control and Information Sciences

Edited by A.V. Balakrishnan and M. Thoma

39

Feedback Control of Linear and Nonlinear Systems

Proceedings of the Joint Workshop on Feedback and Synthesis of Linear and Nonlinear Systems
Bielefeld/Rom

Edited by D. Hinrichsen and A. Isidori

Springer-Verlag
Berlin Heidelberg GmbH 1982

Authors
Prof. Dr. D. Hinrichsen
Forschungsschwerpunkt Dynamische Systeme
Universität Bremen, Postfach 330 440
2800 Bremen, FRG

Prof. Dr. A. Isidori
Istituto di Automatica
Universitá di Roma, Via Eudossiana 18
00184 Roma, Italy

ISBN 978-3-540-11749-0 ISBN 978-3-540-39479-2 (eBook)
DOI 10.1007/978-3-540-39479-2

Library of Congress Cataloging in Publication Data
Joint Workshop on Feedback and Synthesis of Linear and Nonlinear Systems
(1981 : University of Bielefeld and University of Rome)
Feedback control of linear and nonlinear systems.
(Lecture notes in control and information sciences ; 39)
Bibliography: p.
Includes index.
1. Control theory--Congresses. 2. Feedback control systems--Congresses.
I. Hinrichsen, Diederich, 1939- .
II. Isidori, Alberto. III. Title.
IV. Series.
QA402.3.J64 1981 629.8'312 82-10428

Originally published by Springer-Verlag Berlin Heidelberg New York in 1982

2061/3020-543210

PREFACE

This volume comprises the invited papers presented at the Joint Workshop on Feedback and Synthesis of Linear and Nonlinear Systems, held at the Centre of Interdisciplinary Research in Bielefeld, June 22 - 26, and the University of Rome, June 29 - July 3, 1981.

The workshop was supported by grants of the *Centre of Interdisciplinary Research* (ZiF) of the University of Bielefeld, the *University of Rome* and the *University of Bremen*.
We greatly appreciate the generous financial support rendered by these institutions.

The organization of the workshop was a joint initiative of the linear systems group of the *Forschungsschwerpunkt Dynamische Systeme* (U. Bremen) and the nonlinear systems group of the *Istituto di Automatica* (U. Rome) together with *A. J. Krener* (Davis) The organizing committee consisted of *D. Hinrichsen* (Bremen), *A. Isidori* (Rome), *A. J. Krener* (Davis), *H. F. Münzner* (Bremen) and *D. Prätzel-Wolters* (Bremen).

The aim of the workshop was to stimulate the exchange of ideas between linear and nonlinear system theory and to assess recent advances in both research areas. The list of lectures (below) offers a rough picture of the variety of subjects and problems treated. Nearly all the invited speakers and many other contributors to the workshop were present at both the linear part in Bielefeld and the nonlinear part in Rome. In total more than 90 participants from 18 countries attended the workshop.

The morning sessions were occupied by comprehensive one-hour-talks of the invited speakers while more specialized research papers were presented in the afternoon seminars. Many of the contributed papers are meanwhile available in preprint form or have already been published in *System & Control Letters*. The present volume contains only the invited lectures (which have been submitted). We take the opportunity to express our appreciation to all of the authors who have contributed to this volume.

Special thanks go to *Uwe Helmke* and *Arno Linnemann* for their continual assistance in preparing the workshop, to *Eva Sieber* for her careful secretarial work and to the staff of the Centre for Interdisciplinary Research (ZiF), in particular *Mrs. Niemeier* and *Dr. Sprenger* for their efficient organizational support.

Finally, we would like to thank *Art Krener*, *Hans-Friedrich Münzner* and *Dieter Prätzel-Wolters* for their advice and cooperation.

March 1982

D. Hinrichsen
A. Isidori

PARTICIPANTS

ACKERMANN, J.	Oberpfaffenhofen
ANTOULAS, A. C.	Zürich
BABALOLA, V. A.	Ibadan
BACCIOTTI, A.	Firenze
BALESTRINO, A.	Napoli
BARTOSIEWICZ, Z.	Białystok
BOSGRA, O. H.	Delft
BROCKETT, R. W.	Cambridge, Mass.
BYRNES, C. I.	Cambridge, Mass.
CATTANEO GASPARINI, I.	Roma
CLAUDE, D.	Gif-Sur-Yvette
COLONIUS, F.	Bremen
COMMAULT, C.	Grenoble
CONTE, G.	Genova
COPPEL, W. A	Canberra
DESCUSSE, J.	Nantes
DION, J. M.	Grenoble
DRÜCKE, P.	Bremen
DUNCAN, T.	Lawrence
EISING, R.	Eindhoven
FLIESS, M.	Gif-Sur-Yvette
FORNASINI, E.	Padova
FRANKE, D.	Hamburg
FUHRMANN, P.	Beer Sheva
FUSARO, B.	Gainesville
VAN GELDEREN, J.A.	Delft
HANZON, B.	Rotterdam
HAUTUS, M. L. J.	Eindhoven
HAZEWINKEL, M.	Rotterdam
HEYMANN, M.	Haifa
HINRICHSEN, D.	Bremen
IRVING, M.	Warwick
ISIDORI, A.	Roma
JAKUBCZYK, B.	Warsaw
KAESBAUER, D.	Oberpfaffenhofen
KALOUPTSIDIS, N.	Athen
KARCANIAS, N.	London
KLIEMANN, W.	Bremen
KÖHNE, M.	Siegen
KOKOTOVIC, P. V.	Urbana
KRENER, A. J.	Davis
LAUB, A. J.	Los Angeles
LINDQUIST, A.	Lexington
LINNEMANN, A.	Bremen
LOBRY, C.	Bordeaux
MARCHESINI, G.	Padova
MICHEL, M.	Nantes
MITTER, S. K.	Cambridge, Mass.
MOOG, C.	Nantes
MORSE, A. S.	New Haven
MOURA, J. M. F.	Lisboa
MÜNZNER, H. F.	Bremen
NIHTILÄ, M.	Helsinki
NIJMEIJER, H.	Amsterdam
NOMURA, T.	Warwick
NORMAND-CYROT, D.	Gif-Sur-Yvette
OLBROT, A. W.	Warsaw
OWENS, D. H.	Sheffield
PANDOLFI, L.	Torino
PERDON, A.	Padova
PERNEBO, L.	Lund
PICCI, G.	Padova
PRÄTZEL-WOLTERS, D.	Bremen
PRZYŁUSKI, K. M.	Warsaw
QUADRAT, J. P.	Le Chesnay
RESPONDEK, W.	Warsaw
ROSENBROCK, H. H.	Manchester
SALAMON, D.	Bremen
VAN DER SCHAFT, A.	Groningen
SCHMALE, W.	Oldenburg
SCHULZ, R.	Bielefeld
SCHUMACHER, J. M.	Amsterdam
SILVA LEITE, M. F.	Warwick
SILVERMAN, L. M.	Los Angeles

SMITH, M. C.	Cambridge
SOLAK, M. K.	Warsaw
SONDERGELD, K. P.	Oberpfaffenhofen
SONTAG, E. D.	New Brunswick
STEFANI, G.	Firenze
STEVENS, P. K.	Cambridge
SUSSMANN, H. J.	New Brunswick
TARN, T. J.	St. Louis
TRENTELMAN, H. L.	Groningen
VARDULAKIS, A.. I. G.	Cambridge
VORST, A. C. F.	Rotterdam
VAN DER WEIDEN, A.J.J.	Delft
WILLEMS, J. C.	Groningen
WILLEMS, J. L.	Gent
WIMMER, H.	Würzburg
WINTER, A. D.	Copenhagen
WONHAM, W. M.	Toronto
YANNAKOUDADIS, A.	Grenoble

LECTURES

Linear part (Bielefeld)

ACKERMANN, J.	Robust flight control system design
ANTOULAS, A. C.	The minimality problem of generalized invariant subspaces with applications to linear systems
BARTOSIEWICZ, Z.	Completability of neutral systems
BOSGRA, O. H.	On invariants and the partial realization problem for linear multivariable systems
BYRNES, C. I.	Root loci in one and in several variables, with applications to problems of output feedback
CONTE, C.	Generalized state space realizations of non proper rational transfer functions
COPPEL, W. A.	Polynomial lattices
DION, J. M.	Some factorizations at infinity of rational matrix functions and their control interpretation
EISING, R.	Polynomial matrices and feedback
FUHRMANN, P. A.	On the application of polynomial models to some classical stability criteria
HAUTUS, M. L. J.	Controlled invariance in systems over rings
HEYMANN, M.	System factorization: Feedback and stability
HINRICHSEN, D.	Parametrization of (C,A)-invariant subspaces I

ROSENBROCK, H. H. — Automation and society

SCHUMACHER, J. M. — Stabilizing a delay system by integral control

SILVERMAN, M. — Spectral theory of the linear quadratic control problem for continuous time systems

SOLAK, M. K. — A differential representation for multivariable linear systems with disturbances

SONDERGELD, K. P. — A generalization of the Routh-Hurwitz stability criteria and applications to linear system theory

TRENTELMAN, H. L. — Multivariable root loci, high gain feedback and (almost) controlled-invariant subspaces

VARDULAKIS, A. I. G. — On certain connections between: Infinite zeros of proper rational matrices, dynamic equivalence and the "Interactor"

VAN DER WEIDEN, A.J.J. — On decoupling zeros at infinity

WILLEMS, J. C. — Almost disturbance rejection by measurement feedback

WILLEMS, J. L. — Criteria for stabilization of stochastic systems and for robust stabilization of deterministic systems

WIMMER, H. — Polynomial matrices and dualities

YANNAKOUDADIS, A. — Output feedback equivalence for linear multivariable control systems

Nonlinear part (Rome)

BACCIOTTI, A. — Poisson stabilizability via nonlinear feedback

BALESTRINO, A.	Hyperstable adaptive model following control of nonlinear plants
BROCKETT, R. W.	Linear and nonlinear systems on flat spaces
CATTANEO GASPARINI, I.	Group action and differential operators
CLAUDE, D.	Sur le decouplage des systemes non lineaires
DUNCAN, T.	Some topological properties of systems with symmetries
FLIESS, M.	Syntactic Lie algebras and nonlinear realizations of regular (or bilinear) systems
FRANKE, D.	Synthesis of variable structure feedback control of distributed parameter systems
HAZEWINKEL, M.	Topics in nonlinear filtering and Lie algebras
IRVING, M.	On the equivalence of the Lagrange and gradient formulations of the nonlinear network problem
JAKUBCZYK, B.	Construction of formal and analytic realizations of nonlinear systems
KALOUPTSIDIS, N.	On equivalence and stability of nonlinear systems
KOKOTOVIC, P. V.	A two stage Lyapunov-Bellman feedback design of a class of nonlinear systems
KRENER, A. J.	Topics on nonlinear decoupling
LINDQUIST, A.	Some topics in stochastic realization theory

CONTENTS

THE GLOBAL DESCRIPTION OF LOCALLY LINEAR SYSTEMS

R. W. Brockett*

Division of Applied Sciences
Harvard University
Cambridge, Massachusetts 02138

ABSTRACT

Let $\pi : E \to X$ be a vector bundle with a flat affine connection ∇ and let $\gamma \in \Gamma(E,\pi^*TX)$ be a control system which has, in the neighborhood of any point in X , a description of the form

$$\dot{x} = Ax + Bu + \xi$$

with $\nabla\left(\frac{\partial}{\partial u_i}\right) = 0$. We show that under a mild hypothesis that X must then admit the structure of a complete flat affine space. Using the standard representation of such spaces we investigate the relationship between the holomony group and the Kronecker indices of the system and establish the appropriate canonical form with respect to state feedback.

1. INTRODUCTION

Let X be a connected Hausdorff manifold and let $\pi : E \to X$ denote a rank m vector bundle over X . Let TX be the tangent bundle of X and let π^*TX be the pullback of TX over E . The notation $\Gamma(A,B)$ for B a vector bundle over A indicates the space of all sections of B . The elements of $\Gamma(E,\pi^*TX)$ are called <u>control systems</u> and the elements of $\Gamma(X,E)$ are called <u>feedback control laws</u>. Notice that for a pair $\gamma \in \Gamma(E,\pi^*TX)$ and $\alpha \in \Gamma(X,E)$ there is an element $\gamma^{\alpha} \in \Gamma(E,\pi^*TX)$ which is obtained from γ by shifting the zero section of E to α . Each $\gamma \in \Gamma(E,\pi^*TX)$ also defines a vector field γ_0 which is obtained by restricting $\gamma : E \to \pi^*TX$ to the zero section of E . If F is a subset of $\Gamma(X,E)$ we say that $\gamma \in \Gamma(X,\pi^*TX)$ is F-<u>complete</u> if γ_0^{α} is a complete vector field for each $\alpha \in F$. Finally, if Y is a second manifold then a pair (γ,h) with $h : X \to Y$ and γ as above is called an <u>input-output system</u>. We say that (γ,h) is <u>externally consistent</u> if E is the pullback by h of some vector bundle over Y .

Suppose now that Y is a flat affine space and that $\pi : E \to X$ is a flat vector bundle with respect to the connection ∇ . An input-output system (γ,h) will be said to be <u>locally linear</u> with respect to the connection ∇ if at each point in E there is a local trivialization $\nu : \pi^{-1}(V) \to \mathbb{R}^m \times V'$; $V' \subset \mathbb{R}^n$ such that γ and h are described by

*This work was supported in part by the Army Research Office under Grant DAAG29-76-C-0139, the U.S. Office of Naval Research under the Joint Services Electronics Program Contract N00014-75-C-0648, the National Science Foundation under Grant ENG-79-09459, and the Air Force Office of Scientific Research under Grant AFSOR-81-0054.

$$\dot{x} = Ax + Bu + n \ ; \ y = Cx + p$$

with A and B constant and u being ∇-adapted in the sense that $\nabla\left(\frac{\partial}{\partial u_i}\right) = 0$.

In section 2 of this paper we show that under a mild hypothesis the state space of a locally linear system admits the structure of a complete flat affine space. This is the geometric part of the paper. In section 3 we construct controllable and observable linear systems on complete flat affine spaces and establish an analog of Brunovsky's normal form appropriate to this setting. Section 4 is devoted to the discussion of input-output models.

Earlier work on locally linear systems appears in [1-3]. I would like to thank Chris Byrnes for his help in studying these questions.

2. LOCALLY LINEAR SYSTEMS

Let $\pi : E \to X$ be a vector bundle with a flat connection. We say that a feedback control law $\alpha \in \Gamma(X,E)$ is locally linear if γ^α is locally linear for every locally linear system $\gamma \in \Gamma(E,\pi^*TX)$. A given γ is said to be complete if for every choice of a locally linear feedback $\alpha \in \Gamma(X,E)$ the vector field γ_0^α is a complete vector field on X . A differentiable manifold is said to admit the structure of a flat affine space if there exists a subset of its atlas which covers the space and has the property that the transition maps are all affine transformations. Such spaces are called complete if every straight line segment can be continued indefinitely.

Theorem 1: Let $\pi : E \to X$ be a vector bundle having a flat connection ∇ . Suppose that $\gamma \in \Gamma(E,\pi^*TX)$ is locally linear with respect to ∇ and reachable. Then X admits the structure of a flat affine space. If γ is complete then X is a complete flat affine space.

Proof: Given $e \in E$ there exists a neighborhood of V of $\pi(e) \in X$ such that we have a local trivialization given by $\nu : \pi^{-1}(V) \to \mathbb{R}^m \times V'$ with $V' \subset \mathbb{R}^n$. Because E is flat we can choose coordinates $(u_1, u_2, \ldots, u_m)$ for $\mathbb{R}^m$ such that $\nabla\left(\frac{\partial}{\partial u_i}\right) = 0$. In terms of these coordinates we can describe γ by

$$\dot{x} = Ax + Bu + n$$

with A and B constant matrices.

Suppose that (w,z) are a second set of coordinates for E with γ be given by

$$\dot{z} = Fz + Gw + n'$$

If the domain of definition of the x-system and the z-system overlap then there must be a mapping $\overline{\psi} : (u,x) \mapsto (w,z) = (Mu, \psi(x))$. However, by solving the equations of motion we have in the overlap

$$\psi\left(e^{At}x(o) + \int_o^t e^{A(t-\sigma)}(Bu(\sigma) + n)d\sigma\right) = e^{Ft}\psi(x(o)) + \int_o^t e^{F(t-\sigma)}(GMu(\sigma) + n^1)d\sigma$$

Since this is to hold for all u , and since the system is assumed to be reachable, it follows that ψ must be an affine mapping. Thus we can find an atlas for X such that the transition maps are all affine. That is to say, X admits the structure of a flat affine space. To show that X is complete it is enough to show that every straight line segment in X can be continued indefinitely. If not there exists a coordinate chart in the affine atlas of X such that some line $\{x \mid x = \alpha a + \beta\}$ can not be continued to some $\alpha = \alpha o$. Consider the description of γ in this chart

$$\dot{x} = Ax + Bu + \xi \ .$$

By completeness of γ for locally linear feedbacks we can assert that

$$\dot{x} = (A+BK)x + Bu_o + \xi$$

is a complete vector field. By virtue of the controllability assumption there are locally linear control laws which steer the system to any point in a neighborhood $\alpha_o a + \beta$ and therefore there are free motion trajectories which pass through $\alpha_o a + b$, contradicting the assumed incompleteness.

3. THE ALGEBRAIC STRUCTURE OF LOCALLY LINEAR SYSTEMS

We now describe a useful representation for locally linear systems. In effect, the results here identify locally linear systems with ordinary linear systems having a suitable group of symmetries.

Let Af(n) denote the group of affine transformations on $\mathbb{R}^n$ and let $P \subset Af(n)$ be a subgroup. We say that P acts freely on $\mathbb{R}^n$ if for an affine transformation $P_i x + p_i \in P$ the equations $P_i x + p_i = x$ implies that $p_i = 0$ and P_i is the identity. We say that P acts <u>properly discontinuously</u> on $\mathbb{R}^n$ if for any sequence without repeated elements $(P_i, p_i) \in P$ we can assert that $\{P_i x + p_i\}$ has no accumulation points. Subject to these two hypotheses the space $\mathbb{R}^n/P$, i.e. the set of equivalence classes in $\mathbb{R}^n$ defined by $x \sim x'$ iff $x' = P_\alpha x + p_\alpha$ for some $(P_\alpha, p_\alpha) \in P$, admits the structure of a differentiable manifold. In fact, it has considerably more structure since there exists for this manifold, a covering by open balls and coordinates for these balls, such that the maps relating the coordinate description of points in the intersection of two open balls are affine maps. It is always possible to think of those manifolds as being identified with a <u>fundamental domain</u> D , i.e. a connected subset of $\mathbb{R}^n$ such that each equivalence class contains exactly one point in D . Such manifolds are complete flat affine spaces and <u>every complete flat affine space arises in this way</u> [4].

Observe that if $\mathbb{R}^n/N$ is a complete flat affine space with $N = \{(N_i, n_i)\}$ and if $F \subset \mathbb{R}^n$ is any invariant subspace for the set $\{N_i\}$ then F defines a vector bundle over $\mathbb{R}^n/N$. To see this notice that the tangent bundle of $\mathbb{R}^n/N$ is itself a flat space with the equivalence relation on $\mathbb{R}^{2n}$ being given by

$$(\dot{x},x) \approx (N_i\dot{x}, N_i x+n_i)$$

But if we restrict $\dot{x}$ to range F and if range F is invariant under N , it is clear that it singles out a smooth subbundle of the tangent bundle. This vector bundle need not be trivial, i.e. it need not be equivalent to a product of $\mathbb{R}^m$ and $\mathbb{R}^n/N$. It does admit a flat connection.

Theorem 2: Let $P = \{(P_i,p_i)\} \subset Af(p)$ and $N = \{(N_i,n_i)\} \subset Af(n)$ act freely and properly discontinuously on $\mathbb{R}^p$ and $\mathbb{R}^n$, respectively. Suppose that there exist group homomorphisms $\phi_x : P \to N$ and $\phi_u : P \to G\ell(m)$ such that $\phi_x : (P_i,p_i) \mapsto (N_i,n_i)$ and $\phi_u : (P_i,p_i) \to M_i$. Then any triple (A,B,C) which appears in the description of a standard linear system

$$x = Ax + Bu \; ; \; y = Cx$$

on $E = \mathbb{R}^m \times \mathbb{R}^n$ and $Y = \mathbb{R}^p$ defines a locally linear system on $(E, \mathbb{R}^p/P)$ with E being defined by range B , provided that the following compatability conditions are met.

(i) $N_iB = BM_i$

(ii) $N_iA = AN_i \quad ; \quad An_i = 0$

(iii) $CN_i^{-1} = P_iC \; ; \quad Cn_i = p_i$

Proof: If x is to be identified with N_ix+n_i then computing x in the two different descriptions yields

$$\begin{aligned} N_ix &= A(N_ix + n_i) + BM_iu \\ &= N_iAx + N_iB \end{aligned}$$

which imply (i) and (ii) whereas $Cx = y$ and

$$P_iy + p_i = C(N_ix + n_i)$$

implies (iii).

There are two groups of interest here in connection with the spaces $\mathbb{R}^n/N$. One is the fundamental group of the space, which can be identified with N itself, and the other is the holonomy group which can be identified with the linear transformations $\{N_i\}$ in N . In theorem 3 we describe the interplay between the controllability indices and the structure of the holonomy group.

We state the following theorem using the notation for block matrices

$$\sum_{i=1}^{r} \oplus M = \begin{bmatrix} M & 0\ldots0 \\ 0 & M\ldots0 \\ \cdots & \cdots \\ 0 & 0\ldots M \end{bmatrix}$$

Theorem 3: Suppose that $\dot{x} = Ax + Bu$ defines a controllable system on the flat affine space $\mathbb{R}^n/N$. Suppose that rank B = dim U . Let $k_1 > k_2 > \cdots > k_r$ be the distinct Kronecker indices and suppose that $\alpha_1,\alpha_2,\ldots,\alpha_r$ are the multiplicities of these indices. Then after a suitable change of basis in $\mathbb{R}^m$ and $\mathbb{R}^n$ we have (all entries below the diagonal blocks are zero)

$$N_i = \begin{bmatrix} \sum_{j=1}^{k_r} \oplus N_j^{11} & \sim & \cdots & \sim \\ & \sum_{j=k_r+1}^{k_{r-1}} \oplus N_j^{22} & & \vdots \\ & & \ddots & \sim \\ & & & \sum_{j=k_2+1}^{k_1} \oplus N_j^{rr} \end{bmatrix}$$

With $M_i = N_i^{11}$ and N_i^{pp} being of the form

$$N_i^{p+1,p+1} = \begin{bmatrix} N_i^{pp} & \sim \\ & \\ 0 & \sim \end{bmatrix}$$

<u>Proof</u>: By a suitable change of basis in the control space we can arrange matters so that $b_1, b_2, \ldots, b_m, Ab_1, Ab_2, \ldots, Ab_m, \ldots, A^{k_r}b_m$ is a basis for $\mathbb{R}^n$ having the additional properties that (i) if $A^i b_j$ belongs to this list and $A^{i+1}b_j$ does not then $A^{i+1}b_j$ is expressible in terms of those elements of the list $A^\ell b_j$ which have $\ell \leqslant i$ and (ii) if $A^i b_j$ belongs to the list so does $A^i b_\ell$ for $\ell \geqslant j$. In terms of such a basis A takes the form

$$A = \begin{bmatrix} \sim & 0 & \sim & 0 \ldots \sim & 0 & \sim \\ 0 & I & \sim & 0 \ldots \sim & 0 & \sim \\ 0 & 0 & 0 & I \ldots \sim & 0 & \sim \\ \cdot & \cdot & \cdot & \cdot \cdots \cdot & \cdot & \cdot \\ & & & 0 & I & 0 \end{bmatrix}$$

With the identity matrix which appears in the ith row being of dimension rank $(B, AB, \ldots, A^{i-1}B)$ - rank $(B, AB, \ldots, A^{i-2}B)$. The structure of the N_i then follows from the fact that $N_i A = AN_i$. The last remark of the theorem is self explanatory since, as we have already remarked, the kth order tangent bundle of a complete flat affine space can be viewed as $(x,x^{(1)},\dots,x^{(k)})$ with the equivalence relations

$$(x,x^{(1)},\dots,x^{(k)}) \approx (N_i x + n_i, N_i x^{(1)},\dots,N_i x^{(k)}) .$$

It is to be observed that the pairs (N_i,M_i) can be thought of as elements of the stabilizer of the feedback group acting on controllable pairs (A,B) as in [5]. The results of this reference suggest the form of the preceding theorem.

There are a number of interesting corollaries, of which we mention two. The proof of the first is immediate.

Corollary 1: If $\mathbb{R}^n/N$ is the state space of a complete, reachable, locally linear system and if

$$N_i = \begin{bmatrix} A_i^{11} & A_i^{12}\dots A_i^{1\ell} \\ 0 & A_i^{22}\dots A_i^{2\ell} \\ \dots & \dots \\ 0 & 0 \dots A_i^{\ell\ell} \end{bmatrix}$$

indicates a Jordan-Hölder decomposition of the holonomy group then $\dim u \geq \sum \dim A_i^{jj}$ where the sum is taken over the distinct diagonal blocks.

Corollary 2: If $x = Ax + Bu$ defines a complete reachable linear system on $\mathbb{R}^n/N$ with B being injective then there exists a locally linear feedback control law $\alpha \in \Gamma(X,E)$ such that the closed-loop system has all its eigenvalues at zero.

Proof: Notice that $u = Kx$ is a well defined element of $\Gamma(X,E)$ provided $Kn_i = 0$ and $KN_i = M_i K$ for all (N_i,n_i) in N. According to Bronovsky [6] we can express A and B as

$$A = H(A_o + B_o K)H^{-1} ,$$

$$B = HB_o L$$

with A_o and B_o in Brunovsky normal form. In terms of $X' = Hx$ and $u' = L^{-1}u$ we know that the N_i take the form given by theorem 3. Since $N_i A_o = A_o N_i$ and $N_i B_o = B_o M_i$ we know that $N_i B_o K = B_o K N_i$ which, because $N_i B = BM_i$, implies $M_i K = KN_i$. Thus there exists a feedback which is locally linear which reduces (A,B) to a pair which is Brunovsky normal form modulo the change of coordinates in x and u given above.

This last result admits an interpretation in terms of the feedback group action appropriate in the present context.

Theorem 4: If $\dot{x} = Ax + Bu$ defines a complete, reachable, locally linear system on $\mathbb{R}^n/N$ with B being injective and if F_N denotes the set of invertible transformations on (x,u) which are of the form

$$\begin{bmatrix} x' \\ u' \end{bmatrix} = \begin{bmatrix} H & 0 \\ K & L \end{bmatrix} \begin{bmatrix} x \\ u \end{bmatrix}$$

with

$$\begin{bmatrix} N_i & 0 \\ 0 & M_i \end{bmatrix}\begin{bmatrix} H & 0 \\ K & L \end{bmatrix} = \begin{bmatrix} H & 0 \\ K & L \end{bmatrix}\begin{bmatrix} N_i & 0 \\ 0 & M_i \end{bmatrix}$$

then F_N acts on pairs (A,B) with a finite number of orbits, distinct orbits corresponding to distinct sets of Kronecker indices.

4. INPUT-OUTPUT SYSTEMS

Let $Y = \mathbb{R}^m/P$ be a complete flat affine space and let $\pi' : E' \to Y$ be a rank m vector bundle over Y . Let ∇' be a flat connection on E' and let $(u_1,u_2,\ldots,u_m)$ be coordinates for the fiber which are adapted to ∇ . Let $(y_1,y_2,\ldots y_p)$ be coordinates in an affine chart of the manifold Y . Consider curves in E' of the form $(t \geq o)$

$$y(t) = \int_0^t W(t-\sigma)u(\sigma)\,d\sigma + \int_a^0 W(t-\sigma)u_o(\sigma)\,d\sigma$$

where u_o is some function of time and $a < 0$ is arbitrary. To be more explicit, this is a description of a set of curves in terms of one local trivialization of E' . If the patching data on E' is given, for example, by

$$y' = P_i y + p_i$$

$$u' = M_i u$$

then we must have, for consistency

$$P_i W(t) = W(t) M_i$$

and for $t \geq 0$ and some u_i

$$p_i = \int_a^0 W(t-\sigma)u_i(\sigma)\,d\sigma \quad .$$

Given these two conditions the integral equation will be said to define an input-output system on $\pi' : E' \to Y$.

<u>Theorem 5</u>: If $Y = \mathbb{R}^n/P$ is a complete flat affine space and if

$$y(t) = \int_0^t W(t-\sigma)u(\sigma)\,d\sigma + n(t)$$

defines an input-output system on $\pi' : E' \to Y$ then there exists a finite dimensional, complete, flat affine space X , a vector bundle $\pi : E \to X$ with a flat connection ∇ , and a locally linear input-output system (γ,h) such that its input-output pairs all satisfy the given integral equation provided

(i) $W(\cdot)$ is C^∞ with exponential growth and has a rational Laplace transform

(ii) $P_i W(t) = W(t) M_i$

where $(y',u') = (P_i y + p_i, M_i u)$ is the patching relation for E' .

<u>Proof</u>: This is almost an immediate consequence of the standard linear theory and theorem 2. In view of (i) we can express W as $Ce^{At}B$ with (A,B) reachable and (C,A)

observable. If $P_iCe^{At}BM_i^{-1} = Ce^{At}B$ then the state space isomorphism theorem tells us that there exists N_i such that

$$\begin{aligned} P_iC &= CN_i^{-1} \\ A &= N_iAN_i^{-1} \\ BM_i^{-1} &= N_iB \end{aligned}$$

Moreover, if for $t > 0$

$$p_i = \int_a^0 Ce^{A(t-\sigma)}u_o(\sigma)\,d\sigma$$

then there exists n_i in the kernel of A such that $Cn_i = p_i$. In view of theorem 2 we can take X to be $\mathbb{R}^n/N$ where $N = \{(N_i,n_i)\}$ and the patching data for the vector bundle $\pi : E \to X$ is $(x',u') = (N_ix + n_i\ ,\ M_iu)$.

REFERENCES

1. R.W. Brockett, A Geometrical Framework for Nonlinear Control and Estimation (Notes for a CBMS Conference, to appear).

2. R. Hermann, Cartanian Geometry, Nonlinear Waves and Control Theory, Math Sci. Press, Brookline, MA, 1980.

3. R.W. Brockett, "On the Asymptotic Properties of Solutions of Differential Equations with Multiple Equilibria," J. of Differential Equations, (to appear).

4. J. Wolf, Spaces of Constant Curvature, McGraw-Hill, New York, 1967.

5. R.W. Brockett, "The Geometry of the Set of Controllable Linear Systems," Research Reports of Automatic Control Laboratory, Faculty of Engineering, Nagoya University, Vol. 24, (June 1977) pp. 1-7.

6. P. Brunovsky, "A Classification of Linear Controllable Systems," Kibernetika, Vol. 6, (1970) pp. 173-188.

GLOBAL PROPERTIES OF THE ROOT-LOCUS MAP

Christopher I. Byrnes*
Department of Mathematics and
Division of Applied Science
Harvard University
Cambridge, Massachusetts 02138

Peter K. Stevens**
Division of Applied Science
Harvard University
Cambridge, Massachusetts 02138

ABSTRACT

In this paper we use a recently proven "general position lemma" for transfer functions to derive several important qualitative properties, some new, of the root-locus map for multivariable systems. Among the immediate applications which we derive is that it is not in general possible to develop a formula, involving rational operations and the extraction of r-th roots, for an output feedback gain (either complex or real, should a real solution exist) which places a given set of closed-loop poles. This is in sharp contrast to the state feedback situation [2] and is a partial affirmation of a conjecture made in [3]. The technique is to reduce the problem, via global reasoning, to a tractible problem in Galois theory. Having proved this result, the prerequisite global analysis is applied to give the positive result that the pole-placement equations can, however, be solved numerically by the homotopy continuation method. Since this global analysis of the root-locus map also plays a vital role in recent work on generic stabilizability ([3], [10]) and pole placement by output feedback ([6], [9]) and has never yet appeared in its full generality, we thought it would be useful to collect these basic topological and geometric results and derive them in a coherent fashion based on the "general position lemma."

1. INTRODUCTION

In this paper, we analyze the qualitative, quantitative, and numerical behaviour of the root-locus map. That is, if $G(s)$ is a fixed $p \times m$ transfer function representing the input-output system,

$$y(s) = G(s)u(s), \tag{1.1}$$

we study the behaviour of the map

$$\chi(K) = (s_1, \ldots s_n) \tag{1.2}$$

where K is a constant output feedback law,

$$u(s) = Ky(s), \tag{1.3}$$

leading to the closed-loop feedback system

$$y(s) = G(s)(I - KG(s))^{-1}u(s) \tag{1.4}$$

---

*Research partially supported by the National Science Foundation under Grant ENG-79-09459, the National Aero and Space Administration under Grant NSG-2265, and the Air Force Office of Scientific Research under Grant AFSOR 81-0054.

**Research partially supported by the National Science Foundation under Grant ENG-79-09459.

with poles at $(s_1,\dots,s_n)$. As explained in Section 2, we coordinatize the unordered n-tuple $(s_1,\dots,s_n)$ via the coefficients $c_i(K)$ of

$$\prod_{i=1}^{n} (s-s_i) = s^n + c_1(K)s^{n-1} + \cdots + c_n(K). \tag{1.2'}$$

One question which we study here (following [2],[3]) is the problem of finding explicit formulae for solutions to (1.2) which involve only rational preprocessing of the $c_i(K)$, the coefficients of $G(s)$, and the extraction of r-th roots for arbitrary r. We begin with an easy example, suppose $m = 1$ while p is arbitrary. Given $K = (k_1,\dots,k_p)$ we can calculate (1.2)' from (1.4) as $(1+KG(s))\,\Delta(s) = s^n + c_1(K)s^{n-1} + \dots + c_n(K)$, where $\Delta(s)$ is the open-loop characteristic polynomial, $s^n + c_1(0)s^{n-1} + \dots + c_n(0)$. Setting $G(s)^t = [\Delta_1(s)/\Delta(s),\dots,\Delta_p(s)/\Delta(s)]$ with $\Delta_i(s)$ a polynomial of degree $\leq n-1$, one has

$$s^n + c_1(K)s^{n-1} + \dots + c_n(K) = \Delta(s) + k_1\Delta_1(s) + \dots + k_p\Delta_p(s). \tag{1.5}$$

Now (1.5) can be made arbitrary provided the linear mapping

$$(k_i,\dots,k_p) \mapsto \sum_{i=1}^{p} k_i\Delta_i(s) \tag{1.6}$$

is onto the vector space of polynomials of degree $\leq n-1$. If $\rho = n$, (1.6) is surjective if, and only if, it is injective, i.e. provided

$$\sum_{i=1}^{n} k_i\Delta_i(s) = 0 \Rightarrow k_i = 0. \tag{1.7}$$

It is well worth remarking that (1.7) in this example is precisely the condition referred to in ([6], [7]) as **nondegeneracy**. Thus, if $G(s)$ is nondegenerate, (1.2)' is always solvable, with unique solution K given by inverting (1.5). This involves only rational preprocessing in the c_i's and the coefficients of $G(s)$.

In [2], the important question of whether one could always solve (1.2) rationally, as in the case of state feedback, was raised. That this is not always possible follows from more subtle arguments in the case $m = p = 2$ and $n = 4$. In this case Willems and Hesselink showed (1.2) could not always be solved over $\mathbb{R}$ but their proof [19] shows much more, indeed after eliminating all but one variable in the equations

$$c_i^o = c_i(k) \tag{1.2'}$$

they arrive at a quadratic formula for the entries (k_{ij}). This has since been derived from many other points of view ([6] , [7]). Indeed, following the derivation in [7], with the notation

$$G(s) = \frac{1}{\Delta(s)}\begin{pmatrix} p_1(s) & p_3(s) \\ p_2(s) & p_4(s) \end{pmatrix},$$

$$p_5(s) = \Delta(s)\,.\,\det G(s),$$

$$K = \begin{pmatrix} k_1 & k_2 \\ k_3 & k_4 \end{pmatrix},$$

$$k_5 = \det K = k_1 k_4 - k_2 k_3 \tag{1.8}$$

the pole placement map (1.2) becomes $\Delta_c(s) = \prod_{i=1}^{4} (s-s_i) = \Delta(s) + \sum_{j=1}^{5} p_j(s)k_j$. Expressing the polynomials

$$p_j = \sum_i p_{ji} s^i \tag{1.9}$$

(1.2') becomes

$$c_i^o = \Delta_i + \sum_j p_{ji} k_j \tag{1.10}$$

together with the quadratic relation (1.8) defining k_5. Denote a general solution of (1.10) by $\qquad k_i = k_i^o + \gamma e_i \tag{1.11}$

where (k_i^o) is a particular solution and (e_i) an element of the kernel of (1.10). The quadratic relation (1.8) becomes

$$\alpha \gamma^2 - \beta \gamma + \sigma = 0 \tag{1.8'}$$

where

$$\alpha = e_1 e_4 - e_2 e_3$$

$$\beta = k_2^o e_3 + k_3^o e_2 - e_1 k_4^o - e_4 k_1^o + e_5$$

$$\sigma = k_1^o k_4^o - k_2^o k_3^o - k_5^o$$

The explicit form of the solution, provided $\alpha \neq 0$, is

$$k_i = k_1^o + \frac{\beta \pm \sqrt{\beta^2 - 4\alpha\sigma}}{2\alpha} e_i \tag{1.12}$$

which clearly requires the use of radicals provided $\sigma \neq 0$. It is also here well worth remarking that the condition $\alpha \neq 0$ is again precisely the nondegeneracy condition of ([6], [7]).

As an application (Theorem 5.1) of the qualitative results derived below is a proof that, in general, the output feedback equations cannot be solved by rational preprocessing and the extraction of r-th roots. Indeed, we show that in the next simplest cast, $\min(m,p) = 2$ and $\max(m,p) = 3$, the Galois group of the system of equations $X(K) = (c_i)$ is the full symmetric group S_5 for generic (A,B,C) and generic (c_i). In particular (1.2)' is not solvable by radicals and one must turn to numerical or transcendental methods. However, as another application of our analysis of the global behaviour of the root locus map X we can show (Corollary 3.5) that for arbitrary m,p and $n = mp$ that the homotopy continuation method provides a convergent numerical scheme for solving the output feedback equations.

Another area of application for this global analysis is the study of generic stabilizability, which is taken up in joint work with B.D.O. Anderson in [3]. For example, it is shown in [3] that for m, n, and p fixed that a necessary condition for generic stabilizability by constant gain output feedback is $mp \geq n$. The proof begins by noting that generic stabilizability with respect to the left-half plane (continuous-time) is equivalent to generic stabilizability with respect to the unit disc (discrete-time). These are, in turn, equivalent to generic stabilizability with a prescribed degree of stability, i.e. with respect to a disc of arbitrarily small radius. By invoking one of the main theorems proved here -- that the image of the root-locus map is generically closed -- Anderson and Byrnes proved that generic stabilizability is equivalent to pole-positioning at the origin, for the generic system. This replaces the inequalities arising in, for example, the Routh-Hurwitz criteria by algebraic equalities, thus rendering the generic stabilizability question as a problem in algebraic geometry. For example, a dimension count yields the necessary condition

$$mp \geq n$$

for generic stabilizability.

That image $\mathcal{X}$ is generically closed also provides a sharpening of previous results on generic pole-assignability. For example, we give a proof that if $mp \geq n$ than for the generic (A,B,C) arbitrary pole placement is possible over $\mathbb{C}$, refining the almost onto result proved by Hermann and Martin [14] who used the dominant morphism theorem. That image $(\mathcal{X})$ is generically closed plays a similar role in the more subtle questions of the pole assignability using real feedback and also plays a major role in a recent derivation of sufficient conditions for generic stabilizability [10]. For these reasons, we thought that it would be useful to give a detailed derivation of the qualitative theory of the root-locus map.

2. THE GENERAL POSITION LEMMA

If $k = \mathbb{R}$ or $\mathbb{C}$, then one may consider a control system

$$\dot{x} = Ax + Bu , \qquad (2.1)$$

$$y = Cx$$

(where $u \in k^m$, $y \in k^p$, $x \in k^n$), as a point $(A,B,C) \in k^N$ where $N = n^2 + nm + mp$. In this language, a property P(A,B,C) is generic if it is enjoyed by all (A,B,C) except perhaps those (A,B,C) lying on a proper algebraic subset of k^N. Similarly, a feedback law $u = Ky$ can be thought of as a point, $K \in k^{mp}$ and, for (A,B,C) fixed, the roots of (1.2)' define n algebraic functions $s_1(K),\ldots,s_n(K)$ of K. More precisely, one could consider the function

$$\mathcal{X} : K \to (s_1(K),\ldots,s_n(K))$$

which assigns to K the <u>unordered</u> <u>n-tuple</u> of closed-loop roots or poles.

If $m = p = 1$, then the (oriented) image of

$$X : \mathbb{R} \to \{\text{unordered n-tuples of points on } S^2\}$$

is the classical root-locus, hence we refer to X as <u>the root-locus map</u>. However, one must be quite careful in topologizing the range of X, i.e. if S^2 denotes the 2-sphere $\mathbb{C} \cup \{\infty\}$ then

$$X : k^{mp} \to S^2 \times \ldots \times S^2/\sim \text{ , where}$$

$$(p_1,\ldots,p_n) \sim (q_1,\ldots,q_n) \iff p_i = q_{\Pi(i)}$$

for Π some permutation.

As it turns out, this space is well known in classical algebraic geometry and this observation will greatly facilitate analysis of the qualitative and quantitative topological properties of X; e.g. is X surjective over $\mathbb{R}$ (pole-placement), what is the degree of X (over $\mathbb{R}$) if $mp = n$? Or, how can one invert X if X is onto?

We turn first to the question of identifying $S^2 \times \ldots \times S^2/\sim$. Let $\mathbb{C}\mathbb{P}^n$ denote the space of lines, i.e. complex 1-dimensional subspaces, in $\mathbb{C}^{n+1}$ One can for example identify S^2 with $\mathbb{C}\mathbb{P}^1$, for $\mathbb{C}\mathbb{P}^1$ is a one point compactification of $\mathbb{C}$, regarded as the line $z_2 = 1$ in $\mathbb{C}^2$. Moreover, an unordered n-tuple on S^2 is then an unordered collection of n lines in $\mathbb{C}^2$. Now, if $p(z_1,z_2)$ is a nonzero homogeneous polynomial of degree n, then the zeroes of p, denoted $V(p)$, is a collection of n lines in $\mathbb{C}^2$. And, if $V(p) = V(q)$ then $p = cq$, for some constant $c \in \mathbb{C} - \{0\}$. Conversely, any unordered collection of n lines in $\mathbb{C}$ corresponds, qua $V(p)$, to a line $\{cp : c \in \mathbb{C} - \{0\}\}$ of homogeneous polynomials of degree n. Since the vector space V_n of $p(z_1,z_2)$ homogeneous of degree n has dimension $n+1$, one has a bijection

$$S^2 \times \ldots \times S^2/\sim \;\simeq\; \mathbb{C}\mathbb{P}^1 \times \ldots \times \mathbb{C}\mathbb{P}^1/\sim \;\simeq\; \mathbb{C}\mathbb{P}^n.$$

If each of the s_i are finite, then $p(z_1,z_2)$ is not zero on the line $z_2 = 0$, so that this correspondence reduces to

$$\mathbb{C} \times \ldots \times \mathbb{C}/\sim \;\simeq\; \mathbb{C}^n$$

given by

$$[s_1,\ldots,s_n] = (c_1,\ldots,c_n) = \text{coeff's of } \prod_{i=1}^{n} (s-s_i).$$

Thus, we shall consider the root locus map as the map

$$X : k^{mp} \to k^n \subset \mathbb{P}^n_k$$
$$X(K) = (c_1(K),\ldots,c_n(K)) \tag{2.3}$$

but we have compactified $k^n \subset \mathbb{P}^n_k$ in order to examine the behaviour of X in the high gain limit.

Our methods will be based on a geometric "General Position Lemma" [10] and a classical geometric interpretation of the pole-placement problem ([6], [7]). Passing to the transfer function

$$G(s) = C(sI - A)^{-1}B \tag{2.4}$$

one may reinterpret the condition that $s_1, \ldots, s_n$ be roots of (2.2) for a particular choice of $-K$ by asserting that, modulo pole-zero cancellations, $(s_i, -K)$ must satisfy

$$0 = \det(I - KG(s_i)) = \frac{\det(s_i I - A + BKC)}{\det(s_i I - A)} \tag{2.5}$$

However, (2.5) has a spectacular reinterpretation in Euclidean geometry, based on the observation [6]:

$$0 = \det(I - KG(s)) \Longleftrightarrow 0 = \det \begin{bmatrix} I & G(s) \\ K & I \end{bmatrix} \tag{2.5)'$$

Thus, to say s is a root of (2.2) is to say

$$\dim \left(\text{col.span} \begin{bmatrix} I \\ K \end{bmatrix} \cap \text{col.span} \begin{bmatrix} G(s) \\ I \end{bmatrix} \right) \geq 1, \tag{2.5)''$$

where we have assumed s is not a pole of $G(s)$. We shall consider the general case shortly, but first we remark that

$$\text{col.span} \begin{bmatrix} I \\ K \end{bmatrix} = \text{graph}(K) \subset k^p \oplus k^m \tag{2.6}$$

is a p-plane, hence a point in the compact manifold $\text{Grass}_k(p, m+p)$. Similarly,

$$\text{col.span} \begin{bmatrix} G(s) \\ I \end{bmatrix} = \text{graph}G(s) \tag{2.7}$$

is a m-plane in $k^p \oplus k^m$, which we take to be

$$\text{graph } G(s) = \text{col.span} \begin{bmatrix} N(s) \\ D(s) \end{bmatrix}, \quad (N,D) \text{ a coprime factorization of } G$$

if s is a pole of G.

Our point of view is that, first, $\text{Grass}_k(p, m+p)$ contains the space of feedback laws K, qua graph(K), as an open dense subspace. Of course, not every p-plane V is of the form graph(K), for such a V must be complementary to the subspace $k^m \subset k^p \oplus k^m$. Indeed, one can (see [6], [7], [8]) interpret those p-planes V such that

$$\dim (V \cap k^m) \geq 1 \tag{2.8}$$

as infinite gains or as high gain limits.

The fundamental geometric ingredient is to express the condition that $(s,-K)$ be a root of (2.2) in terms of $(s, \mathrm{graph}(K))$. Define $\sigma(s_i) \subset \mathrm{Grass}_k(p,m+p)$, for $s_i \in \mathbb{C} \cup \{\infty\}$, via

$$\sigma(s_i) = \{V : \dim_k(V \cap \mathrm{graph}G(s_i)) \geq 1\}. \tag{2.9}$$

$\sigma(s_i)$ is classically referred to as a Schubert hypersurface on $\mathrm{Grass}_k(p,m+p)$.

Lemma 2.1: To say $-K$ places the roots of (2.2) at the distinct complex frequencies $\{s_i\}$ is to say (2.5), or equivalently (2.5)", holds. That is,

$$\mathrm{graph}(K) \in \bigcap_{i=1}^{n} \sigma(s_i) \subset \mathrm{Grass}_k(p,m+p),$$

and in particular,

$$\bigcap_{i=1}^{n} \sigma(s_i) \neq \emptyset.$$

This tautology does not imply, conversely, that if

$$\emptyset \neq \bigcap_{i=1}^{n} \sigma(s_i) \subset \mathrm{Grass}_k(p,m+p)$$

one has pole-placement at the s_i by a finite gain. For, a priori, none of the points V of $\bigcap_{i=1}^{n}\sigma(s_i)$ might be of the form $\mathrm{graph}(K)$. That this is an equivalence is a much deeper statement, viz.

General Position Lemma [10]: If $mp \geq n$, then for generic $(A,B,C) \in \mathbb{C}^N$ and generic points $s_1,\ldots,s_n \in \mathbb{C}$, $s_{n+1} = \infty$, one has in $\mathrm{Grass}_{\mathbb{C}}(p,m+p)$

(i) $\dim \bigcap_{i=1}^{n+1} \sigma(s_i) = mp - n - 1$

(ii) $\dim \bigcap_{i=1}^{n} \sigma(s_i) = mp - n.$

Moreover, for $(A,B,C) \in \mathbb{R}^N$ and $\{s_1,\ldots,s_n\}$ a self-conjugate set, if

$$\emptyset \neq \bigcap_{i=1}^{n} \sigma(s_i) \subset \mathrm{Grass}_{\mathbb{R}}(p,m+p)$$

then this intersection contains a finite point.

3. BASIC TOPOLOGICAL PROPERTIES OF THE ROOT-LOCUS MAP

The first result which we present plays a sizable role in recent work on generic stabilizability ([3],[10]) as well as in the numerical aspects of pole-placement. This theorem holds for any fixed values of m, n, and p.

Theorem 3.1: (i) If $k = \mathbb{C}$ then, for generic (A,B,C), image (X) is a (closed) subvariety of $\mathbb{C}^n$.

(ii) If $k = \mathbb{R}$ then, for generic (A,B,C), image (X) is a Euclidean closed semialgebraic subset of $\mathbb{R}^n$.

If $mp \leq n$, we can prove a much stronger result, Theorem 3.3, which gives an explicit characterization (see [6], [7]) of a particular generic class of systems for which (i) and (ii) hold.

Proof: Suppose $mp \geq n$, $k = \mathbb{C}$ and that (A,B,C) is fixed. Consider the problem of extending $X : \mathbb{C}^{mp} \to \mathbb{C}^n$ to a map $\overline{X} : \mathrm{Grass}_{\mathbb{C}}(p,m+p) \to \mathbb{C}\mathbb{P}^n$. If $mp > n$, such an extension will not exist, but one can consider

$$\Gamma = \overline{\mathrm{graph}(X)} \subset \mathrm{Grass}_{\mathbb{C}}(p,m+p) \times \mathbb{C}\mathbb{P}^n .$$

Γ is a subvariety and is therefore defined by algebraic conditions. These are easily defined, if one regards $\mathbb{C}\mathbb{P}^n$ as the space of unordered n-tuples $(s_1,\ldots,s_n)$ of points $s_i \in \mathbb{C}\mathbb{P}^1 = \mathbb{C} \cup \{\infty\}$, as in Section 2.

Lemma 3.2: $\Gamma = \{(V,(s_1,\ldots s_n)) : V \in \bigcap_{i=1}^{n} \sigma(s_i)\}$.

Proof: According to Lemma 2.1,

$$\mathrm{graph}(X) = \{(K,(s_1,\ldots,s_n)) : \mathrm{graph}(K) \in \bigcap_{i=1}^{n} \sigma(s_i)\}$$

from which Lemma 3.2 follows by taking closures.

Now consider $\mathrm{proj}_2((V,(s_1,\ldots,s_n))) = (s_1,\ldots,s_n)$, restricted to Γ. Thus,

$$\mathrm{proj}_2 : \Gamma \to \mathbb{C}\mathbb{P}^n$$

is an extension of X to a compact, algebraic subset Γ of $\mathrm{Grass}_{\mathbb{C}}(p,m+p) \times \mathbb{C}\mathbb{P}^n$. By compactness of Γ and regularity of proj_2, $\mathrm{proj}_2(\Gamma) \subset \mathbb{C}\mathbb{P}^n$ is an algebraic set. Moreover, since $\overline{\mathbb{C}^{mp}} = \Gamma$, Γ and hence $\mathrm{proj}_2(\Gamma)$ is irreducible. Hence, $\mathrm{proj}_2(\Gamma)$ is a subvariety of $\mathbb{C}\mathbb{P}^n$. Assertion (i) then follows from the identify

$$\mathrm{proj}_2(\Gamma) \cap \mathbb{C}^n = X(\mathbb{C}^{mp}). \tag{3.1}$$

which in turn follows from the General Position Lemma. That is, since proj_2 extends X one clearly has $X(\mathbb{C}^{mp}) \subset \mathrm{proj}_2(\Gamma) \cap \mathbb{C}^n$. If $(s_1,\ldots,s_n) \in \mathrm{proj}_2(\Gamma) \cap \mathbb{C}^n$, then each $s_i \in \mathbb{C}$ and

$$\mathrm{proj}_2^{-1}(s_1,\ldots,s_n) \cap \Gamma \neq \emptyset. \tag{3.2}$$

But (3.2) is, by Lemma (3.2), simply

$$\mathrm{proj}_2^{-1}(s_1,\ldots,s_n) \cap \Gamma = \bigcap_{i=1}^{n} \sigma(s_i). \tag{3.3}$$

If (3.3) contains a finite point, $\mathrm{graph}(K)$, then by Lemma 2.1 we see $X(K) = (c_1,\ldots,c_n)$, where $s^n + c_1 s^{n-1} + \ldots + c_n = \prod_{i=1}^{n} (s-s_i)$, and therefore $(s_1,\ldots,s_n) \in \mathrm{image}(X)$. By the

General Position Lemma for generic(A,B,C) we have

$$\dim(\mathrm{proj}_2^{-1}(s_1,\ldots,s_n) \cap \Gamma \cap \sigma(\infty)) < \dim(\mathrm{proj}_2^{-1}(s_1,\ldots,s_n) \cap \Gamma). \qquad (3.4)$$

But, to say $V \in \mathrm{Grass}_{\mathbb{C}}(p,m+p)$ is infinite is to say $\dim(V \cap \mathbb{C}^m) \geq 1$ in $\mathbb{C}^p \oplus \mathbb{C}^m$. Since $\mathrm{gr}G(\infty) = \mathrm{gr}(0) = \mathbb{C}^m$ in $\mathbb{C}^p \oplus \mathbb{C}^m$, to say V is infinite is to say

$$\dim(V \cap \mathrm{gr}G(\infty)) \geq 1 \text{ , i.e. } V \in \sigma(\infty).$$

Thus, the meaning of (3.4) is that there exists V,

$$V \in \mathrm{proj}_2^{-1}(s_1,\ldots s_n) \cap \Gamma,$$

which satisfy $V \notin \sigma(\infty)$, i.e. V which are of the form $\mathrm{gr}(K)$.

If $k = \mathbb{C}$ and $mp < n$, the theorem is proved in [7]. If $k = \mathbb{R}$ and the unordered set $(s_1,\ldots,s_n)$ is self-conjugate, the assertion (ii) follows from the General Position Lemma as above. Q.E.D.

In the course of the above proof, we introduced a compactification Γ of the space k^{mp} of gains, such that there exists a map

$$\mathrm{proj}_2 : \Gamma \to \mathbb{P}_k^n$$

extending the root-locus map

$$\chi : k^{mp} \to k^n.$$

If $mp \leq n$, then one can show that for generic(A,B,C)

$$\mathrm{proj}_1 : \Gamma \to \mathrm{Grass}_k(p,m+p)$$

is 1-1, and therefore over $\mathbb{C}$ (and hence $\mathbb{R}$)

$$\mathrm{proj}_1 : \Gamma \simeq \mathrm{Grass}_k(p,m+p)$$

by Zariski's Main Theorem [19]. In this case, χ extends to a continuous (in fact regular) map defined on $\mathrm{Grass}_k(p,m+p)$:

<u>Theorem 3.3 [7]</u>: (i) If $k = \mathbb{C}$, and $mp \leq n$, for generic(A,B,C) the root-locus map extends to a map

$$\overline{\chi} : \mathrm{Grass}_{\mathbb{C}}(p,m+p) \to \mathbb{C}\mathbb{P}^n$$

of algebraic varieties.

(ii) If $k = \mathbb{R}$ and $mp \leq n$, for generic(A,B,C) the root locus map extends to a differentiable map

$$\overline{\chi} : \mathrm{Grass}_{\mathbb{R}}(p,m+p) \to \mathbb{R}\mathbb{P}^n$$

of differentiable manifolds.

(iii) In both cases, if $(s_1,\ldots,s_n) \in \mathbb{P}_k^n$ is a finite point, i.e. each s_i is finite, then $\overline{\chi}^{-1}(s_1,\ldots,s_n)$ consists entirely of finite points V, i.e. $V = \mathrm{graph}(K)$ for some feedback transformation K.

In particular, (iii) asserts that for such an (A,B,C) the root locus map χ has the property that if a 1-parameter family of gains K_λ becomes infinite as $\lambda \to \infty$, then at least one of the roots of $\mathcal{X}(K_\lambda)$ tends to infinity as $\lambda \to \infty$. The generic class of systems (A,B,C) referred to in this theorem are precisely the nondegenerate systems in the sense of [6] and [7]. This statement, and Theorem 3.3, will be proved at the end of this section. It should be remarked that for $mp > n$, this asymptotic property of root-loci is not generically satisfied, see [8] for counterexamples.

Assuming Theorem 3.3, one can deduce several important properties of $\mathcal{X}$. For example, if $X \subset k^n$ is compact, then $\mathcal{X}^{-1}(X)$ is compact; i.e. $\mathcal{X}$ is proper. This gives a second proof, based on the General Position Lemma, of

Corollary 3.4 [7]: If $mp \leq n$, then for generic (A,B,C)

$$\mathcal{X} : k^{mp} \to k^n$$

is proper. In particular, if $mp = n$, for generic (A,B,C)

$$\text{image}(\mathcal{X}) \subset k^n$$

is a connected, closed subset of k^n with nonempty interior and infinite Lebesgue measure.

Proof of the Corollary: Since k^{mp} is connected, all that remains to be shown is that if $mp = n$ image($\mathcal{X}$) has nonempty interior and $\mu(\text{image}(\mathcal{X})) = \infty$, where μ is Lebesgue measure on k^n. If image($\mathcal{X}$) has empty interior, then the Lebesgue covering dimension $\dim(\text{image}(\mathcal{X})) \leq n-1$. By a standard theorem of dimension theory,

$$mp = \dim(k^{mp}) \leq \dim(\text{image}(\mathcal{X})) + \dim(\mathcal{X}^{-1}(y))$$

where $y \in \text{image}(\mathcal{X})$ is arbitrary. Since $\mathcal{X}$ is polynomial and proper, $\dim(\mathcal{X}^{-1}(y)) = 0$, so one obtains the inequality $mp \leq n-1$ contradicting our hypothesis. Since $\mathcal{X}$ is proper and finite-to-one, one easily sees that

$$\mu(\text{image}(\mathcal{X})) < \infty \iff \mu(k^n) < \infty. \qquad \text{Q.E.D.}$$

Remark: One also knows, for generic(A,B,C), that whenever $mp \geq n$ image($\mathcal{X}$) contains an open set by a Jacobian calculation, see [14], [20], or [7]. Over $\mathbb{C}$, this calculation also implies that image($\mathcal{X}$) is open and dense [13]. Alternatively, it follows from statement (ii) of the General Position Lemma and Lemma 2.1 that image($\mathcal{X}$) is open and dense in $\mathbb{C}^n$, for generic(A,B,C), whenever $mp \geq n$. By either route, one knows by Theorem 3.1 that image$\mathcal{X}$ is also closed, for generic(A,B,C). Therefore,

Corollary 3.5: If $mp \geq n$, then for generic(A,B,C) arbitrary pole-placement over $\mathbb{C}$ is always possible. Moreover, if $mp = n$, then the system of algebraic equations

$$\mathcal{X}(K) = c$$

can be solved numerically by the homotopy continuation method.

For, as in [1] and [12], the homotopy continuation method which allows one to deform a solution to a nominal problem, e.g. for (A_0,B_0,C_0) one takes the solution

$$\chi(0) = p_0(s) = \text{open-loop characteristic polynomial}$$

and continues it to a solution, for (A_1,B_1,C_1),

$$\chi(K) = p_1(s)$$

along paths from (A_o,B_o,C_o) to (A_1,B_1,C_1) and $p_0(s)$ to $p_1(s)$ -- will work, without a bifurcation analysis at the branch points, provided there is a path from (A_o,B_o,C_o) to (A_1,B_1,C_1) along which χ remains proper. Since over $\mathbb{C}$ the generic set of nondegenerate (A,B,C) is necessarily connected and since χ is always proper for (A,B,C) in this set, by Corollary 3.4, the homotopy continuation method applies.

We conclude this section with a proof of Theorem 3.3 and a discussion of non-degeneracy.

<u>Proof</u>: Consider $\Gamma \subset \text{Grass}_{\mathbb{C}}(p,m+p) \times \mathbb{C}\mathbb{P}^n$ and the map

$$\text{proj}_1 : \Gamma \to \text{Grass}_{\mathbb{C}}(p,m+p)$$

given by projection on the first factor. The claim is that proj_1 is 1-1 for generic (A,B,C), provided $mp \leq n$. Supposing this is false, one has for some $V \in \text{Grass}(p,m+p)$ at least two distinct points in the fiber

$$(\text{proj}_1)^{-1}(V) \simeq \{(s_1,\ldots,s_n) : V \in \bigcap_{i=1}^{n} \sigma(s_i)\}.$$

Since $(s_1,\ldots,s_n)$, $(s_1',\ldots,s_n')$ are distinct unordered n-tuples, one must have an unordered $(n+1)$-tuple $(s_1,\ldots,s_n,s_{n+1})$ between these two collections. That is, $V \in \bigcap_{i=1}^{n+1} \sigma(s_i)$ with $s_1,\ldots,s_{n+1}$ distinct. Consider the Hermann-Martin representation [15] of the transfer function, i.e. $G(s) = C(sI-A)^{-1}B$ gives rise to a holomorphic map

$$\underline{G} : \mathbb{C}\mathbb{P}^1 \to \text{Grass}_{\mathbb{C}}(m,m+p).$$

In [7], it is explicitly shown that to say $V \in \sigma(s_i)$ is to say $\underline{G}(s_i) \in \sigma(V)$, so that $\underline{G}(s_1),\ldots,\underline{G}(s_{n+1}) \in \sigma(V)$. Therefore, see [11],

$$\underline{G}(s) \in \sigma(V), \quad \text{for all } s \in \mathbb{C}\mathbb{P}^1 \tag{3.5}$$

and, in particular, for $s_1,\ldots,s_n$ generic points on $\mathbb{C}$ and $s_{n+1} = \infty$

$$V \in \bigcap_{i=1}^{n+1} \sigma(s_i). \tag{3.5'}$$

Therefore, by the General Position Lemma, $mp - n - 1 = \dim \bigcap_{i=1}^{n+1} \sigma(s_i) \geq 0$, contradicting $mp \leq n$. Therefore, $\text{proj}_1 : \Gamma \to \text{Grass}_{\mathbb{C}}(p,m+p)$ is 1-1; i.e. $\Gamma = \text{graph}(\overline{\chi})$ for $\overline{\chi}$ a mapping

$$\overline{\chi} : \text{Grass}_{\mathbb{C}}(p,m+p) \to \mathbb{C}\mathbb{P}^n$$

of sets. Since $\text{graph}(\overline{\chi})$ is a subvariety of $\text{Grass}_{\mathbb{C}}(p,m+p) \times \mathbb{C}\mathbb{P}^n$, $\overline{\chi}$ is a morphism of varieties. If (A,B,C) is real, then one can restrict to the sets of real points to obtain $\overline{X} : \mathrm{Grass}_{\mathbb{R}}(p,m+p) \to \mathbb{R}\mathbb{P}^n$, an algebraic and hence smooth map of real algebraic manifolds. Thus, assertions (i) and (ii) are proved. As for (iii), suppose $s_1,\ldots,s_n \in \mathbb{C}$ and that $V \in \overline{X}^{-1}(s_1,\ldots,s_n)$. If $V \neq \mathrm{graph}(K)$, for some K then $V \in \sigma(\infty)$ as well. Therefore, again by the General Position Lemma

$$mp - n - 1 = \dim \bigcap_{i=1}^{n+1} \sigma(s_i) \geq 0$$

contradicting $mp \leq n$. Q.E.D.

In [6] and [7], systems which satisfy (3.5) or equivalently (3.5)' are called degenerate, and in [6] several explicit algebraic criteria are derived for the non-degeneracy of a system. We note that the above proof, and statement (iii) of the Theorem, give a new proof of the following result, which is quite basic to the geometric theory of pole placement and stabilizability of outnut feedback.

Corollary 3.6: If $mp \leq n$, then the generic system (A,B,C) is nondegenerate and in this case (i), (ii), (iii) of Theorem 3.3 hold. In particular, if K_λ is a 1-parameter family of output gains for which at least one entry $(k_{ij})_\lambda$ tends to ∞ as $\lambda \to \infty$, then at least one of the closed loop poles tends to ∞ as $\lambda \to \infty$.

4. THE BROUWER DEGREE OF THE ROOT-LOCUS MAP

According to Corollary 3.5, if $mp = n$, then for generic (A,B,C) the map

$$X : \mathbb{C}^{mp} \to \mathbb{C}^n \tag{4.1}$$

is a proper surjection and the equation

$$X(K) = c \tag{4.2}$$

can be solved numerically by continuation methods. This is important for, as we shall show in the next section, for $m = 2$ and $p = 3$ (4.2) cannot be solved using rational operations and the extraction of rth roots. Thus, one must turn to numerical or transcendental techniques for the solution of pole-placement problems.

Now, in order to carry out a similar analysis over $\mathbb{R}$, one must understand when the map

$$X : \mathbb{R}^{mp} \to \mathbb{R}^n$$

is proper and surjective. By Sard's Theorem, for a generic point $(s_1,\ldots,s_n) \in \mathbb{R}^n$ the fiber $X^{-1}(s_1,\ldots,s_n)$ in $\mathbb{R}^{mp}$ will be an $(mp-n)$-dimensional manifold. If $mp = n$, since $X^{-1}(s_1,\ldots,s_n)$ is compact for generic (A,B,C), one knows that

$$X^{-1}(s_1,\ldots,s_n) = \{K_1,\ldots,K_d\} \in \mathbb{R}^{mp} .$$

In this setting, the real pole-placement problem is to determine whether $d \geq 1$. One defines [17] the degree of X at K_i to be ± 1 according to whether or not

$\det(\mathrm{Jac}X)$ is positive at K_i, and then

$$\deg_{\mathbb{R}}(X) = \sum_{i=1}^{d} \{\deg(X) \text{ at } K_i\}.$$

If X is proper [17], $\deg_{\mathbb{R}}(X)$ is independent of the (regular) value $(s_1,\ldots,s_n)$ and satisfies the important property

$$\deg_{\mathbb{R}}(X) \neq 0 \Rightarrow X^{-1}(c) \neq \emptyset \text{ for all } c.$$

Since image(X) is closed and the set of regular values is dense, one has

<u>Lemma 4.1</u>: If $mp = n$, for generic real (A,B,C), $\deg_{\mathbb{R}}(X)$ is defined and

$$\deg_{\mathbb{R}}(X) \neq 0 \Rightarrow (A,B,C) \text{ is pole-assignable by real output feedback.} \tag{4.3}$$

<u>Conjecture 4.2</u>: The condition $\deg_{\mathbb{R}}(X) \neq 0$ is necessary and sufficient for pole-placement by real output feedback, for generic real (A,B,C).

In any case, in light of Lemma 4.1 it is interesting to compute $\deg_{\mathbb{R}}(X)$ explicitly. Brockett and Byrnes [6] showed, using the Schubert calculus, that

$$\deg_{\mathbb{C}}(X) = \frac{1!\ldots(p-1)!(mp)!}{m!\ldots(m+p-1)!} \tag{4.4}$$

Moreover, it is known [5] that this number is odd if, and only if,

$$\min(m,p) = 1 \quad \text{or} \quad \min(m,p) = 2 \quad \text{and} \quad \max(m,p) = 2^r-1 \tag{4.5}$$

Since $\deg_{\mathbb{R}}(X) \equiv \deg_{\mathbb{C}}(X) \bmod 2$ one deduces

<u>Theorem 4.3 (Brockett-Byrnes)</u>: Arbitrary pole placement by real constant output gain feedback is possible, for nondegenerate (A,B,C), whenever (4.5) holds and $n = mp$. In fact, pole placement holds for generic (A,B,C) whenever (4.5) holds and $n \leq mp$.

In [9], it was announced that these calculations may be considerably refined, i.e.

<u>Theorem 4.4</u>: $\deg_{\mathbb{R}}(X) \neq 0$ if, and only if, (4.5) holds.

Returning to the conjecture, it was noted in the Introduction that if $\min(m,p) = 1$, X is an affine map so that $\deg X = 1$ provided (A,B,C) is non-degenerate, and if $m = p = 2$ pole placement is not possible, and therefore $\deg X = 0$, according to the Willems-Hesselink calculation [21]. On the other hand, if $\min(m,p) = 2$, and $\max(m,p) = 3$, then

$$\deg_{\mathbb{C}}(X) = 5$$

according to (4.4) and hence $\deg_{\mathbb{R}}(X) \neq 0$. One can in fact show that

$$|\deg_{\mathbb{R}}(X)| = 1.$$

Preliminary calculations indicate that, in the case $\min(m,p) = 2$, $\max(m,p) = 4$ not only is $\deg_{\mathbb{R}}(X) = 0$ but that for generic (A,B,C) the map

$$X : S^8 \to S^8$$

does not contain a neighborhood of ∞. There is at present, however, no published proof of this fact.

5. THE GALOIS GROUP OF THE POLE-PLACEMENT EQUATIONS

Consider the fixed, but generic (indeed, a nondegenerate) system (A,B,C), where A is a 6×6, B is a 6×2, and C is a 3×6 real matrix. In this section, we shall prove that the pole-placement equations

$$X_{(A,B,C)}(K) = (c_1,\ldots,c_n) = c \tag{5.1}$$

cannot be solved by radicals. That is, we first consider the problem of finding a formula, involving only combinations of rational expressions and arbitrary n-th roots, of the c_i which is valid for all choices of c.

Complexifying, one obtains the map

$$X : \mathbb{C}^6 \to \mathbb{C}^6$$

and if $E = \mathbb{C}(c_i)$ and $F = \mathbb{C}(k_{ij})$ are the fields of rational functions, then composition with X gives a map

$$X^* : E \to F,$$

$$X^*(f) = f \circ X.$$

Since X is surjective (Corollary 3.5) if (A,B,C) is nondegenerate, X^* is injective, so one can regard

$$E \simeq X^*E \subset F$$

as an extension of fields. For example, to say that there exists rational formulae for the entries k_{ij} in terms of the c_i is to say $E = F$. In [3] it was shown that, in contrast to pole placement by state feedback, such a formula exists if, and only if, $\min(m,p) = 1$ -- thereby answering in the negative a question raised in [2].

F is, of course, a vector space over the subfield E, and from Theorems 6-7 on pp. 116-117 of [20] it follows that

$$[F : E] = \dim_E(F) = \deg_{\mathbb{C}}(X).$$

And, from the calculation (4.4) made by Brockett and Byrnes, one has in general

$$[F : E] = \frac{1!\ldots(p-1)!(mp)!}{m!\ldots(m+p-1)!}. \tag{5.3}$$

If $m = 2$, $p = 3$ then (5.3) reduces to $[F : E] = 5$, therefore the minimal polynomial over $\mathbb{C}(c_i)$ of $k_{ij}(c)$, where $X(K(c)) = c$, has degree 5 for generic c ([20], pp. 116-117). And, since X extends to a globally defined map $\overline{X}$ on $\mathrm{Grass}_{\mathbb{C}}(p,m+p)$, the minimal polynomial has its coefficients in $\mathbb{C}[c_i]$. Moreover, if $c_i \in \mathbb{R}$ then the coefficients of the minimal polynomial of $k_{ij}(c)$ are real polynomials in the c_i.

Theorem 5.1: If $\min(m,p) = 2$, $\max(m,p) = 3$ and $n = 6$, then for generic (A,B,C) and for generic $(c_i) \in \mathbb{R}^6$, the equation

$$\chi(K) = (c_i)$$

is not solvable by radicals.

Proof: To say that $\chi(K) = (c_i)$ is solvable by radicals, is to say that the minimal polynomial of $k_{ij}(c)$ is solvable by radicals. Since this is an equation of prime order defined over a subfield of $\mathbb{R}$, by Galois theory [4] one has a dichotomy provided the Galois group is in fact solvable: either

(i) all 5 roots $k_{ij}(c)$ are real; or

(ii) just 1 root $k_{ij}(c)$ is real.

In terms of the extended map, which is globally defined if (A,B,C) is nondegenerate,

$$\overline{\chi} : \text{Grass}_{\mathbb{R}}(3,5) \to \mathbb{RP}^6$$

this is the assertion:

(i) $\overline{\chi}$ is 5 to 1 on an open subset, 1 to 1 on its complement,

(ii) $\overline{\chi}$ is $1-1$ everywhere.

Lemma 5.2: If for an open set of (A,B,C) the equation $\chi(K) = (c_i)$ is not solvable by radicals for an open set of (c_i) of (A,B,C), then this equation is not solvable by radicals for the generic choice of (c_i) and (A,B,C).

Proof: Denote by $V \subset \mathbb{C}^{66}$ the open, dense subset of nondegenerate (A,B,C) and consider the map

$$\overline{\chi} : V \times \text{Grass}_{\mathbb{C}}(3,5) \to V \times \mathbb{CP}^6$$

defined by $\overline{\chi}((A,B,C),\Pi) = \overline{\chi}_{(A,B,C)}(\Pi)$ for a 3-plane $\Pi \subset \mathbb{C}^5$. If K_1 denotes the field of rational functions on $V \times \mathbb{CP}^6$ and K_2 denotes the field of rational functions on $V \times \text{Grass}_{\mathbb{C}}(3,5)$ then $\overline{\chi}^* K_1 \subset K_2$ and it follows from the formula (5.3) that $\deg[K_2 : \chi^* K_1] = 5$. Moreover, the extension $K_2/\chi^* K_1$ is solvable if, and only if, the extension F/E defined in (5.2) is solvable for generic (A,B,C) by elementary Galois theory ([16] pp. 244-249). This, in turn, is solvable if, and only if, the extension field associated to the equation $\chi(K) = (c_i)$ is solvable for generic (c_i), again by Galois theory. From these statements, the assertion in the Lemma follows by taking contrapositives. **Q.E.D.**

Turning to the proof of the Theorem, one can see for purely topological reasons that (ii) can never occur for a nondegenerate system. That is, if $\overline{\chi}$ were $1-1$, then since $\overline{X}$ is continuous and $\text{Grass}_{\mathbb{R}}(3,5)$ is compact

$$\overline{X} : \text{Grass}_{\mathbb{R}}(3,5) \simeq \mathbb{RP}^6$$

would be a homeomorphism. But this is well known to be false, for example one may compute the homotopy groups

$$\Pi_2(\text{Grass}_{\mathbb{R}}(3,5)) \simeq \mathbb{Z} \;, \quad \Pi_2(\mathbb{RP}^6) = \{0\}$$

which would be isomorphic were $\overline{X}$ a homeomorphism.

We next give an example of a nondegenerate system for which (i) is false. Consider

$$A = \begin{bmatrix} 1 & 2 & -1 & 2 & 1 & 1 \\ 1 & 3 & 1 & 2 & 3 & 1 \\ -1 & 1 & 2 & 3 & -1 & 1 \\ 3 & 2 & 1 & -3 & -1 & -2 \\ -1 & -3 & -2 & -1 & 1 & -3 \\ -2 & -1 & 1 & 3 & 2 & 1 \end{bmatrix} , \quad B = \begin{bmatrix} 1 & 0 \\ 0 & 1 \\ 0 & 0 \\ 0 & 0 \\ 0 & 0 \\ 0 & 0 \end{bmatrix} \tag{5.5a}$$

$$C = \begin{bmatrix} 1 & 0 & 0 & 0 & 0 & 0 \\ 0 & 1 & 0 & 0 & 0 & 0 \\ 0 & 0 & 1 & 0 & 0 & 0 \end{bmatrix} \tag{5.5b}$$

Lemma 5.3: (A,B,C) is nondegenerate.

Proof: In order to prove that a linear system is nondegenerate, it was shown in [6] that it suffices to demonstrate that there do not exist 2 independent linear functionals ϕ_1, ϕ_2 on $\mathbb{C}^5$, operating on the columns $g_j(s)$ of the 5×2 matrix defined in (2.7) and satisfying

$$\det \left[\begin{pmatrix} \phi_1 \\ \phi_2 \end{pmatrix} \cdot (g_1 g_2) \right] \equiv 0. \tag{5.6}$$

Going over to Plucker coordinates:

$$m_1 = \begin{pmatrix} [\phi]_{23} \\ [\phi]_{24} \\ [\phi]_{25} \end{pmatrix} \quad m_2 = \begin{pmatrix} [\phi]_{13} \\ [\phi]_{14} \\ [\phi]_{15} \end{pmatrix} \quad m_3 = \begin{pmatrix} [\phi]_{45} \\ -[\phi]_{35} \\ [\phi]_{34} \end{pmatrix}$$

where $[\phi]_{ij}$ denotes the 2×2 minor constructed from the ith and jth column of $(\phi_1^T,\phi_2^T)^T$, the equation (5.6) becomes equivalent to the linear equation

$$L(m_1,m_2,m_3) = 0 \tag{5.6)'}$$

together with the quadratic constraints

$$\begin{cases} m_1 \wedge m_2 = 0 \\ m_1^T m_3 = 0 \\ m_2^T m_3 = 0 \end{cases} \tag{5.6)''}$$

The solutions to the linear part (5.6)' can be expressed as

$$\begin{cases} m_1 = M_1 x \\ m_2 = M_2 x \\ m_3 = M_3 x \qquad x \in \mathbb{R}^3 \end{cases}$$

whereas the condition (5.6)" reduce to the following two cases

1. $m_2 = 0, \qquad m_1^T m_3 = 0$

2. $m_2 \neq 0, \qquad m_1 \wedge m_2 = 0, \qquad m_2^T m_3 = 0$

For the example above one obtains

$$M_1 = \begin{pmatrix} 0.012 & 1.744 & -0.982 \\ 2.887 & 0.107 & 0.029 \\ 1.577 & -0.655 & -0.703 \end{pmatrix}$$

$$M_2 = \begin{pmatrix} -3.404 & 2.621 & -1.602 \\ -0.012 & -1.744 & 0.982 \\ -2.803 & -2.275 & 0.578 \end{pmatrix}$$

$$M_3 = \begin{pmatrix} 8.325 & -0.242 & -0.064 \\ 0.594 & 9.012 & 0.510 \\ -0.258 & 0.398 & 9.716 \end{pmatrix}$$

Since

$$\det M_2 \neq 0$$

only the second case above needs to be further examined. Here it suffices to compute the generalized eigenvalues and eigenvectors of M_1, according to

$$\det(\alpha M_2 + M_1) = 0.$$

It turns out that for these solutions the last condition $m_2^T m_3$ is not satisfied, so that the system is indeed nondegenerate. □

Lemma 5.4: The inverse image of the pole placement map of the above system, at the open loop poles, has 3 simple real roots and 1 simple complex pair of roots.

Proof: The feedback gains, placing the poles at the open loop values are

$$\mathbb{R}1 \quad : \quad \begin{pmatrix} 0 & 0 & 0 \\ 0 & 0 & 0 \end{pmatrix}$$

$$\mathbb{R}2 \quad : \quad \begin{pmatrix} -3.86 & -6.52 & 43.56 \\ 2.47 & 3.86 & 36.24 \end{pmatrix}$$

$$\mathbb{R}3 \quad : \quad \begin{pmatrix} 27.60 & -12.36 & -10.35 \\ 58.29 & -27.60 & -18.18 \end{pmatrix}$$

$$\mathbb{C}4,5 \quad : \quad \begin{pmatrix} -0.91 \pm j2.52 & 2.30 \pm j4.50 & -2.33 \pm j2.94 \\ -1.76 \pm j1.36 & 0.91 \pm j2.52 & -0.65 \pm j6.70 \end{pmatrix}$$

with maximal errors ± 0.01. This shows that indeed the roots are simple.

Corollary 5.5: In an open neighborhood of $((A,B,C),0)$ there exist 3 real roots.

Q.E.D.

As above, we fix $\min(m,p) = 2$, $\max(m,p) = 3$ and $n = 6$.

Theorem 5.6: For generic (A,B,C) and generic (c_i), the Galois group of the equation

$$X_i(K) = (c_i)$$

is the full symmetric group S_5 on 5 letters.

Proof: It follows from the above argument for the generic (A,B,C) and a generic choice of (c_i), that the equation

$$X(K) = (c_i)$$

is not solvable by radicals. Moreover, the minimal polynomial of the entries k_{ij} of K has degree 5 so that the Galois group G is a nonsolvable subgroup of S_5. It is a well known and straightforward proposition that the only such subgroups are A_5, the alternating subgroup, and S_5. Thus, we shall have $G = S_5$ if we can prove that G contains a simple transposition. Now, by elementary Galois theory ([16] pp. 244-]49), it suffices to find a particular choice of nondegenerate (A,B,C) and c_i such that $G \simeq S_5$, and for this example we return to (5.5a)-(5.5b), leading to the map

$$\overline{X} : \mathrm{Grass}_{\mathbb{C}}(3,5) \to \mathbb{CP}^6 .$$

By Lemma 5.5, the Galois group of the equation

$$\overline{X}(K) = (c_i) \in \mathbb{CP}^6$$

is nonsolvable for generic (c_i). We prove that G contains a simple transposition by using a result due to Joe Harris:

Lemma 5.7 ([13] p. 698): Let $\Pi : Y \to X$ be a holomorphic map of degree n. If there exists a point $p \in X$ such that the fiber of Y over p consists exactly of $n-1$ distinct points -- i.e. $n-2$ simple points $q_1,\ldots,q_{n-2}$ and one double point q_{n-1} -- and if Y is locally irreducible at q_{n-1}, then the monodromy group M of Π contains a simple transposition.

Lemma 5.8 ([13] p. 689): If X,Y are irreducible algebraic varieties of the same dimension over the complex numbers C, and $\Pi : Y \to X$ is a map of degree $d > 0$, the monodromy group equals the Galois group.

Using numerical techniques, we have proved

Lemma 5.9: For (A,B,C) defined in (5.5a)-(5.5b) there exists a branch point $(c_i) \in \mathbb{R}^6$ at which there are three distinct solutions K_1, K_2, K_3 -- three simple real solutions -- and one real double solution K_4 to the equation

$$\mathcal{X}(K) = (c_i)$$

Assuming Lemma 5.9, we have thus shown that the Galois Group of the equation

$$\mathcal{X}(K) = (c_i),$$

and thus of the extension field $\mathcal{X}^*E \subset F$, is

$$\mathcal{Gal}(F/E) = S_5$$

For generic (A,B,C) and generic (c_i) the Galois group G of the pole-placement equation is a subgroup

$$G \subset S_5,$$

while for fixed nondegenerate (A,B,C) and (c_i) the Galois group G' is a homomorphic image of G. In particular

$$G \to S_5$$

is surjective and therefore, by a counting argument, one has

$$G = S_5$$

for generic (A,B,C) and (c_i).

We now complete the argument by proving Lemma 5.9.

Proof: Consider the following path linking the open loop characteristic polynomial of (5.5) with the polynomial s^6

$$s^6 - 5ts^5 + 4ts^4 + 12ts^3 - 87ts^2 + 623ts - 246t, \quad t \in [0,1].$$

Then the solutions to the pole placement problem can schematically be represented as follows:

It follows that for $t \cong 0.603$, there is a unique branch point. The solutions at this point are

$$\mathbb{R}1 = \begin{pmatrix} 1.25 & 3.80 & 0.98 \\ -1.42 & 0.75 & -1.33 \end{pmatrix}$$

$$\mathbb{R}2 = \begin{pmatrix} -3.14 & -5.06 & 41.68 \\ 3.14 & 5.05 & 36.70 \end{pmatrix}$$

$$\mathbb{R}3 = \begin{pmatrix} 2.81 & 3.89 & -1.91 \\ 5.09 & -0.84 & -0.19 \end{pmatrix}$$

$$\mathbb{R}4 = \mathbb{R}5 = \begin{pmatrix} -1.24 & 1.34 & -0.81 \\ -2.40 & 3.24 & 3.37 \end{pmatrix}$$

with error ± 0.01. Q.E.D.

REFERENCES

[1] J.C. Alexander, "The Topological Theory of an Embedding Method," Continuation Methods (H. Wacker, ed.), Academic Press, NY, 1978.

[2] B.D.O. Anderson, N.K. Bose and E.J. Jury, "Output Feedback Stabilization and Related Problems - Solutions via Decision Algebra Methods," IEEE Trans. Aut. Control, AC-20 (1975), pp. 53-66.

[3] B.D.O. Anderson and C.I. Byrnes, "Output Feedback and Generic Stabilizability," submitted to SIAM J. Control.

[4] E. Artin, "Galois Theory," University of Notre Dame Press, Notre Dame, 1971.

[5] I. Berstein, "On the Ljusternick-Šnirel'mann Category of Grassmannians," Proc. Camb. Phil. Soc. 79 (1976), pp. 129-134.

[6] R.W. Brockett and C.I. Byrnes, "Multivariable Nyquist Criteria, Root-Loci and Pole Placement: A Geometric Viewpoint," IEEE Trans. Aut. Control, AC-26 (1981), pp. 271-284.

[7] C.I. Byrnes, "Algebraic and Geometric Aspects of the Analysis of Feedback Systems," in Geometric Methods in Control Theory (C.I. Byrnes and C.F. Martin, eds.), D. Reidel, Dordrecht, Holland, 1980.

[8] C.I. Byrnes, "Root Loci in Several Variables: Continuity in the High Gain Limit," Systems and Control Letters 1 (1981), pp. 69-73.

[9] C.I. Byrnes, "On the Topology and Arithmetic of Real Algebraic Sets," submitted to Bull. Amer. Math. Soc.

[10] C.I. Byrnes, "Stabilizability of Multivariable Systems and the Ljusternick-Snirel'mann Category of Real Grassmanians," submitted to Systems and Control Letters.

[11] S.S. Chern, Complex Manifolds without Potential Theory, Springer-Verlag, NY, 1979.

[12] F.J. Drexler, "A Homotopy Method for the Calculation of all Zeroes of Zero Dimensional Polynomial Ideals," in Continuation Methods (H. Wacker, ed.), pp. 69-93, Academic Press, NY, 1978.

[13] J. Harris, "Galois Groups of Enumerative Problems," Duke Math. J. 46 (1979), pp. 685-724.

[14] R. Hermann and C. Martin, "Applications of Algebraic-Geometry to System Theory - Part I," IEEE Trans. Aut. Control 22 (1977), pp. 19-25.

[15] R. Hermann and C. Martin, "Applications of Algebraic Geometry to Systems Theory: The McMillan Degree and Kronecker Indices of Transfer Functions as Topological and Holomorphic Invariants," SIAM J. Control 16 (1978), pp.743-755.

[16] S. Lang, Algebra, Addison-Wesley, Reading, MA, 1971.

[17] J. Milnor, Topology from the Differentiable Viewpoint, Univ. of Virginia Press, 1965.

[18] A.S. Morse, W.A. Wolovich, and B.D.O. Anderson, "Generic Pole Assignment: Preliminary Results," Proc. 20th IEEE Conf. Dec. and Control, San Diego, 1981.

[19] D. Mumford, Algebraic Geometry I: Complex Projective Varieties, Springer-Verlag, NY, 1976.

[20] I.R. Shafarevich, Basic Algebraic Geometry, Springer-Verlag, NY, 1974.

[21] J.C. Willems and W.H. Hesselink, "Generic Properties of the Pole-Placement Problem," Proc. of 7th IFAC Congress (1978), pp. 1725-1729.

GROUP ACTION AND DIFFERENTIAL OPERATORS

I. Cattaneo Gasparini
Institute of Applied Mathematics
University of Rome
Rome (Italy)

In this talk we shall speak of a geometric tool and results which can be useful in some problems of controllability and observability.
The geometric tool is a flat connection which we called "Lie connection" and which we introduced in a previous paper of 1969 [4]. It is a global operator for almost parallelizable manifold and, for foliated manifolds, it allows in a natural way to define a transport along the leaves of a foliation of the transverse bundle of the foliation; the leaves are geodesic submanifolds relatively to this connection. In control theory a particular case of state manifold in which it is possible to define this connection is a manifold having the accessibility property and not the strong accessibility property.
This operator is particularly suitable when there is a group acting differentiably on the manifold, or more generally when there are vector fields with particular physical or geometric meaning such as symplectic automorphism on a symplectic manifold or isometric vector field on a riemannian manifold. In our case on the manifold there is a Lie algebra of vector fields associated to the control system.
The holonomy of this connection is linked with the holonomy of the leaves and therefore in some cases it gives informations on the space of the leaves; this is of great importance in many problems of realization of the state manifold relatively to controllability and observability. Some results on this subject are obtained in theorems 4-1, 4-2, 4-3, 4-4. Other topological results are obtained as a consequence of the existence on the state manifold of different distributions $E_1 \ldots E_k$ of dimension n_i $(i=1,2,\ldots,k)$ (vanishing of certain Pontryagin classes). Finally we describe an invariant of a q-foliation, namely the Godbillon and Vey class [10]. It is a cohomology class of dimension 2q + 1 which represents a sort of global twisting of the leaves.

1 - PRELIMINARIES

Let M be a C^∞ connected, compact differentiable manifold of dimension n, representing the state manifold of a system. Suppose that the evolution of the states is represented by the differential equation

$$\frac{dx(t)}{dt} = X(x(t),u(t)) \tag{1}$$

where the function u(t), control function, is piece-wise constant from $[0,\infty)$ to a subset Ω of $\mathbb{R}^n$, $(x^1,\dots,x^n)$ are local coordinates in a chart (U,φ) of M and X is a smooth vector field on M for each $u \in \Omega$. As a consequence of the assumption that M is compact for each $u \in \Omega$, $X(\ ,u)$ is a complete vector field, i.e. for all u the corresponding solution to (1) with $x(0)=x$ exists for all $t \in R$.

Let E be the set of the vector fields associated to the system and $\underline{E}$ the Lie algebra of the system, i.e. the Lie algebra over R generated by the set E.

Following important papers of D.L. Elliott, R. Hermann, V. Jurdievic, C. Lobry, H.J. Sussmann, the definition of accessibility and strong accessibility leads to the consideration of an ideal $\underline{E}_0$ of the Lie algebra $\underline{E}$. If the dimension of $\underline{E}$ is k, the dimension of $\underline{E}_0$ is k - 1 or k; namely if the system has the accessibility property and not the strong accessibility property the dimension of $\underline{E}$ is n and the dimension of $\underline{E}_0$ is n-1 [22]. We have then a foliation of dimension n-1 and the Lie algebra of the vector fields associated to it is an ideal of the Lie algebra of the system. This is the situation we will generalize and study by means of the introduction of a suitable operator of connection.

2 - LIE CONNECTION

It is classically known that there is a very important link between the flow of a vector field and a fundamental operator called Lie derivative. If X is a vector field on M, its flow determines a local transformation φ_t mapping differentiably a neighborhood U(x) of M into M.

The differential $(\varphi_t)_*$ of this map induces a map on contravariant tensors. If S is such a tensor field, the Lie derivative L(X)S of S relative to X is

$$L(X)S \underset{=}{\text{def}} \lim_{t \to 0} \frac{(\varphi_t)_*^{-1} S(\varphi_t x) - S(x)}{t}$$

A similar definition with some adaptation is valid for the Lie derivative of covariant tensor fields.

It is also classically known that we do not have on a manifold an *intrinsic way* of comparing two tangent vectors in two different points. The geometric object that permits the comparison of vectors and permits then to define the parallelism of vectors and an absolute differential is a connection operator or linear connection, the most important of which is the parallelism of Levi-Civita in a riemannian manifold.

A linear connection on M associates to each piece-wise smooth path $\sigma : I = (0,1) \to M$ a linear map of tangent spaces in $\sigma(0)$ and $\sigma(1)$ *depending on* σ. It can be defined on the principal fibre bundle of the linear frames of M as a field of *horizontal spaces* (i.e. supplementary to the tangent spaces to the fibres) satisfying conditions of differentiability and of invariance by the right action of the linear group (see for instance [15]).

From an algebraic point of view a linear connection ∇ can equivalently be defined as a rule which assigns to each $X \in \mathcal{X}(M)$ (smooth vector fields of M) a map of $\mathcal{X}(M)$ into itself called covariant differentiation satisfying the following axioms

i) $\nabla_X(Y' + Y'') = \nabla_X Y' + \nabla_X Y''$

ii) $\nabla_X fY = Xf.Y + f\nabla_X Y$

iii) $\nabla_{fX+gZ} Y = f\nabla_X Y + g\nabla_Z Y$

$X, Y, Z \in \mathcal{X}(M)$ and $f, g \in F(M)$ (ring of smooth functions on M). The notion of Lie derivative could seem to be a good candidate to define a connection operator, but it satisfies the first two axioms and not the third, because if X, Y are vector fields of M then

$$L(fX)Y \neq fL(X)Y \qquad \text{namely}$$

$$L(fX)Y = fL(X)Y - (Yf)X$$

i.e. the Lie derivative is not F-linear with respect to the vector field X. On the other hand if we have on our manifold some distinguished vectors fields we would like to use them to define a connection operator by means of the Lie derivative associated to their flow. In that case we could construct a connection without assigning any extra data.

For this purpose we have given in [4] the following

Definition 2.1. If $\{X_i\}$ $i = 1,2,\ldots,n$ is a field of frames on a coordinate domain U of M, for any vector field Z with $Z = Z^r X_r$ we define a differential operator D_Z by

$$D_Z \ \underline{\underline{\mathrm{def}}}\ Z^r L(X_r)$$

As it is easily verified D_Z satisfies all the axioms of a connection operator.
We have called D_Z: "Lie connection".
D_Z depends naturally on the local parallelization $\{X_i\}$ but we have proved in [4] the following

Proposition 2.2. The Lie connection associated to the frame field $\{X_i\}$ on U does not change if the frame field changes by linear transformations with constant coefficients.
We have therefore that if the manifold M admits an almost parallelization, i.e. if the structural group of the frame bundle can be reduced to a discrete subgroup then D_Z is a global operator.
We have also proved that D_Z characterizes the connections with zero curvature.
The advantages of this operator are:
1) It allows to define a covariant derivative of a connection, notion which is not possible to define with an ordinary connection.
2) If on the manifold there are some vector fields with particular geometrical or physical meaning this operator can be intrinsically linked to them.
3) It allows to define in a natural way a transport along curves on a submanifold N of M of tensor fields defined on M but not necessary tangent to N. In this case of

foliated manifold, it allows then to define a natural parallelism along the leaves. This, as we have previously remarked, is the situation we have in some control problems.

3 - TRANSVERSE BUNDLE TO A FOLIATION OF G ORBITS

If E is an integral subbundle of the tangent bundle TM defining a foliation F, we can have some geometric informations on the leaves by the study of the transverse bundle $Q = TM/E$ to F, called also normal bundle to F (supposed to have chosen a riemannian metric).
We shall suppose in the following that the leaves are the G-orbits of an almost free action of a Lie group G. We have then

<u>Proposition</u> 3.1. If φ is an almost free action of a Lie group (i.e. the isotropy group is discrete) on M, then the spaces E_x of vectors tangent to the orbits of G through $x \in M$ form a trivial subbundle of TM.

Proof. Let φ be a differentiable action of a Lie group G on a differentiable manifold M. The tangent space to GxM at a point (e,x) ("e" unit of G) is identified in a natural way to the direct sum $T_e(G) \oplus T_x(M)$.
The map $\varphi_*(e,x)\colon T_e(G) \oplus T_x(M) \to T_x(M)$ at the fixed point $x \in M$ is <u>injective</u> on $T_e(G)$. In fact to each vector field X of the Lie algebra $g \equiv T_e(G)$ of G there corresponds a vector field X_M of M associated to the one-parameter subgroup exp(tX) of G, (for a fixed X) i.e. to an action or flow of R on M. As the action of φ is almost free, $\exp(tX)(x) = x$ for $X \in \underline{g} \Rightarrow X = 0$ i.e. $\varphi_*(e,x)X = 0$ for $X \in T_e(G) \Rightarrow X = 0$.

The image of $T_e(G)$ is the subspace E_x of $T_x(M)$ consisting of the vectors tangent to the orbit of G through x. The map φ_* gives therefore a global trivialization of the subbundle E. We have then a global parallelization of the leaves and the operator D can, in that case, be defined globally on the leaves of the foliation.

<u>Definition</u> 3.2. If $\{X_\alpha\}\alpha = 1,\dots,n-q$, $\{X_r\} r = n-q+1,\dots,n$ are local trivializations of the fibre bundles E and Q over $U \subset M$ then for any $Z \in E/U$ and $Y \in Q/U$ we have

$$D_Z Y \underset{\equiv}{\text{def}} Z^\alpha L(X_\alpha)(Y^r X_r) \equiv Z^\alpha\{(L(X_\alpha)(Y^r X_r))_E + (L(X_\alpha)(Y^r X_r))_Q\}.$$

Let [Y] be the equivalence class of any $Y \in TM$, then

$$[D_Z Y] \underset{\equiv}{\text{def}} Z^\alpha(L(X_\alpha)(Y^r X_r))_Q.$$

<u>Remark</u>. If the structural group of the frame bundle R(Q) of Q can be reduced to a discrete subgroup, the definition of $[D_Z Y]$ is globally valid for $Z \in E$ and $Y \in Q$. We can examine the different cases

a) $\qquad (L(X_\alpha)X_r)_E \neq 0 \qquad\qquad (L(X_\alpha)X_r)_Q = 0$

this is equivalent to

$$[D_Z Y] = 0 \qquad \text{for any } Z \in F \qquad \text{and any } Y \in Q$$

i.e. the vector field Y is parallel along the leaves.

Proposition 3.3. The leaves are totally geodesic submanifolds of M relatively to the connection D.

Proof. From the defintion of the operator D, we have

$$D_{X_\alpha} X_\alpha = 0 \qquad \alpha = 1,\dots,n-q;$$

the leaves are then totally geodesic submanifolds of M.
For a codimension one foliation we have the following well known result

Proposition 3.4. If F is a codimension one orientable foliation and if $Y \in Q$ is parallel along the leaves then the foliation F is invariant by the flow generated by Y.

Proof. A foliation is orientable if its normal bundle is orientable; there exists then a global 1-form w on M such that $w(X_\alpha) = 0$ for $X_\alpha \in F$ and a global vector field $Y \in Q$ such that $w(Y) = 1$. Assuming the vector field Y as a base of the 1-dimensional vector space Q from $[D_Z Y] = 0$ $\forall Z \in F$ we have $L(X_\alpha)Y = 0$ $\alpha = 1,\dots,n-q$. By the definition of the Lie derivative $(\varphi_t)_* E_x = E_{\varphi_t x}$ i.e. the distribution E_x is invariant by the flow φ_t generated by the complete vector field Y. The leaves are all diffeomorphic and if $\varphi_t p$ and $\varphi_t q$ are two integral curves of the vector field Y with the points p and q belonging to the same leave L_0 and $\varphi_{t_0} p = p$, $\varphi_{t_0} q = q$, then at time t_1 we have that $\varphi_{t_1} p$ and $\varphi_{t_1} q$ belong to the same leave L_1. This is a very well known result in the case of systems satisfying accessibility property and not strong accessibility property.

b) $\qquad (L(X_\alpha)X_r)_E = 0 \qquad (L(X_\alpha)X_r)_Q = 0$

i.e. the fields X_r commute with the fields X_α.
In this case we have topological conditions on the manifold as a consequence of the existence of commuting vector fields (see F. Lima. Ann. of Math. (2) 81 (1965))

c) $\qquad (L(X_\alpha)X_r)_E = 0 \qquad (L(X_\alpha)X_r)_Q \neq 0$

i.e. the trivialization of Q by the vector fields $\{X_r\}$defines a distribution Q which is invariant by the distribution E.

4 - LINK BETWEEN FIBRE BUNDLES WITH DISCRETE STRUCTURAL GROUP, FUNDAMENTAL GROUP AND COVER SPACE OF THE BASE SPACE

Essential to what we developped is to have a foliated manifold M whose transverse frame bundle F(Q) (Q = TM/E) has a discrete structural group. The link of that with the homotopy group of M will give some global informations on the manifold. In fact a manifold whose principal fibre bundle has a discrete structural group is an almost parallelizable manifold and it admits then a flat connection.

From the definition of curvature in terms of the connection form w and from the classical formula $dw(X,Y)=Xw(Y)-Yw(X)-w([X,Y])$,we see that the horizontal distribution on the principal fibre bundle defining the connection is integrable. Such a foliation can be constructed through the fundamental group of the base manifold F and the universal space of F; namely if F is the base space, Q is the fibre and M is the total space

$$M = \frac{\tilde{F} \times Q}{\pi_1(F)}$$

where $\tilde{F}$ is the universal covering space of F and $\pi_1(F)$ is the first homotopy group of F (see Lawson [16]).

In fact the trivial foliation $\tilde{F} \times Q \to Q$ pass to the quotient as $\pi_1(F)$ acts on both the factors: on F in a natural way from the definition of universal covering space and on Q through a given homomorphism

$$\varphi : \pi_1(F) \to \text{Diff }(Q)$$

The study of the representation φ of the fundamental group $\pi_1(F)$ on Diff(Q) is essential in the study of many foliations.

For instance if M is the Moebius band, φ is the map: $Z \to Z_2$.

Associated to a connection is the notion of holonomy group, notion that we shall briefly recall. If M is a manifold with connection, for each $x \in M$, consider the set of all closed curves starting and ending at x. The parallel displacement along such closed curves is an isomorphism of the fibre $\pi^{-1}(x)$ into itself. The set of all such isomorphism forms a group called the holonomy group $\phi(x)$ of the connection with reference point x. The subgroup of the holonomy group consisting of the parallel displacement along curves homotopic to zero is denoted $\phi_0(x)$ and is called the restricted holonomy group. If $\pi_1(M)$ is the homotopy group of M, it can be defined an homomorphism

$$f: \ \pi_1(M) \to \frac{\phi(M)}{\phi_o(M)}$$

If the structure group of F(Q) can be reduced to a discrete group, then the holonomy group of the connection is a discrete group.

The holonomy associated to transport along a leaf is strongly connected with the notion introduced by Ehresmann of "holonomy of a leaf L" of a codimension q foliation F on M which is a representation $\phi : \pi_1(L,p) \to \Gamma_q$ where $p \in L$ and Γ_q is the group of the germs of the local diffeomorphism of R which leave 0 fixed.

We prove then a result which we do not think to be known, at least under this form:

Theorem 4.1. Let the compact manifold M have a Lie connection of the transverse bundle on the leaves of a q foliation F, if M admits a compact leaf L, and if the holonomy group of D along the leaves is finite then M is a fibre bundle, M/F is the base space and the leaves are the fibres.

Proof. If M admits a Lie connection D of Q = TM/F and the holonomy group is finite, then Q is almost parallelizable and foliated with horizontal integrable distributions. The leaves are then diffeomorphic by the flow associated to any vector field of Q, so that if one leaf is compact all the leaves are compact.

Consider now a point $p \in L$ and a small transverse q-ball Bp centered in p. As the holonomy group of the connection D is finite the holonomy of the leave is finite and the ball Bp can be chosen so that it intersects any leaf at most in a finite number of points.

The vectors in p belonging to B_p are vectors of the bundle Q, so that B_p parallel transport along L and if p is another point of L the ball $B'_{p'}$ (in a neighborhood of p) obtained from B_p by parallel transport meets the same leaves as B_p. There is then a neighborhood of L diffeomorphic to the foliation $B_p \times L$ obtained from the projection $B_p \times L \to B_p$ so that locally M is a product.

Let L and L' be two leaves of F, p and q points respectively of L and L'. If $\bar{B}_p$ is the closure of the transverse ball B_p, if h(p) indicates the points of the leaf L in B_p, as the holonomy group of the leaves is finite, h(p) consist of a finite number of points. In the compact $\bar{B}_p$ we can take $\inf(h(p),q) = \varepsilon > 0$. If $U_p = (p,\varepsilon/2)$ we obtain then two neighborhoods U_p, V_q of M/F such that $U_p \cap V_q = \emptyset$.

The leaf space is then Hausdorff, M is locally a product and $(M,\pi,M/F,F)$ is a fibre bundle on the base space M/F with fibre F and canonical projection π.

Theorem 4.2. If M has a compact leaf, is simply connected and has the accessibility property, but not the strong accessibility property then M is a trivial fibre bundle.

Proof. As M is simply connected $\pi_1(M) = 0$. From the assumption that M has the accessibility property and not the strong accessibility property, we have a codimension one foliation on M and a Lie connection on the leaves. The holonomy group of this flat connection must be the identity; the linear holonomy of all the leaves is then the identity. It follows that the leaves are all compact; M is then a fibre bundle and as the fibres are (n-1)-dimensional, the base space must be S_1 or R. Since $\pi_1(M)=0$ the base space must be R, but R is contractible then the fibre bundle is trivial, i.e. $M \cong L \times R$.

These techniques can be useful also in problems of observability under the additional condition that the foliation of indistinguishable sets admits a transverse bundle Q parallel along the leaves. We have then the theorems:

Theorem 4.3. If the foliation F formed by the indistinguishable sets in the compact manifold M admits a compact leave and a Lie connection D of the transverse bundle Q along the leaves, then if the holonomy group of D along the leaves is finite, M is a fibre bundle, M/F is the base space and the leaves are the fibres.

For the proof see theorem 4.1.

Theorem 4.4. If $\pi_1(M)$ is finite and M is compact and of dimension 3, and the foliation of indistinguishable sets admits a Lie connection of the transverse 2 codimensional bundle Q, along the leaves, then the leaves cannot be circles.

Proof. If one leave is a circle, all the leaves are circles by a previous result. Now on M there exists by hypothesis a Lie connection D and as $\pi_1(M)$ is finite the holonomy group of D is finite. M/F is then the base space of a fibration whose fibres are circles, but as the base space must be parallelizable the base space must be a torus which is the only compact, connected parallelizable 2-manifold. By the hypothesis that $\pi_1(M)$ is finite this cannot be; the hypotesis that one leaf is circle is then absurd.
In the next section we shall consider the more general situation in which the state manifold admits complementary distributions (not necessary integrable) and we give some topological conditions for the existence of such a structure.

5 - TOPOLOGICAL CONDITIONS

Naturally in order that a manifold M admits a foliation of codimension q it must admit a continuous field of (n - q)-planes.
In a previous paper by an application of Chern-Weil homomorhism we have proved the following vanishing theorem which gives necessary conditions for the existence of k complementary distributions on M [5].

Theorem 5.1. Let M be a compact smooth orientable manifold of dimension n which admits k complementary, $(k > 2)$ smooth, distributions of oriented n_i-planes $(i = 1,\dots,k)$. Then the real Pontryagin classes $P_r(M)$ are null for $2r > \max(n_1,\dots,n_k)$.
For k = 2, using the generalized Gauss-Bonnet formula, we have a result of Samelson and Willemore [20], [23] which we found independently in [6].

Theorem 5.2. Let M be a smooth compact orientable n dimensional manifold (n even) and suppose that it has a distribution of oriented q-planes with q odd $(1 \leq q \leq n)$. Then the Eulero-Poincaré characteristic of M is null.
We have also proved the following [7].

Theorem 5.3. Let M be a riemannian smooth orientable manifold of dimension n. Furthermore let M have k complementary distribution $E_i \subset TM$ and let the bundle $Q_i = TM/E_i$ have fibre of dimension q_i with $q_i = n - n_i$ and $n_1 +,\dots,n_k = n$. Finally let $P_r(E) \in H^{4r}(M,R)$ denote the r-th real Pontryagin class of the bundle E.
Then if

$$P_h(Q_j)\, P_s(E_j) = 0 \qquad h,s \geq 1 \text{ and } h + s = r$$

one has

$$P_r(Q_j) = 0 \qquad 2r > \max(n_1,\dots,n_k)$$

This is not the only obstruction to the existence of foliations. An important problem formulated by G. Reeb in [19], was the following: given a distribution on M of codimension q, under which conditions is this distribution homotopic to an integrable distribution of codimension q, i.e. to a codimension q foliation?
Bott gave an answer to this problem by proving the important "vanishing theorem".

Theorem. (Bott 1971) [2]. If a (n - q) distribution on M is homotopic to a foliation, then the Pontryagin classes of the normal bundle Q satisfy the conditions

$$P_k(Q) = 0 \qquad \text{for } k > 2q$$

The result is very important and from it Bott gave the first examples of plane fields non homotopic to integrable fields. For instance the complex projective n-space $P^n(C)$ for n odd admits a plane field of codimension two, but such plane field cannot be integrable as a consequence of the above conditions.
This theorem is proved by an application of Chern-Weil homomorphism using a connection ∇ called sometimes in the literature "Bott's connection" introduced by Bott in [2,1971] and defined by

$$\nabla_{X_E} Y = [X_E, Y]_Q$$

This is exacly our "Lie connection" introduced in [4,1969]. In fact by the projection of the Lie derivative on Q, the F(M)-linearity property with respect to X_E is satisfied. A different operator of partial connection has been introduced previously (1959) by C. Cattaneo for not necessary integrable distributions (C. Cattaneo "Proiezioni naturali e derivazione trasversa in una varieta riemanniana a metrica iperbolica normale", Ann. di Mat. Pura ed Appl. (IV) vol. XLVIII (1959)).

6 - INVARIANTS OF FOLIATIONS

C. Godbillon and J. Vey have introduced in [10] certain cohomology classes associated to orientable foliations (an orientable foliation is a foliation whose normal bundle is orientable) and which are invariants of the foliation.
The construction of the Godbillon-Vey class is obtained in the following way.
Let F be an orientable foliation of codimension one on a manifold M, and let w be the 1-form on M whose zero set in each tangent space T_xM is the (n-1)-dimensional vector space E_x tangent to the leaf in x.
The condition of integrability for the subbundle E_x is $w \wedge dw = 0$.
This condition equivalently means that there exists a 1-form w_1 with $dw = w_1 \wedge w$.

Theorem . (Godbillon-Vey) [10]. The form $\Omega = w_1 \wedge dw_1$ is closed and its cohomology class belonging to $H^3(M,R)$ is determined by the foliation F.

We denote it by $\Omega(F)$ and its cohomology class by $[\Omega(F)]$.

The construction has been then generalized to codimension q foliations; the Godbillon-Vey class is in that case an element of $H^{2q+1}(M,R)$. We give here an elementary proof of an important property of $[\Omega(F)]$.

Theorem 6.1. If F is a codimension one orientable foliation which is a fibration then its Godbillon-Vey class is zero.

Proof. If (M,F,π) is a fibration of codimension one, on the base space M/F there exists a global closed 1-form $\bar{w}$. If $\pi : M \to M/F$ is the canonical projection, then $\operatorname{Ker} \pi^*\bar{w} = F$. The 1-form $w = \pi^*\bar{w}$ is defined globally on M, is different from zero in any point and defines the foliation F. Moreover such a form is closed on M as $dw = d\pi^*w = \pi^*dw = 0$.
The result is valuable also in codimension q. It is justified then to consider $[\Omega(F)]$ as a "measure" of a global twisting of the leaves. We see from this the interest of the construction of $[\Omega(F)]$ in problems of observable non linear realization.

REFERENCES

[1] W.M. BOOTHBY: Transversally complete e-foliations of codimension one... The 1976 Ames Research Center (NASA). Edited by C. Martin, R. HERMAN.

[2] R. BOTT: On a topological obstruction to integrability. Proc. International Congress Math. (Nice, 1970) vol. 1, 27-36. Gouthier-Villars Paris, 1971.

[3] R. BOTT: Lectures on characteristic classes and foliations (Notes by Lawrence Conlon). Lecture Notes in Mathematics n. 279, 1-94. Springer-Verlag, New York 1972.

[4] I. CATTANEO GASPARINI: Operatori intrinseci di derivazione su una varietà parallelizzabile. Rend. Acc. Naz. Lincei, serie VIII, fasc. 6, giugno 1969.

[5] I. CATTANEO GASPARINI: Curvature e classi caratteristiche; Anno 1968-69. Rend. Sem. Matem. Università e Politecnico Torino vol. 28.

[6] I. CATTANEO GASPARINI: Su una condizione necessaria per l'esistenza di un campo di r-piani. Rend. Acc. Naz. Lincei. Serie VIII, vol. XLII. Maggio 1967.

[7] I. CATTANEO GASPARINI e G. DE CECCO: Complementary distributions and Pontryagin classes. Rend. Accademia Naz. Lincei. Serie VIII, vol. LXIX. 2° sem. 1980.

[8] L. CONLON: Transversally parallelizable foliations of codimension two. Trans. of the Amer. Math. Soc. vol. 194, 1974.

[9] D. L. ELLIOTT: A consequence of controllability. J. Diff. Equations 10 (1971) 364-376.

[10] C. GODBILLON e J. VEY: Un invariant des feuilletages de codimension un. C.R. Acad. Sc. Pris, 273 (1971).

[11] A. HAEFLIGER: Variétès Feuilletées. Ann. Scuola Normale Sup. Pisa (3) 16 (1962) 367-397.

[12] R. HERMANN: Differential geometry of foliations, J of Math. and Mech., vol. II pp. 303-316, 1962.

[13] F.W. KAMBER and P. TONDEUR: Foliated bundles and characteristic classes. Lecture Notes in Math. Springer-Verlag. n° 493 (1975).

[14] N.KALOUPTOIDIS and D.L. ELLIOTT: Accessibility properties of smooth nonlinear control systems. The 1976 Ames Research Center (NASA) ed. by C. Martin, R. Hermann.

[15] S. KOBAYASHI and K. NOMIZU: Foundations of differential geometry, Interscience Tracts in Pure and Appl. Math. vol. I and II, New York, 1963 and 1969.

[16] H.B.Jr. LAWSON: Foliations, Bull. AMS, 80, 369-418, 1974.

[17] C. LOBRY: Controllabilité des systemes non linéaires. SIAM J. on Control, 8, 1970, pp. 573-605.

[18] S.P. NOVIKOV: Topology of foliations, Trudy Moskov. Mat. Obsc. 14 (1965) 248-éè^. Trans. Moscov. Math. Soc. (1965) 268-304.

[19] G. REEB: Sur certaines propriétés topologiques des variétés feuilletées, Actualités Sc. Indust., n. 1183. Publ. Inst. Math. Univ. Strasbourg 11, Hermann, Pris, 1952 pp. 91-154.

[20] H. SAMELSON: A theorem on differentiable manifolds. Portugaliae Math 10. 129-133 (1951).

[21] H. SUSSMANN: Orbits of families of vector fields and integrability of systems with singularities Trans. Amer. Math. Soc. vol. 180, (1973) pp. 171-188.

[22] H. HUSSMANN and V.JURDIEVIC: Controllability of Nonlinear Systems. Journal of Differential Equations, vol. 12 (1972) pp. 95-116.

[23] T.J. WILLEMORE: Les plans parallèles dans les espaces riemanniens globaux. C.R. Acad. Sc. Paris 232. (1951).

SOME FACTORIZATIONS AT INFINITY OF RATIONAL MATRIX FUNCTIONS AND THEIR CONTROL INTERPRETATION

J.M. DION and C. COMMAULT
Laboratoire d'Automatique de Grenoble
B.P. 46
38402 Saint Martin d'Hères - FRANCE

ABSTRACT

Polynomial Matrix Description of transfer matrices received a great deal of attention during the last decade. This formulation exhibits the finite pole and zero structure generalizing the monovariable case. In the last few years, there has been an increasing interest in factorizations at infinity. The present paper focuses these factorizations, which permits the pointing out of invariant structures under some groups of transformations. The paper is organized as follows. First, left and right Wiener-Hopf factorizations at infinity are presented. Basic properties and some control interpretations are recalled. A characterization of dynamic equivalence is given in terms of Wiener-Hopf factorizations. In the second part we study the Smith Mc Millan factorization at infinity of a transfer function and propose some control interpretations of this factorization. A characterization of the stabilizer of Morse group at (A, B, C) is given for irreducible systems.

1 . INTRODUCTION

In this paper, we study the structural properties of linear systems and their transfer matrices which remain invariant under transformation groups such as the feedback group or the Morse group. This group includes output injection transformations. For a number of years polynomial Matrix Description of rational matrix functions received a great deal of attention. This formulation clearly exhibits the finite pole and zero structure generalizing the monovariable case. For complete study of these factorizations, see [1]. In the last few years, there has been an increasing interest in factorizations at infinity [2] - [6]. The interest of these factorizations is to point out invariant structures under some groups of transformations. Wolowich and Falb [5] studying invariants and canonical forms under dynamic compensation, associate with any rational matrix function T(s), a polynomial matrix called the interactor. This interactor which is obtained from the infinite behaviour of T(s) characterizes together with the rank of T(s) the set of transfer matrices which are equivalent to T(s) under dynamic compensation. Considering a Hermite s form over a particular principal ideal domain, Morse [4] generalizes the concept of interactor which allows us to deal with stability. Verghese [2] studies the infinite structure of rational matrices using a generalized Smith McMillan form at infinity which turns out to be the Smith form over the P.I.D. of the proper rational functions. Fuhrmann and Willems [3] study the Wiener-Hopf factorization at infinity of a transfer function and relate the corresponding factorization indices to the reachability indices of a feedback equivalent feedback irreducible system.

Pernebo [6] defines Λ-generalized polynomials as the set of rational functions with no poles in Λ and uses Λ-generalized polynomial matrices to describe transfer functions. Many authors have defined the infinite zeros of a rational matrix [7], [2], [8]. Infinite zeros turn out to be crucially important in solving problems such as decoupling [9], disturbance decoupling [10], root locus theory [11] and singular optimal control [12]. The factorizations at infinity permit us to point out invariant structures under some groups of transformations, as the infinite zeros or as the reachability indices of a reduced system.

In this paper, we present the control interpretation of these factorizations at infinity and some applications thereof. The paper is organized as follows. Some background and preliminaries are recalled in § 2. In § 3, the Wiener-Hopf factorizations at infinity are presented. As an application, we provide a characterization of equivalent systems under dynamic compensation. In § 4, the Smith McMillan factorizations at infinity are presented and basic properties are studied. In § 5, a control interpretation of the Smith-McMillan form at infinity is given. A connection is made with Morse's canonical form. As an application some results in dynamic equivalence and feedback are given. The stabilizer of Morse's group at (A, B, C) is given for an irreducible system.

2 . PRELIMINARIES AND NOTATIONS

Script letters $\mathcal{X}$, $\mathcal{Y}$, ... denote real vector spaces with elements x, y, ...
Let $R^{nxn}[s]$ be the ring of (nxn) polynomial matrices. An invertible element A(s) in $R^{nxn}[s]$ is called unimodular. A(s) is unimodular iff det (A(s)) is a non zero constant. Write R(s) for the field of fractions of R[s]. Let $R_p^{nxn}(s)$ be the ring of (nxn) proper rational matrices. An invertible element B(s) in $R_p^{nxn}(s)$ is called bicausal isomorphism. B(s) is a bicausal isomorphism iff $\det (\lim_{s \to \infty} B(s)) \neq 0$

As in [4], let $\# : R_p(s) \to$ {non-negative integers} denote the function defined by $\#(n(s)/d(s)) = \deg(d(s)) - \deg(n(s))$ and $\#(0) = \infty$; $\#(n(s)/d(s))$ is called the size of n(s)/d(s).

Consider the linear system (Σ).

$$\dot{x} = Ax + Bu \qquad x \in R^n = \mathcal{X},\ u \in R^m = U$$
$$y = Cx \qquad y \in R^p = \mathcal{Y}$$

The system is standard if B is monic and C is epic. In this paper, we shall restrict our attention to such systems.

(Σ) is said to be minimal if (A, B) is controllable and (C, A) observable.

Let $G(s) = C(sI-A)^{-1}B$ be the transfer matrix of (Σ). There exist two unimodular matrices U(s) and V(s) such that $G(s) = U(s)\,\Lambda(s)\,V(s)$ where :

$$\Lambda(s) = \begin{bmatrix} \Delta(s) & 0 \\ 0 & 0 \end{bmatrix} \qquad \Delta(s) = \text{diag}\left(\frac{n_1(s)}{d_1(s)}, \frac{n_2(s)}{d_2(s)}, \ldots, \frac{n_r(s)}{d_r(s)}\right)$$

$r = \text{rank}\,(G(s))$, $n_i(s)$ divides $n_{i+1}(s)$ and $d_{i+1}(s)$ divides $d_i(s)$ for $i = 1, \ldots, r-1$.
$\Lambda(s)$ is called the Smith McMillan form of G(s).

For minimal systems, the non trivial numerators of the Smith McMillan form of G(s) are called the transmission polynomials of the system. The roots of these polynomials are called finite transmission zeros of (Σ).

For left prime factorizations $D_1^{-1}(s)\,N_1(s)$ and for right prime factorizations $N_2(s)$ $D_2^{-1}(s)$ of a transfer matrix G(s), [1], transmission zeros are the roots of the invariant factors of $N_1(s)$ (or $N_2(s)$).

A classical definition of the infinite structure of (Σ) is the following: [7], [8], The (pxm) rational matrix G(s) possesses an infinite zero of order k when $G(\frac{1}{\omega})$ has a finite zero of precisely that order at $\omega = 0$.

$\mathcal{G}$ denotes the transformation group including input, state and output changes of coordinates, state feedback and output injections [13].
By feedback group, we mean state and input changes of coordinates and state feedback.
By output injection group, we mean state and output changes of coordinates and output injection. This group is the dual of the feedback group.

We will use in the sequel a result of Hautus and Heymann [17] which gives a complete characterization of the dynamic precompensators that can be implemented by action of the feedback group.

<u>Theorem 2.1</u> : Let G(s) be a proper rational function, the transfer function $G_1(s)$ is feedback equivalent to a possibly non minimal realization of G(s) if and only if $G_1(s) = G(s)\,B(s)$ where B(s) is a bicausal isomorphism.

3 . WIENER HOPF FACTORIZATIONS AT INFINITY

Definition 3.1 : Let G(s) be a (pxm) rational matrix. A left Wiener Hopf factorization at infinity is a factorization of G(s) of the form :

$$G(s) = B(s)\ \Lambda(s)\ U(s) \tag{3.1}$$

with U(s) is a (mxm) unimodular matrix, B(s) is a (pxp) bicausal isomorphism and $\Lambda(s) = \begin{bmatrix} \Delta(s) & 0 \\ 0 & 0 \end{bmatrix}$ where $\Delta(s) = \text{diag}\ (s^{q_1}, \ldots, s^{q_r})$.

The integers q_i assumed to be decreasingly ordered are called the left factorization indices at infinity. In the same way, we can define a right Wiener Hopf factorization permuting the roles of B(s) and U(s). Basic properties and some control interpretations may be found in [3]. Let us recall the main result of this paper. For a proper rational matrix, the left (right) factorization indices at infinity are uniquely defined and are non positive. These factorizations are non unique and we can state the following theorem [3] :

Theorem 3.1 : Let G(s) be a (mxm) non singular rational matrix and let $G(s) = B_1(s)\ \Lambda_1(s)\ U_1(s) = B_2(s)\ \Lambda_2(s)\ U_2(s)$ be two left factorizations then $\Lambda_1(s) = \Lambda_2(s)$ and there exists a (mxm) unimodular matrix U(s) satisfying :

$$\begin{aligned} &u_{ij}(s) = 0 \text{ if } q_i > q_j \\ &\deg\ (u_{ij}(s)) \leq q_j - q_i \text{ if } q_j \geq q_i \end{aligned} \tag{3.2}$$

for which $U_2(s) = U(s)\ U_1(s)$ and $B_2(s) = B_1(s)\ \Lambda_1(s)\ U^{-1}(s)\ \Lambda_1^{-1}(s)$.

The set of all such unimodular matrices U(s) form a group called the left factorization group of G(s), which depends only on Λ(s). Note that these unimodular matrices are those for which there exists a bicausal isomorphism B(s) such that :

$$B(s)\ \Lambda(s) = \Lambda(s)\ U(s) \tag{3.3}$$

Notice that analogous results hold for right factorizations.

If G(s) is singular the left factorization group is characterized as follows, (3.3) becomes :

$$\begin{bmatrix} B_{11}(s) & B_{12}(s) \\ B_{21}(s) & B_{22}(s) \end{bmatrix} \begin{bmatrix} \Delta(s) & 0 \\ 0 & 0 \end{bmatrix} = \begin{bmatrix} \Delta(s) & 0 \\ 0 & 0 \end{bmatrix} \begin{bmatrix} U_{11}(s) & U_{12}(s) \\ U_{21}(s) & U_{22}(s) \end{bmatrix} \tag{3.4}$$

which implies in turn :

$B_{11}(s)\ \Delta(s) = \Delta(s)\ U_{11}(s)$ then $U_{11}(s)$ is defined by theorem 3.1 .

$U_{12}(s) = 0$

$U_{22}(s)$ is any unimodular $(m-r)\times(m-r)$ matrix

$U_{21}(s)$ is any polynomial $(m-r)\times(r)$ matrix.

Fuhrmann and Willems [3] obtained the following control characterization of the factorization indices :

Theorem (3.2) : Let G(s) be a (pxm) proper rational matrix. Then the right factorization indices at infinity are equal to the negatives of the reachability indices of any canonical realization of any feedback irreducible system feedback equivalent to G(s).

We illustrate this theorem with the following example used in [5].

Example (3.1)

$$G(s) = \begin{bmatrix} \frac{1}{s+1} & \frac{1}{s+2} \\ \frac{1}{s+3} & \frac{1}{s+4} \end{bmatrix}$$,a right Wiener Hopf factorization at infinity is :

$$G(s) = \begin{bmatrix} 1+s/2 & 1 \\ s/2 & 1 \end{bmatrix} \begin{bmatrix} s^{-2} & 0 \\ 0 & s^{-2} \end{bmatrix} \begin{bmatrix} \frac{2s^2+12s+16}{s^2} & \frac{2s^2+8s+6}{s^2} \\ \frac{s^2+6s+8}{s^2} & \frac{2s^2+8s+6}{s^2} \end{bmatrix}$$

In this case, the right factorization group is composed of all the non singular (2x2) constant matrices.

Let (A, B, C) be a minimal realization of G(s) in Luenberger controllable form :

$$A = \begin{bmatrix} 0 & 1 & 0 & 0 \\ -3 & -4 & 0 & 0 \\ 0 & 0 & 0 & 1 \\ 0 & 0 & -8 & -6 \end{bmatrix}, \quad B = \begin{bmatrix} 0 & 0 \\ 1 & 0 \\ 0 & 0 \\ 0 & 1 \end{bmatrix}, \quad C = \begin{bmatrix} 3 & 1 & 4 & 1 \\ 1 & 1 & 2 & 1 \end{bmatrix}$$

G(s) is full rank and has no finite zeros then this system is feedback irreducible. The reachability indices are 2,2 and are equal to the negatives of the right factorization indices of G(s).

Now, we study the equivalence of transfer matrices under dynamic compensation.

Definition 3.2 : Two (pxm) proper rational matrices $G_1(s)$ and $G_2(s)$ are dynamically equivalent [5] if there exist two (mxm) proper matrices $Q_1(s)$ and $Q_2(s)$ such that :

$$\begin{aligned} G_1(s)\, Q_1(s) &= G_2(s) \\ G_2(s)\, Q_2(s) &= G_1(s) \end{aligned} \tag{3.5}$$

In [4] is proven the following.

Theorem (3.3) : $G_1(s)$ and $G_2(s)$ two (pxm) proper rational matrices are dynamically equivalent if and only if there exists a bicausal isomorphism $B(s)$ such that :

$$G_1(s)\ B(s) = G_2(s) \tag{3.6}$$

Thus by theorem (2.1) the dynamic compensator may be achieved by feedback.
We can now characterize dynamic equivalence in terms of Wiener Hopf factorizations.

Theorem (3.4) : Let $G_1(s) = U_1(s)\ \Lambda_1(s)\ B_1(s)$ and $G_2(s) = U_2(s)\ \Lambda_2(s)\ B_2(s)$ two (pxm) proper rational matrices and their right Wiener Hopf factorizations at infinity. $G_1(s)$ and $G_2(s)$ are dynamically equivalent if and only if :

$$\begin{aligned} \Lambda_1(s) &= \Lambda_2(s) \\ U_1(s) &= U_2(s)\ U(s) \end{aligned} \tag{3.7}$$

where $U(s)$ belongs to the right factorization group of $G_1(s)$ (or $G_2(s)$).

Proof : Assume that $G_1(s)$ and $G_2(s)$ are dynamically equivalent. From theorem (3.3), there exists a bicausal isomorphism $B(s)$ such that $G_1(s)\ B(s) = G_2(s)$ then $U_1(s)\ \Lambda_1(s)\ B_1(s)\ B(s)$ is a right factorization of $G_2(s)$. The part "if" of the theorem then follows from theorem (3.1) applied to right factorizations. Reciprocally, let $G_1(s) = U_1(s)\ \Lambda_1(s)\ B_1(s)$, $G_2(s) = U_2(s)\ \Lambda_2(s)\ B_2(s)$ with $\Lambda_1(s) = \Lambda_2(s)$ and $U_1(s) = U_2(s)\ U(s)$. $U(s)$ belongs to the left factorization group of $G_2(s)$ then there exists $B(s)$ such that $U(s)\ \Lambda_2(s) = \Lambda_2(s)\ B(s)$. So $G_1(s) = U_2(s)\ U(s)\ \Lambda_2(s)\ B_1(s) = U_2(s)\ \Lambda_2(s)\ B(s)\ B_1(s)$ which is feedback equivalent to $G_2(s)$ with the bicausal isomorphism $B_1^{-1}(s) B(s) B_2(s)$ ∎

Remarks

* $U_1(s)\ \Lambda_1(s)$ represents a feedback irreducible system which can be obtained by feedback from a minimal realization of $G_1(s)$.

* In [4], Morse solves the problem of feedback equivalence on minimal realizations For left invertible systems. In this case, it is necessary to add to the condition of dynamic equivalence the identity of the transmission zeros. In the case of non-left invertible systems, these conditions are not sufficient. One must add at least the identity of the controllable structures of the maximally non-observable parts of canonical realizations.

4 . SMITH MCMILLAN FACTORIZATIONS AT INFINITY, BASIC PROPERTIES

Let us introduce now a factorization at infinity which was used before in [2].

Definition 4.1 : Let G(s) be a (pxm) proper rational matrix, a factorization of G(s) of the form :

$$G(s) = B_1(s)\ \Lambda(s)\ B_2(s) \tag{4.1}$$

is called a **Smith McMillan factorization at infinity of G(s).**

$B_1(s)$ is a (pxp) bicausal isomorphism and $B_2(s)$ is a (mxm) bicausal isomorphism.

$$\Lambda(s) = \begin{bmatrix} \Delta(s) & 0 \\ 0 & 0 \end{bmatrix} \quad \text{where } \Delta(s) = \text{diag}\ (s^{n_1}, \ldots, s^{n_r})$$

The integers n_i are assumed to be decreasingly ordered.

The following theorem insures the existence of such factorizations and gives some of their basic properties.

Theorem 4.1 : Let G(s) be a (pxm) proper rational matrix. Then there exist Smith McMillan factorizations at infinity of G(s), moreover $\Lambda(s)$ is uniquely defined and the negatives of the n_i's are the infinite zero orders of G(s) and r = rank (G(s)).

Proof : Following Morse [4] one can consider G(s) as a matrix over the Principal Ideal Domain of all the proper rational functions. Then the Smith form of G(s) over the P.I.D. admits the uniquely determined structure $\Lambda(s)$ where r = rank (G(s)). Then G(s) may be factorized as $G(s) = Q_1(s)\ \Lambda(s)\ Q_2(s)$ where $Q_1(s)$ and $Q_2(s)$ are two invertible matrices over the P.I.D. These are bicausal isomorphisms. Bicausal isomorphisms have neither poles nor zeros at infinity, thus clearly the structure at infinity of G(s) is contained in $\Lambda(s)$ and from the definition of infinite zeros it follows that the n_i's are the negatives of the infinite zero orders of G(s) ■

For a geometric interpretation of the infinite structure see [14]. We do not have uniqueness of these factorizations and more precisely we can state :

Proposition 4.1 : Let $G(s) = B_1'(s)\ \Lambda'(s)\ B_2'(s) = B_1(s)\ \Lambda(s)\ B_2(s)$ be two Smith McMillan factorizations at infinity of a proper rational matrix G(s). Then :

$$\Lambda(s) = \Lambda'(s) \text{ and } B_1(s) = B_1'(s)\ B_L^{-1}(s),\ B_2(s) = B_R(s)\ B_2'(s)$$

where $B_L(s)$ and $B_R(s)$ are bicausal isomorphisms such that :

$$B_L(s)\ \Lambda(s) = \Lambda(s)\ B_R(s) \tag{4.2}$$

The $B_L(s)$ (resp. $B_R(s)$) form a multiplicative group denoted by G_L (resp. G_R).

Proof : From theorem 4.1, $\Lambda(s)$ is uniquely determined so $\Lambda(s) = \Lambda'(s)$.

Then $B_1'(s)\ \Lambda(s)\ B_2'(s) = B_1(s)\ \Lambda(s)\ B_2(s)$ it follows that :

$$B_1^{-1}(s)\ B_1'(s)\ \Lambda(s) = \Lambda(s)\ B_2(s)\ B_2'^{-1}(s)$$

pose $\quad B_1^{-1}(s)\ B_1'(s) = B_L(s)$ and $B_2(s)\ B_2'^{-1}(s) = B_R(s)$

such matrices $B_L(s)$ are bicausal isomorphisms such that there exists $B_R(s)$ bicausal isomorphism with $B_L(s)\ \Lambda(s) = \Lambda(s)\ B_R(s)$ ■

Clearly such $B_L(s)$ form a multiplicative group denoted by G_L. Roughly speaking, we have uniqueness of the Smith McMillan factorization at infinity of G(s) modulo the bicausal isomorphisms which can cross the diagonal.

Now, let us characterize such bicausal isomorphisms.

Lemma 4.1 : Let $B_1(s)\ \Lambda(s)\ B_2(s)$ be a Smith McMillan factorization at infinity of a (mxm) proper rational non singular matrix $G(s)$, where $\Lambda(s) = \text{diag}\ (s^{n_1} \ldots s^{n_m})$. The elements of G_L are bicausal isomorphisms such that :

$$\# b_{ij}(s) \geq n_j - n_i \qquad i = 1, \ldots, m,\ j = 1, \ldots, m \tag{4.3}$$

Proof : Let $B(s)$ be a bicausal isomorphism of G_L, there exists $B'(s)$ bicausal isomorphism such that $B(s)\ \Lambda(s) = \Lambda(s)\ B'(s)$. This implies :

$$b_{ij}(s)\ s^{n_j} = b'_{ij}(s)\, s^{n_i} \quad i = 1, \ldots, m,\ j = 1, \ldots, m \tag{4.4}$$

Then $\# b_{ij}(s) = n_j - n_i + (\# b'_{ij}(s))$

Since $\# b'_{ij}(s) \geq 0$, it follows that $\# b_{ij}(s) \geq n_j - n_i$, $i = 1, \ldots, m$, $j = 1, \ldots, m$.

Now, suppose that $B(s)$ is a (mxm) bicausal isomorphism verifying condition (4.3). Relation (4.4) determines uniquely a proper rational matrix $B'(s)$ such that $B(s)\ \Lambda(s) = \Lambda(s)\ B'(s)$. It just remains to prove that $B'(s)$ is a bicausal isomorphism.

(4.3) and (4.4) imply that $B(s)$ and $B'(s)$ are as follows :

$B(s) =$ 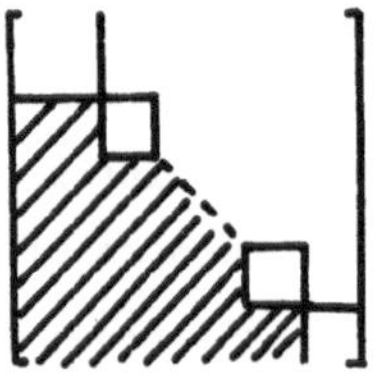 $\qquad B'(s) =$ 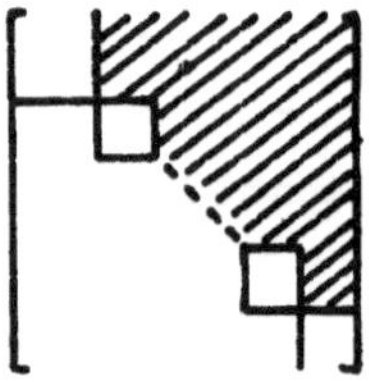 (4.5)

The elements of the shaded parts are strictly proper and the block diagonal parts of $B(s)$ and $B'(s)$ are identical. Blocks of dimension greater than 1 appear in case of equality of n_i' s. The bicausality of $B(s)$ implies that the blocks of the diagonal have full rank at infinity, which in turn together with the structure of $B'(s)$, imply the bicausality of $B'(s)$. Note that in this case (4.2) establishes a bijective correspondence between G_L and G_R ■

We can now characterize G_L in the general case.

Theorem 4.2 : Let $G(s)$ be a (pxm) rational proper matrix whose Smith McMillan factorization at infinity is $G(s) = B_1(s)\ \Lambda(s)\ B_2(s)$ with $\Lambda(s) = \begin{bmatrix} \Delta(s) & 0 \\ 0 & 0 \end{bmatrix}$ and where $\Delta(s) = \text{diag}\ (s^{n_1}, \ldots, s^{n_r})$.

Then, the elements of the group G_L have the following form :

$$\begin{bmatrix} B(s) & P(s) \\ 0 & \bar{B}(s) \end{bmatrix} \tag{4.6}$$

where $B(s)$ is any (rxr) bicausal isomorphism such that :

$$\# b_{ij}(s) \geq n_j - n_i \qquad i = 1 \ldots r,\ j = 1 \ldots r$$

$P(s)$ is any (rx(p-r)) proper matrix.

$\bar{B}(s)$ is any $((p-r)\times(p-r))$ bicausal isomorphism.
Proof is immediate by implementing (4.2) and using lemma (4.1).

Remarks :

* An analogous result holds for G_R.
* In the non injective case for a given $B_L(s)$ of G_L, we obtain several solutions for $B_R(s)$ satisfying (4.2).
* The constant matrices of G_L are upper block triangular in case of lemma (4.1).

Considering again example (3.1) :

$$G(s) = \begin{bmatrix} \frac{1}{s+1} & \frac{1}{s+2} \\ \frac{1}{s+3} & \frac{1}{s+4} \end{bmatrix}$$

Let us take two different Smith McMillan factorizations at infinity of G(s).

$$G(s) = B_1(s)\ \Lambda(s)\ B_2(s) = \begin{bmatrix} 1 & 0 \\ \frac{s-2}{s} & 1 \end{bmatrix} \begin{bmatrix} s^{-1} & 0 \\ 0 & s^{-3} \end{bmatrix} \begin{bmatrix} \frac{s}{s+1} & \frac{s}{s+2} \\ \frac{6\ s^2}{(s+1)(s+3)} & \frac{8\ s^2}{(s+2)(s+4)} \end{bmatrix}$$

$$= B_1'(s)\ \Lambda(s)\ B_2'(s) = \begin{bmatrix} 1 & \frac{-4s}{s+4} \\ \frac{s+1}{s+3} & \frac{-2s(s+2)}{(s+3)(s+4)} \end{bmatrix} \begin{bmatrix} s^{-1} & 0 \\ 0 & s^{-3} \end{bmatrix} \begin{bmatrix} \frac{s}{s+1} & \frac{s+2}{s+4} \\ 0 & \frac{s}{s+2} \end{bmatrix}$$

$$B_1^{-1}(s)\ B_1'(s) = \begin{bmatrix} 1 & \frac{-4s}{s+4} \\ \frac{6}{s(s+3)} & \frac{2s^2-24}{(s+3)(s+4)} \end{bmatrix},\ B_2(s)\ B_2'^{-1}(s) = \begin{bmatrix} 1 & \frac{-4}{s(s+4)} \\ \frac{6\ s}{s+3} & \frac{2s^2-24}{(s+3)(s+4)} \end{bmatrix}$$

Clearly $B_1^{-1}(s)\ B_1'(s)$ belongs to G_L, $B_2(s)\ B_2'^{-1}(s)$ belongs to G_R and furthermore $B_1^{-1}(s)\ B_1'(s)\ \Lambda(s) = \Lambda(s)\ B_2(s)\ B_2'^{-1}(s)$.

5 . SMITH MCMILLAN FACTORIZATIONS AT INFINITY, CONTROL INTERPRETATION

Recall some tools of geometric control theory [15], which will be necessary in the following. Consider the linear multivariable standard system Σ. A subspace $\mathcal{V} \subset \mathcal{X}$ is said to be (A,B)-invariant if there exists a feedback F such that $(A+BF)\mathcal{V} \subset \mathcal{V}$. The set of all (A,B)-invariants contained in a given subspace is closed under addition, so there exists a uniquely defined supremal element in this set. Let $\mathcal{V}^*$ be the largest (A,B)-invariant contained in kernel C. $\mathcal{V}^*$ is the largest unobservable subspace which can be obtained by feedback.

In a dual way, we can define (C,A)-invariant subspaces. Denote by T_* the smallest (C,A)-invariant subspace containing the image of B. T_* is then the smallest controllable subspace which can be obtained by output injection. In [13], Morse studies the structural invariants of linear multivariable systems under the transformation group G of all input, output and state changes of coordinate state feedback and output injections.

An element (T, F, G, K, H) of G transforms a triple (A, B, C) in :

$$(T^{-1}(A+BF+KC)T,\ T^{-1}BG,\ HCT) \tag{5.1}$$

It is shown that the orbit of any standard triple (A, B, C) under G is uniquely characterized by three lists of positive integers I_1, I_2, I_3 and a list I_4 of monic polynomials called the transmission polynomials of (A, B, C). These lists determine a canonical form $(\hat{A}, \hat{B}, \hat{C})$ under G.

$$\hat{A} = \begin{bmatrix} \hat{A}_1 & 0 & 0 & 0 \\ 0 & \hat{A}_2 & 0 & 0 \\ 0 & 0 & \hat{A}_3 & 0 \\ 0 & 0 & 0 & \hat{A}_4 \end{bmatrix},\quad \hat{B} = \begin{bmatrix} \hat{B}_1 & 0 \\ 0 & 0 \\ 0 & \hat{B}_3 \\ 0 & 0 \end{bmatrix},\quad \hat{C} = \begin{bmatrix} \hat{C}_1 & 0 & 0 & 0 \\ 0 & \hat{C}_2 & 0 & 0 \end{bmatrix}$$

The partition of the above matrices is made according to a decomposition of X : $X_1 \oplus X_2 \oplus X_3 \oplus X_4$ such that :

$(\hat{A}_3, \hat{B}_3)$ is expressed in the controllable Brunovsky form

$(\hat{A}_2, \hat{C}_2)$ is expressed in the observable Brunovsky form

$\hat{A}_4$ is represented in the rational canonical form

$\hat{A}_1$ = block diag $(A_1, A_2, \ldots, A_r)$

$\hat{B}_1$ = block diag $(B_1, B_2, \ldots, B_r)$

C_1 = block diag $(C_1, C_2, \ldots, C_r)$

where $C_i = (1, 0\ 0 \ldots 0)$ is $(1 \times n_i')$.

$$A_i = \begin{bmatrix} 0 & 1 & 0 & \ldots & 0 \\ 0 & & 1 & \ldots & 0 \\ & & & \ddots & 0 \\ & 0 & & \ddots & 1 \\ 0 & \ldots & \ldots & \ldots & 0 \end{bmatrix} \text{ is } (n_i' \times n_i'),\quad B_i = \begin{bmatrix} 0 \\ \vdots \\ 0 \\ 1 \end{bmatrix} \text{ is } (n_i' \times 1)$$

<u>Remarks</u>

* Blocks of the canonical form are not in the same order as in [13].
* The canonical form is obtained firstly by making (A,B,C) maximally non-controllable and maximally non-observable by action of G so $X_3 \oplus X_4 = V^*$, $X_1 \oplus X_3 = T_*$.
* The invariant polynomials of $\hat{A}_4$ are the transmission polynomials (I_4). If (A, B, C) is minimal these polynomials are the non trivial numerators of the Smith McMillan form of $C(sI-A)^{-1}B$ [16]. The roots of these polynomials are called the transmission zeros of the system.
* List (I_3) coincides with the controllability indices of $(\hat{A}_3, \hat{B}_3)$.

* List (I_2) coincides with the observability indices of $(\hat{\hat{A}}_2, \hat{C}_2)$

* List $(I_1) = \{n'_1, n'_2 \dots n'_r\}$ coincides with the controllability indices of $(\hat{A}_1, \hat{B}_1)$ and with the observability indices of $(\hat{A}_1, \hat{C}_1)$.

In the following, we will relate the structural invariants of (A, B, C) defined by Morse to the Smith McMillan form at infinity of $G(s) = C(sI-A)^{-1}B$.

Theorem 5.1 : Let (A,B,C) be a standard triple, the element of G which leads to the Morse canonical form induces a Smith McMillan factorization at infinity of $G(s) = C(sI-A)^{-1}B$. We have $G(s) = B_1(s)\ \Lambda(s)\ B_2(s)$ where $B_2^{-1}(s)$ represents the action of the feedback group and $B_1^{-1}(s)$ represents the action of the output injection group. Furthermore $\Lambda(s) = \hat{C}(sI-\hat{A})^{-1}\hat{B} = \begin{bmatrix}\Delta(s) & 0\\ 0 & 0\end{bmatrix}$ where $\Delta(s) = \hat{C}_1(sI-\hat{A}_1)^{-1}\hat{B}_1 = \text{diag}\ (s^{-n'_1}, \dots, s^{-n'_r})$ then the n'_i's are the infinite zero orders.

Proof : Firstly notice that since we are dealing with transfer functions, the action of the state space basis changes is transparent. Using the main result of [17], the action of an element of the feedback group may be represented by a bicausal precompensator $B_f(s)$. Similarly using the dual version of this result, the action of an element of the output injection group may be represented by a bicausal post compensator $B_o(s)$. So under the action of an element of G, $G(s)$ is transformed in $B_o(s)\ G(s)\ B_f(s)$.

Let (A,B,C) be a standard minimal realisation of G(s). Consider an element of G which transforms (A,B,C) in its canonical form whose transfer function is $\Lambda(s)$. Thus $\Lambda(s) = B_1^{-1}(s)\ G(s)\ B_2^{-1}(s)$ with $B_1^{-1}(s)$ and $B_2^{-1}(s)$ bicausal isomorphisms. A simple calculation gives $\Lambda(s) = \hat{C}(sI-\hat{A})^{-1}\hat{B} = \begin{bmatrix}\Delta(s) & 0\\ 0 & 0\end{bmatrix}$ where $\Delta(s) = \hat{C}_1(sI-\hat{A}_1)^{-1}\hat{B}_1 = \text{diag}\ (s^{-n'_1}, \dots, s^{-n'_r})$.

Then $G(s) = B_1(s)\ \Lambda(s)\ B_2(s)$ where $B_1^{-1}(s)$ and $B_2^{-1}(s)$ represent respectively the action of the feedback group and of the output injection group. Furthermore the infinite structure of G(s) is contained in $\Lambda(s)$, it follows that the n'_i 's are the infinite zero orders∎

Proposition 5.1 : Let $G(s) = B_1(s)\ \Lambda(s)\ B_2(s)$ and $G'(s) = B'_1(s)\ \Lambda'(s)\ B'_2(s)$ be two (pxm) proper rational matrices and two corresponding Smith McMillan factorizations at infinity. A necessary and sufficient condition for the existence of a finite dimensional realization (A,B,C) of G(s) such that G'(s) is obtained from G(s) by action of G on (A, B, C) is that $\Lambda(s) = \Lambda'(s)$.

Proof : Necessity : since G'(s) is obtained from G(s) by action of G, $G'(s) = B_o(s)\ G(s)\ B_f(s)$ where $B_o(s)$ and $B_f(s)$ are bicausal isomorphisms. $G'(s) = B_o(s)\ B_1(s)\ \Lambda(s)\ B_2(s)\ B_f(s) = B'_1(s)\ \Lambda'(s)\ B'_2(s)$. Using theorem 4.1, it follows that $\Lambda(s) = \Lambda'(s)$.

Sufficiency : $G(s) = B_1(s)\ \Lambda(s)\ B_2(s)$, $G'(s) = B'_1(s)\ \Lambda(s)\ B'_2(s)$ thus $G'(s) = B'_1(s)\ B_1^{-1}(s)\ G(s)\ B_2^{-1}(s)\ B'_2(s)$. Using the fact that there exists a finite dimensional realization of G(s) in which the rational bicausal isomorphism $B_2^{-1}(s)\ B'_2(s)$ can be implemented by feedback [17], and using the dual version of this result (i.e. $B'_1(s)\ B_1^{-1}(s)$ implemented by output injection) it is clear that there exists a finite dimensional realization of G(s) such that G'(s) is obtained from G(s) by action of G∎

Remarks : * $\mathcal{G}$ acts on a possibly non minimal realization of G(s).

* Considering $G'(s) = \Lambda(s)$, $B_1^{-1}(s)$ and $B_2^{-1}(s)$ are associated to an element of $\mathcal{G}$ acting on a possibly non minimal realization (A,B,C) of G(s) and such that $\Lambda(s) = HC(sI-A-BF-KC)^{-1}BG$. This result is not the reciprocal of theorem (5.1) because firstly the controller is implemented in a possibly non-minimal realization and secondly no conditions are imposed on the dynamics of the non-controllable or non-observable parts.

* Let (A,B,C) be a realization of G(s) and (T,F,G,K,H) be an element of $\mathcal{G}$ acting on (A,B,C). G(s) is transformed in $G_1(s)$:

$$G_1(s) = H(I-C(sI-A-BF)^{-1}K)^{-1}G(s)(I-F(sI-A)^{-1}B)^{-1}G \tag{5.2}$$

$$= H(I-C(sI-A)^{-1}K)^{-1}G(s)(I-F(sI-A-KC)^{-1}B)^{-1}G \tag{5.3}$$

Note that we got two different factorizations depending on the order in which we apply the transformations.

We can establish an analogous result of theorem (3.4).

Proposition 5.2 : Let $G(s) = B_1(s)\ \Lambda(s)\ B_2(s)$ and $G'(s) = B_1'(s)\ \Lambda'(s)\ B_2'(s)$ be two (pxm) proper rational matrices and their Smith McMillan factorizations at infinity. G(s) and G'(s) are dynamically equivalent if and only if :

$$\Lambda(s) = \Lambda'(s)$$

$$B_1(s) = B_1'(s)\ B(s) \text{ where } B(s) \text{ belongs to } G_L \text{ (of } G(s)).$$

Proof is immediate from theorem 3.3 and from proposition 4.1.

Now, we focus our interest on the elements of $\mathcal{G}$ acting on a minimal realization of the transfer function. Recall briefly a result of [3] which states that the transfer function $G_1(s)$ of a system feedback equivalent to a canonical realization of G(s) is characterized by $G_1(s) = G(s)\ \Gamma_1(s)$ where $\Gamma_1(s)$ is a bicausal isomorphism, furthermore if $N_1(s)\ D_1^{-1}(s)$ is a right coprime factorization of G(s), then in this case, $\Gamma_1(s) = D_1(s)\ (D_1(s)+Q(s))^{-1}G = (I+Q(s)D_1^{-1}(s))^{-1}G$ where G is constant non singular and Q(s) is a (mxm) polynomial matrix such that $Q(s)\ D_1^{-1}(s)$ is strictly proper. We have $G_1(s) = G(s)\ \Gamma_1(s) = N_1(s)\ D_1^{-1}(s)\ D_1(s)\ (D_1(s) + Q(s))^{-1}G = N_1(s)\ (D_1(s) + Q(s))^{-1}G$. The dual result states that the transfer function $G_2(s)$ of a system output injection equivalent to a canonical realization of G(s) has the representation $G_2(s) = \Gamma_2(s)\ G(s)$ where $\Gamma_2(s)$ is a bicausal isomorphism. Let $D_2^{-1}(s)\ N_2(s)$ be a left coprime factorization of G(s), in this case $\Gamma_2(s) = H(D_2(s) + R(s))^{-1}D_2^{-1}(s) = H(I + D_2^{-1}(s)\ R(s))^{-1}$ where H is constant non singular and R(s) is a (pxp) polynomial matrix such that $D_2^{-1}(s)\ R(s)$ is strictly proper. Then $G_2(s) = \Gamma_2(s)\ G(s) = (D_2(s) + R(s))^{-1} N_2(s)$.

Lemma 5.1 : Let $N_1(s)\ D_1^{-1}(s)$ be a right coprime factorization of G(s). The transfer function $G_1(s)$ of a system feedback equivalent to a canonical realization of G(s) is characterized by $G_1(s) = G(s)\ \Gamma_1(s)$ where $\Gamma_1(s) = (I + Q_1(s)\ D_1^{-1}(s))^{-1}$ is a bicausal isomorphism and $Q_1(s)$ is a (mxm) polynomial matrix such that $Q_1(s)\ D_1^{-1}(s)$ is proper.

Proof : Let $G_1(s)$ be the transfer function of a system feedback equivalent to a canonical realization of G(s). From the preceding results, we can then write $G_1(s) =$

$G(s)\ \Gamma_1(s)$ where $\Gamma_1(s)$ is a bicausal isomorphism such that :

$$\Gamma_1(s) = (I+Q(s)D_1^{-1}(s))^{-1}G \text{ with } Q(s)D_1^{-1}(s) \text{ strictly proper}$$

$$\begin{aligned}\Gamma_1(s) &= (G^{-1} + G^{-1}\ Q(s)\ D_1^{-1}(s))^{-1} \\ &= (I + [(G^{-1}-I)D_1(s) + G^{-1}Q(s)]\ D_1^{-1}(s))^{-1} \\ &= (I + Q_1(s)D_1^{-1}(s))^{-1} \text{ where } Q_1(s)\ D_1^{-1}(s) \text{ is proper}\end{aligned}$$

which proves the necessity.

Reciprocally let $\Gamma_1(s)$ be a bicausal isomorphism such that :

$\Gamma_1(s) = (I + Q_1(s)\ D_1^{-1}(s))^{-1}$ where $Q_1(s)\ D_1^{-1}(s)$ is proper.

$Q_1(s)\ D_1^{-1}(s) = G_1 + R(s)\ D_1^{-1}(s)$ where G_1 is constant and $R(s)$ is a polynomial matrix such that $R(s)\ D_1^{-1}(s)$ is strictly proper. Then, $\Gamma_1(s) = (I + G_1 + R(s)\ D_1^{-1}(s))^{-1}$ where $I + G_1$ is invertible. So $\Gamma_1(s) = (I + (I+G_1)^{-1}R(s)\ D_1^{-1}(s))^{-1}(I+G_1)^{-1}$.

$$= (I + Q(s)\ D_1^{-1}(s))^{-1}G$$

Then $G_1(s)$ is feedback equivalent to $G(s)$ ∎ We will deal with irreducible systems defined as follows :

Definition 5.1 : A standard system (Σ) is said to be irreducible if :

$$V^* = 0 \text{ and } T_* = X$$

For another geometric characterization of such systems see [14], now we present a polynomial one :

Lemma 5.2 : A standard system (Σ) is irreducible if and only if there exists a non singular polynomial matrix $D(s)$ such that : $G(s) = C(sI-A)^{-1}B = D(s)^{-1}$.

Proof : Rank $(G(s)) = \dim(C\,T_*) = \dim.\ (\mathcal{B}/\mathcal{B} \cap R^*) = m = p$, this implies that the transfer matrix $G(s) = C(sI-A)^{-1}B$ is invertible. Moreover $V^*/R^* = 0$ then $G(s)$ has no finite transmission zeros. Thus prime factorizations are of the following form $G(s) = N_2(s)\ D_2^{-1}(s) = D_1^{-1}(s)\ N_1(s)$, where $N_1(s)$ and $N_2(s)$ are unimodular. It follows that $G(s) = (D_2(s)\ N_2^{-1}(s))^{-1} = (N_1^{-1}(s)\ D_1(s))^{-1}$. The converse can be proven by reversing the arguments. Note that an irreducible system has the McMillan degree invariance property under the action of G.

Let (A,B,C) be an irreducible system. By lemma (5.2), the transfer matrix is equal to $D^{-1}(s)$. Let us characterize the action of G on (A,B,C). To each element (T,F,G,K,H) of G correspond polynomial matrices $Q(s)$, $R(s)$ and $Q_1(s)$, $R_1(s)$ such that the transformed of (A,B,C) has the transfer :

$$G(D^{-1}(s)) = (I+D^{-1}(s)Q(s))^{-1}\ D^{-1}(s)\,(I+R(s)\,(D(s)+Q(s))^{-1})^{-1} \tag{5.4}$$

$$= (I+(D(s)+R_1(s))^{-1}Q_1(s))^{-1}D^{-1}(s)\,(I+R_1(s)D^{-1}(s))^{-1} \tag{5.5}$$

$$= (D(s)+Q(s)+R(s))^{-1} = (D(s)+Q_1(s)+R_1(s))^{-1} \tag{5.6}$$

where $(I+D^{-1}(s)Q(s))^{-1}$, $(I+R(s)\,(D(s)+Q(s))^{-1})^{-1}$, $(I+(D(s)+R_1(s))^{-1}Q_1(s))^{-1}$ and $(I+R_1(s)D^{-1}(s))^{-1}$ are bicausal isomorphisms, and where $D^{-1}(s)Q(s)$, $R(s)\,(D(s)+Q(s))^{-1}$, $(D(s)+R_1(s))^{-1}Q_1(s)$ and $R_1(s)D^{-1}(s)$ are proper. For an explicit characterization of (5.4), (5.5), see (5.2) and (5.3) ∎ Now, characterize more precisely the feedback group.

Lemma 5.3 : Let (A,B,C) be an irreducible system whose transfer matrix is $D^{-1}(s)$; where $D(s)$ is a (mxm) polynomial matrix. There exists a bijective relation between the elements (I,F,G) of the feedback group and the bicausal isomorphisms $B(s) = (I+Q(s)D^{-1}(s))^{-1}$ where $Q(s)$ is a (mxm) polynomial matrix such that $Q(s)D^{-1}(s)$ is proper.

Proof : Let (I,F,G) be an element of the feedback group by lemma (5.1) there exists $Q(s)$ such that the transformed of (A,B,C) has the transfer matrix $(D(s)+Q(s))^{-1} = D^{-1}(s)(I+Q(s)D^{-1}(s))^{-1} = D^{-1}(s)(I-F(sI-A)^{-1}B)^{-1}G$. This map is clearly onto by lemma 5.1. Let us prove the one to one part. Let (I,F_1,G_1) and (I,F_2,G_2) be two different elements of the feedback group, then $B_1(s) = (I-F_1(sI-A)^{-1}B)^{-1}G_1$ is different from $B_2(s) = (I-F_2(sI-A)^{-1}B)^{-1}G_2$. If not $\lim_{s\to\infty} B_1(s) = \lim_{s\to\infty} B_2(s)$ implies $G_1 = G_2$ and then $(F_1-F_2)(sI-A)^{-1}B = 0$ together with the controllability of (A,B) imply $F_1 = F_2$. $B_1(s)$ different from $B_2(s)$ implies that $D^{-1}(s)B_1(s)$ is different from $D^{-1}(s)\,B_2(s)$. Then there exists $Q_1(s)$ different from $Q_2(s)$ such that $D^{-1}(s)B_1(s) = (D(s)+Q_1(s))^{-1}$ and $D^{-1}(s)\,B_2(s) = (D(s)+Q_2(s))^{-1}$, where $Q_1(s)D^{-1}(s)$ and $Q_2(s)\,D^{-1}(s)$ are proper. We have proved that there exist $Q_1(s)$ different from $Q_2(s)$ such that : $B_1(s) = (I+Q_1(s)D^{-1}(s))^{-1}$ and $B_2(s) = (I+Q_2(s)\,D^{-1}(s))^{-1}$ are associated to (I,F_1,G_1) and (I,F_2,G_2) then the map is one to one ∎ As we have defined G_L, we define :

Definition 5.2 : $G_L(D)$ is the group of all bicausal isomorphisms $B_L(s)$ such that there exists $B_R(s)$ bicausal isomorphism with $B_L(s)D^{-1}(s) = D^{-1}(s)B_R(s)$.

Lemma 5.4 : The elements of $G_L(D)$ associated with elements of the output injection group acting on a minimal realization of $D^{-1}(s)$ are bicausal isomorphisms $B_L(s)$ such that : $B_L(s) = (I+D^{-1}(s)Q(s))^{-1}$, where $D^{-1}(s)\,Q(s)$ and $Q(s)\,D^{-1}(s)$ are proper. Moreover the bicausal isomorphisms $B_R(s)$ such that $B_L(s)D^{-1}(s) = D^{-1}(s)B_R(s)$ represent elements of the feedback group acting on a minimal realization of $D^{-1}(s)$.

Proof : We proved (dual of lemma 5.3) that the bicausal isomorphism $B(s)$ which represents elements of the output injection group acting on a minimal realization of $D^{-1}(s)$ are characterized by : $B(s) = (I+D^{-1}(s)Q(s))^{-1}$ where $D^{-1}(s)Q(s)$ is proper. Then

$$B(s)D^{-1}(s) = (I+D^{-1}(s)Q(s))^{-1}D^{-1}(s)$$
$$= (D(s) + Q(s))^{-1} = D^{-1}(s)(I+Q(s)D^{-1}(s))^{-1} = D^{-1}(s)B_1(s) \qquad (5.7)$$

$B(s)$ belongs to $G_L(D)$ if and only if $B_1(s)$ is a bicausal isomorphism then only if $Q(s)D^{-1}(s)$ is proper. $B_1(s)$ represents an element of the feedback group which can be implemented on a minimal realization of $D^{-1}(s)$, see lemma 5.1. Conversely, let $B(s) = (I+D^{-1}(s)Q(s))^{-1}$ be a bicausal isomorphism where $D^{-1}(s)Q(s)$ and $Q(s)D^{-1}(s)$ are proper. $B(s)$ represents an element of the output injection group acting on a minimal realization of $D^{-1}(s)$. Using (5.7), $\det(B(s)) = \det(B_1(s))$ then $\lim_{s\to\infty}(\det(B_1(s)) = \lim_{s\to\infty}(\det(B_2(s)) = c \neq 0$ since $B(s)$ is a bicausal isomorphism. $Q(s)D^{-1}(s)$ is proper, so is $B_1^{-1}(s) = I+Q(s)D^{-1}(s)$. Then $\lim_{s\to\infty}(\det B_1^{-1}(s)) = \frac{1}{c} \neq 0$ implies that $B_1^{-1}(s)$ and $B_1(s)$ are bicausal isomorphisms. So $B(s)$ belongs to $G_L(D)$ ∎

Remark : This result particularizes lemma 4.1. In effect let $Q(s)$ be a (mxm) polynomial matrix $Q(s)$ such that $Q(s)\,\Lambda(s)$ and $\Lambda(s)Q(s)$ are proper, with $(I+\Lambda(s)Q(s))^{-1}$ bicausal isomorphisms. The properness of $Q(s)\Lambda(s)$ implies $\deg(q_{ij}(s)) \leq -n_j$, similarly one has $\deg(q_{ij}(s)) \leq -n_i$ then the size of the (i-j)th element of $(I+\Lambda(s)Q(s))$ is less or equal to n_j-n_i. So the bicausal isomorphism $(I+\Lambda(s)Q(s))$ belongs to G_L by lemma 4.1, this implies in turn that $(I+\Lambda(s)Q(s))^{-1}$ belongs to G_L. Given a minimal triple (A,B,C), then the set of all elements of G that leave (A,B,C) invariant is a subgroup called the stabilizer of G at (A,B,C).

Given a reachable pair (A,B), the stabilizer of the feedback group at (A,B) is studied in [3], [18]. We study now the stabilizer of G at (A,B,C) for irreducible systems.

<u>Theorem 5.3</u> : Let (A,B,C) be an irreducible system and $D^{-1}(s)$ its transfer matrix. The stabilizer of G at (A,B,C) is in bijective correspondence with the set Q_D of (m×m) polynomial matrices $Q(s)$ such that $(I+D^{-1}(s)Q(s))^{-1}$ is a bicausal isomorphisms and such that $Q(s)D^{-1}(s)$ and $D^{-1}(s)Q(s)$ are proper.

<u>Proof</u> : We will proceed in two steps. In the first one, we study bicausal isomorphisms $B_1(s)$ and $B_2(s)$ associated with elements of G such that $B_1(s)D^{-1}(s)B_2(s) = D^{-1}(s)$. Using (5.6), $B_1(s)D^{-1}(s)B_2(s) = (D(s)+Q(s)-Q(s))^{-1}$ where $B_1(s) = (I+D^{-1}(s)Q(s))^{-1}$, $B_2(s) = (I-Q(s)(D(s)+Q(s))^{-1})^{-1}$ are bicausal isomorphisms and $D^{-1}(s)Q(s)$, $Q(s)(D(s)+Q(s))^{-1}$ are proper. Here we have considered that the output injection is applied first on a minimal realization of $D^{-1}(s)$ and secondly the feedback on a minimal realization of $(D(s)+Q(s))^{-1}$.

$$D^{-1}(s) = (I+D^{-1}(s)Q(s))^{-1}D^{-1}(s)(I-Q(s)(D(s)+Q(s))^{-1})^{-1}$$
$$= (I+D^{-1}(s)Q(s))^{-1}D^{-1}(s)(I+Q(s)D^{-1}(s))$$

which is equivalent to : $(I+D^{-1}(s)Q(s))^{-1}D^{-1}(s) = D^{-1}(s)(I+Q(s)D^{-1}(s))^{-1}$.

Lemma 5.4 proves that $Q(s)D^{-1}(s)$ is proper. Roughly speaking, it follows that the set of all bicausal isomorphisms associated with elements of the output injection which can be compensated by feedback is isomorphic to Q_D.

Now we turn back to the stabilizer of G at (A,B,C). Using the dual of lemma (5.3), we associate in a bijective way with an element $Q(s)$ of Q_D a bicausal isomorphism $B_1(s)$ and an element (I,K,H) of the output injection group. On the other hand to such a $B_1(s)$ is associated in a bijective way a bicausal isomorphism $B_2(s)$ such that $B_1(s) D^{-1}(s)B_2(s) = D^{-1}(s)$, and an element (I,F,G) of the feedback group. The preceding relations are summarized in the following diagram.

$Q(s) \leftrightarrow B_1(s) \leftrightarrow (I,K,H)$ ↘
$\updownarrow$ (I,F,G,K,H) ↔ (T,F,G,K,H)
$B_2(s) \leftrightarrow (I,F,G)$ ↗

So to an element $Q(s)$ of Q_D we associate in a bijective way an element (I,F,G,K,H) of G such that the transformed $(\hat{A},\hat{B},\hat{C})$ of (A,B,C) has the transfer $D^{-1}(s)$. Since (A,B,C) is minimal there exists a unique T such that $T^{-1}\hat{A}T = A$, $T^{-1}\hat{B} = B$ and $\hat{C}T = C$. Since (T,F,G,K,H) belongs to the stabilizer of G at (A,B,C) the proof is complete∎

6 . CONCLUSION

As stated, this paper presents some factorizations at infinity of rational matrix functions. The main interest of these factorizations is to point out the structural properties which remain invariant under transformation groups. The control interpretation of these factorizations permits us to study the problem of dynamic equivalence as a direct application. These factorizations seem attractive for the study of the "model following problem" or for the study of the decoupling problem. For irreducible systems the stabilizer of Morse's group at (A,B,C) is given and it would be interesting to develop a more general result.

R E F E R E N C E S

[1] WOLOWICH W.A., "Linear multivariable systems", Springer Verlag, 1974

[2] VERGHESE G., "Infinite frequency, behaviour of generalized dynamical systems", PhD Thesis, Elect. Eng. Dpt, Stanford University, 1978

[3] FUHRMANN P.A. and WILLEMS J.C., "The factorization indices for rational function matrices", Integral Equations Oper. Theory, vol. 2, pp. 287-301, 1979

[4] MORSE A.S., "System invariants under feedback and cascade control", Proc. International Symp., Udine, Springer, 1975

[5] WOLOWICH W.A. and FALB P.L., "Invariants and canonical forms under dynamic compensation", SIAM J. on Contr. and Opt., vol. 14, pp. 996-1008, 1976

[6] PERNEBO L., "An algebraic theory for the design of controllers for linear multivariable systems", Parts I and II, IEEE Trans. on Auto. Cont., AC 26, pp. 171-194, 1981

[7] ROSENBROCK H.H., "State space and multivariable theory", Nelson, London, 1970

[8] PUGH A.C. and RATCLIFFE P.A., "On the zeros and poles of a rational matrix", Int. J. Control, Vol. 30, pp. 213-226, 1979

[9] VARDULAKIS A.I.G., "On infinite zeros", Int. J. Control, vol. 32, pp. 849-866, 1980

[10] BHATTACHARYYA S.P., "Frequency domain conditions for disturbance rejection", IEEE Trans. Auto. Contr., AC 25, pp. 1211-1213, 1980

[11] OWENS D.H., "On structural invariants and the Root-Loci of linear multivariable systems", Int. J. Contr., vol. 28, pp. 187-196, 1978

[12] FRANCIS B.A., "On totally singular linear quadratic optimal control", IEEE Trans Auto. Cont., AC 24, pp. 616-621, 1979

[13] MORSE A.S., "Structural invariants of linear multivariable systems", SIAM J. Cont. and Opt. , vol.11, pp. 446-465, 1973

[14] COMMAULT C. and DION J.M., "Structure at infinity of linear multivariable systems - A geometric approach", submitted for publication, 1981

[15] WONHAM W.M., "Linear multivariable control : a geometric approach", (2nd Edition Springer Verlag, 1979

[16] ANDERSON B.D.O., "A note on transmission zeros of a transfer matrix", IEEE Trans Auto. Cont., AC 21, pp. 589-591, 1976

[17] HAUTUS M.L.J. and HEYMANN M., "Linear feedback,an algebraic approach", SIAM J. Cont. and Opt., Vol. 16, pp. 83-105, 1978

[18] MUNZNER H. and PRATZEL-WOLTERS O., "Minimal bases of polynomial modules, structural indices and Brunovsky transformations", Int.J.Cont.,vol.30,pp.291-318,1979

SOME TOPOLOGICAL PROPERTIES OF ELECTRICAL MACHINES*

T. E. Duncan**

1. INTRODUCTION

Electrical machines play an important role in many physical systems. In this paper some of the geometry and the topology of electrical machines are studied. Kron [7] was apparently the first person to introduce some geometric ideas in the study of electrical machines and these ideas were expanded by Kondo-Ishizuka [6]. The geometric methods that are used here are more from global differential geometry and are less computational than the work of Kron or Kondo-Ishizuka.

A geometric derivation of the differential equations that describe electrical machines is given. The manner in which the voltage and current are measured on the rotor is shown to affect the geometry and the topology. The nontriviality of the topology of a family of systems formed from an electrical machine is demonstrated.

This topological property has application to the problem of identification of models for electrical machines. The geometric description of electrical machines suggests a natural class of nonlinear systems that are modelled in a vector bundle.

Some of the mathematical techniques that are used to study these nonlinear systems arise from the study of families of linear systems. In particular the families of linear systems that are most directly relevant are those that have symmetric transfer functions. A connected topological component of symmetric transfer functions is denoted Rat(p,q;m) which is the collection of symmetric m x m transfer functions of McMillan degree p+q and Cauchy-Maslov index p-q.

Associated with symmetric transfer functions are special state space realizations called internally symmetric realizations. Geometrically the structure group in the frame bundle is reduced from $GL(p+q;\mathbb{R})$ to $O(p,q)$. A minimal realization (A,B,C) is said to be internally symmetric with respect to $I_{p,q} = I_p \oplus -I_q$ if

$$I_{p,q} A = A^t I_{p,q}$$

$$I_{p,q} B = C^t$$

---

*Research Supported by NSF Grant ECS-8024917

**Department of Mathematics, University of Kansas, Lawrence, KS 66045 USA

In [4] it was shown that $Rat(p,q) \triangleq Rat(p,q;1)$ does not have continuous global internally symmetric realizations if $\min(p,q) > 0$. This was proved by verifying the result for Rat(1,1) and then embedding this construction in Rat(p,q) where $\min(p,q) > 0$. Another proof of this result can be obtained by a method that can be naturally generalized to nonlinear systems.

To accomplish this proof consider the family of transfer functions in Rat(1,1)

$$G(s,\theta) = \frac{s\cos\theta + \sin\theta}{s^2 + 1}$$

where $\theta \in S^1$. These transfer functions were used in [4]. Since the denominators of each of these transfer functions does not depend on θ, each $G(\cdot, \theta)$ is isomorphic to the associated 2 x 2 Hankel matrix

$$H_\theta = \begin{bmatrix} \cos\theta & \sin\theta \\ \sin\theta & -\cos\theta \end{bmatrix}$$

The Cauchy index is the signature of the Hankel matrix by the Hermite-Hurwitz theorem and this is easily computed by the internally symmetric realizations. The positive and the negative eigenspaces are globally trivial if and only if there are continuous global internally symmetric realizations. Restricting the Laplace transform variable to the real line and applying the Cayley map to G we have the map, that by abuse of notation will still be denoted as G,

$$G: S_1^1 \times S_2^1 \longrightarrow S^1$$

If the positive and the negative eigenspaces are globally trivial then the map

$$G: S_2^1 \longrightarrow S^1$$

obtained from the above map G by fixing the first variable is null homotopic. However the degree of this map is nonzero. Thus there do not exist continuous global internally symmetric realizations for Rat(1,1).

In addition to providing a more elementary proof of the nonexistence of continuous global internally symmetric realizations for Rat(1,1) this approach alludes to a well known result in geometry and topology, Bott periodicity for the unitary group. This result states that the stable homotopy of the unitary group is periodic with period 2 [3]. K-theory provides a succinct description of this result

$$K(S^2 \times X) \cong K(S^2) \otimes K(X)$$

where X is a compact space. K-theory naturally appears in questions of stable homotopy of the unitary group because the functor K is represented by homotopy classes of maps from a space into the infinite unitary group. More concretely two vector bundles have the same equivalence class in K-theory if they are stably isomorphic.

The result for Rat(1,1) can be interpreted in this setting by letting $X=S^2$. The isomorphism of line bundles over S^2 is determined by the homotopy class of clutching functions [2] which are maps $S^1 \longrightarrow U(1) \cong S^1$. These clutching functions appear by considering vector bundles on the two hemispheres which are contractible spaces.

2. THE EQUATIONS OF ELECTRICAL MACHINES

To obtain a perspective of electrical machines in electromagnetic theory it is useful to commence with a brief discussion of Maxwell's equations. For electrical machines Maxwell's equations are simplified to include only magnetic fields but additional quantities are included to express the interaction of the magnetic field with a mechanical system.

The geometric description of Maxwell's equations is in terms of a two-form F which is the electromagnetic field.

$$F = 1/2 \sum F_{\mu\upsilon} dx^{\mu} \Lambda\, dx^{\upsilon}$$

This two-form is defined in Minkowski space. Maxwell's equations can be succinctly described as

$$dF = 0$$
$$d{*}F = {*}J$$

where * denotes Hodge duality and J is the current one-form. In physics it is often assumed that J is given so that the equations are linear in the field. If F is exact then there is a one-form A called the potential such that

$$F = dA$$

If F is written as a skew symmetric matrix then the differential form description of Maxwell's equations becomes the following four equations

$$\text{div } H = 0$$

$$\frac{\partial H}{\partial t} + \text{curl } H = 0$$

$$\text{div } E = \rho$$

$$\text{curl } H - \frac{\partial E}{\partial t} = J$$

where H and E are the magnetic and the electric fields respectively and

$$J = \rho dt + \sum_{i=1}^{3} J_i \, dx^i$$

One approach to many equations in physics including Maxwell's equations is by a variational principle, the so-called action principle in physics. The typical action functional A is the integral of a Lagrangian

$$A = \int L \, dx \, dt$$

For an electromagnetic field this expression is

$$A = -\|F\|^2 = \mathrm{Tr} \int F \wedge *F$$

We shall derive the equations of an electrical machine with moving frames by such a variational principle. The space changes from the four dimensional Minkowski space to a manifold with a Riemannian metric. This manifold describes the electrical variables, the charges, and the mechanical variables which for the machines that will be considered here will be the shaft angle. The energy of the system will be the electromagnetic energy and the mechanical energy, the kinetic energy of the shaft. The usual assumptions on the regularity of the magnetic field will be made so that the electromagnetic energy is a quadratic form in the currents through the (effective) inductances. The metric for the manifold from which the energy is computed will be a direct sum of the inductances and the shaft moment of inertia. The inductances will be a function of shaft angle while the shaft moment of inertia will be fixed. Besides the magnetic energy and kinetic energy there are losses due to resistance and friction and there are inputs. Thus the total energy of the system satisfies the equation

$$\int \langle f, \dot{x}_t \rangle \, dt = 1/2 \int (g(\dot{x}_t, \dot{x}_t) + \langle R\dot{x}_t, \dot{x}_t \rangle) dt \qquad (1)$$

where f is the inputs, g is the metric and R is the diagonal form of the resistances and the friction.

__Lemma 1.__ The equations for an electric machine with moving frames on the rotor whose energy satisfies (1) are

$$\sum_j g_{ij} \frac{d^2x^j}{dt^2} + \sum_{jk\ell} g_{ij} \Gamma^j_{k\ell} \frac{dx^k}{dt} \frac{dx^\ell}{dt} + \sum_j R_{ij} \frac{dx^j}{dt} = f_i \qquad (2)$$

__Proof.__ Initially it will be verified that the curves that locally minimize the energy E

$$E = \int_a^b g(\dot{x}_t, \dot{x}_t) dt$$

are the geodesics with respect to the Riemannian connection defined from g. More succinctly the geodesics of the Riemannian connection will locally minimize energy. In differential geometry the minimization is usually of the arc length of the curves rather than the energy. However in both cases the geodesics of the Riemannian connection are the minimizing curves.

Let P be a frame bundle over M. The canonical form θ of P is the $\mathbb{R}^n$-valued 1-form on P defined by

$$\theta(X) = u^{-1}(\pi(X))$$

where $X \in T_uP$, $u \in \mathrm{Hom}\,(R^n, T_{\pi(u)}M)$ and $\pi: P \longrightarrow M$ is the projection [5].

The first structure equation is

$$d\theta = -\omega \wedge \theta + \Theta \tag{3}$$

where ω is the connection form and Θ is the torsion form. If the connection is Riemannian so that the torsion is zero the first structure equation is

$$d\theta = -\omega \wedge \theta \tag{4}$$

This equation will show that geodesics of the Riemannian connection defined from g locally minimize energy. A smooth rectangle in M is defined as a map α

$$\alpha: [a,b] \times [c,d] \longrightarrow M$$

that is smooth. By fixing one of the coordinates of the rectangle we obtain a curve in M. These curves will be described as x-curves or y-curves depending upon whether the second or the first coordinate of the rectangle is fixed. To prove that the geodesics of the Riemannian metric minimize energy the Gauss Lemma will be used. Its statement and a proof will follow. The approach follows that of Ambrose [1].

Let α be a smooth rectangle in M such that all x-curves are geodesics whose tangent vectors have the same length. It is claimed that if the x-tangent vector and the y-tangent vector of α are orthogonal at one point on an x-curve then they are orthogonal at all points. To verify the claim let $\tilde{\alpha}$ be the horizontal lift of α where the curve $\alpha(a, \cdot)$ and the curves $\alpha(\cdot, y)$ where $y \in [c,d]$ are horizontally lifted.

Let $X = \frac{\partial}{\partial x}$ and $Y = \frac{\partial}{\partial y}$ be the canonical vector fields on $\mathbb{R}^2$. The differential forms are pulled back via $\tilde{\alpha}$ to forms on $[a,b] \times [c,d]$. By abuse of notation these pull backs will be denoted by the same symbols as on M. Recall that for a smooth 1-form γ on a manifold we have

$$d\gamma(X,Y) = X\gamma(Y) - Y\gamma(Y) - Y\gamma(X) - \gamma([X,Y])$$

where X,Y are smooth vector fields on the manifold. Applying this result to the first structure equation (4) for a Riemannian connection we have

$$X\theta(Y) - Y\theta(X) = -\omega(X)\theta(Y) + \omega(Y)\theta(X)$$

where X and Y are the canonical vector fields on $\mathbb{R}^2$. Now return to $\tilde{\alpha}$. Since the x-curves are geodesics we have $\omega(X) = 0$. So the previous equation gives

$$X\theta(Y) - Y\theta(X) = \omega(Y)\theta(X)$$

Take the scalar product with $\theta(X)$

$$\langle\theta(X), X\theta(Y)\rangle - \langle\theta(X), Y\theta(X)\rangle = \langle\theta(X), \omega(Y)\theta(X)\rangle \qquad (5)$$

Since ω is skew symmetric

$$\langle\theta(X), \omega(Y)\theta(X)\rangle = 0$$

Now

$$X\langle\theta(X), \theta(Y)\rangle = \langle X\theta(X), \theta(Y)\rangle + \langle\theta(X), X\theta(Y)\rangle$$

Since the x-curves are geodesics, $\theta(X)$ is constant on integral curves of X so that $X\theta(X)=0$. Since $\|\theta(X)\|^2$ is constant because the tangent vectors of the x-curves all have the same length we have

$$\langle Y\theta(X), \theta(X)\rangle = 1/2\, Y\|\theta(X)\|^2 = 0$$

Thus the equation (5) yields

$$X\langle\theta(X), \theta(Y)\rangle = 0$$

or $\langle\theta(X), \theta(Y)\rangle$ is constant on integral curves of X. By the definition of $\tilde{\alpha}$ this proves the claim of the Gauss Lemma.

Now to show that geodesics locally minimize energy recall that the exponential map is locally a diffeomorphism. Let $m \in M$ and $\delta > 0$ such that the ball of radius $\delta, B(\delta)$, centered at the origin in $T_m M$ is mapped diffeomorphically to $C = \exp_m(B(\delta))$ in M. Let τ be a piecewise smooth curve in M from m to n, that is, $\tau:[a,b] \longrightarrow M$, $\tau(a) = m$, $\tau(b) = n$ where $n \in C$. Assume that τ remains in C. Otherwise restrict the subsequent construction to the first exit time of C. Let $\tau_R(u)$ and $\tau_N(u)$ be the radial and normal components of the tangent vector at $\tau(u)$ that are defined in $T_m M$ and mapped to $T_{\tau(u)} M$ by $D \exp_m$. The Gauss lemma shows that τ_R is orthogonal to τ_N so $\|\dot{\tau}(u)\|^2 = \|\tau_R(u)\|^2 + \|\tau_N(u)\|^2$. Comparing τ with the (suitably parameterized) geodesic σ to n we have

$$\|\dot{\sigma}(u)\|^2 = \|\tau_R(u)\|^2 \le \|\dot{\tau}(u)\|^2$$

Integrating the inequality gives the result that the energy is minimum for the geodesics.

For a curve to be a geodesic it means that the derivative vector field $\dot{x}_t$ is parallel along the curve. In terms of the covariant derivative this is

$$\nabla_X X = 0$$

where $X = \dot{x}_t$ or in local coordinates this is

$$\frac{d^2x^i}{dt^2} + \sum_{j,k} \Gamma^i_{j,k} \frac{dx^j}{dt} \frac{dx^k}{dt} = 0$$

However for the subsequent use we want to write the geodesic equation as a first variation of the energy. This follows immediately by considering an orthonormal frame at the initial point $m \in M$. Thus we have

$$\sum_j g_{ij} \frac{d^2x^j}{dt^2} + \sum_{j,k,\ell} g_{ij} \Gamma^j_{k\ell} \frac{dx^k}{dt} \frac{dx^\ell}{dt} = 0 \tag{6}$$

Now consider the equation (1) for the energy of an electrical machine. Since the first variation is the sum of the first variations of each of the energy terms we can restrict consideration to the losses and the inputs. Since R and f do not depend on the manifold M the first variation is obtained by pointwise differentiation. Combining these terms with (6) gives the equation (2). ∎

For the subsequent discussion it is useful to recall the following assumptions that are made on the machines: i) The rotor windings are symmetrically distributed around the smooth rotor, ii) the stator has field poles and asymmetrical windings on it and, iii) the variations of rotor self and mutual inductances with the rotor (shaft) angle are sinusoidal functions.

Machines that satisfy these assumptions can be described by a primitive machine. This idea was emphasized by Kron [7,8] though the ideas for the model can be traced to earlier work on electrical machines. A recent account is given in [9]. The simplest version of it is if there is one (field) winding on the stator and one collection of windings on the rotor (armature). The rotor conductors can be modelled by an inductor that is connected between the brushes and the field can be replaced by an inductance in the "direct" axis. Thus we have the diagram

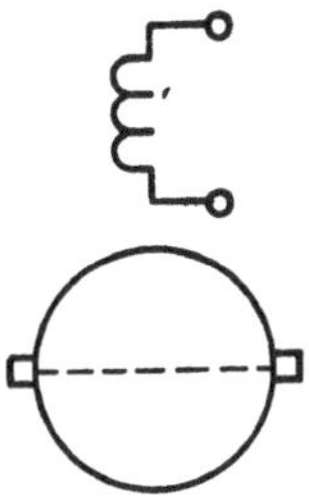

Fig. 1

The antipodal squares on the circle represent the brushes and the dotted line connecting them represents the equivalent armature inductance that is measured at the brushes. More generally one can imagine stator windings in the direct and the quadrature axes and then inductances on the rotor in the direct and the quadrature axes. This 4-coil primitive machine is represented as

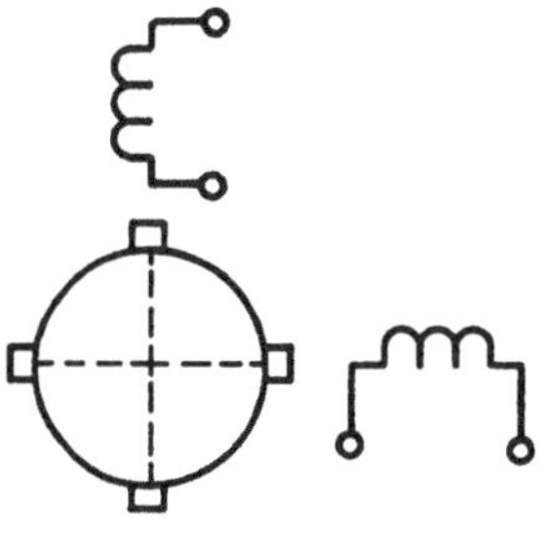

Fig. 2

Kron has shown that all well known machines can be obtained from this primitive machine by adding or subtracting coils along the two axes. For our purposes an electrical machine will be the mathematical data that is necessary to define a model induced from this primitive machine. These data will be the metric which is a direct sum of the inductances and the shaft moment of inertia, a diagonal form that represents the resistances in the windings and the friction in the shaft, and the torque and the voltage inputs to the machine.

Since all machines do not have moving frames on the rotor it is important to understand the effect of other frames on the equations (2) for a machine. For a geometric perspective of this effect some notions from geometry will be recalled.

A symplectic structure on a smooth manifold M is a closed two-form on TM. In the study of variational problems it is natural to preserve the symplectic structure. For example the linear Hamiltonian equations are in the Lie algebra of the symplectic group, the subgroup of the general linear group that preserves the canonical symplectic form J on $\mathbb{R}^{2n}$. For functorial reasons the closed two-form Ω is often put on T*M rather than on TM. More generally a manifold M is said to be symplectic if there is a closed nondegenerate two-form Ω defined on M. If M is finite dimensional then it is even dimensional. Locally Ω can be described by coordinates $q_1, \ldots, q_n, p_1, \ldots, p_n$ as

$$\Omega = \sum dp_i \wedge dq_i$$

The Hamilton-Jacobi equation arises by requiring that the symplectic structure is preserved and that the p variables are expressed in terms of the q variables.

If there is a transformation that preserves the symplectic structure then the variational problem has the same solution expressed in different coordinates. In particular the minimization of energy that was computed before is still the geodesics of the Riemannian connection determined from the metric that is given by the inductances and the moment of inertia. However if a transformation does not preserve the symplectic structure then it is necessary to recompute the equations which describe the curves that minimize the energy integral.

For electrical machines there is one common type of transformation that does not preserve the symplectic structure. It is a transformation from the moving frames given by slip rings on the rotor to fixed frames given by a commutator on the rotor. For a single loop these two measuring devices are indicated in the following diagrams along with the graphs of the voltage that is measured at the brushes.

It will be shown that machines with fixed frames satisfy equations that are different from (2). The most elementary transformation from moving frames in the primitive machine to fixed frames is given by the orthogonal matrix

$$\begin{bmatrix} \cos\phi & \sin\phi \\ -\sin\phi & \cos\phi \end{bmatrix} \tag{7}$$

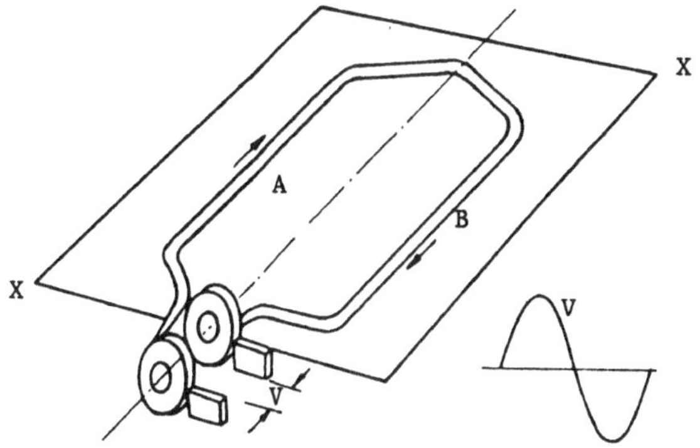

Fig. 3

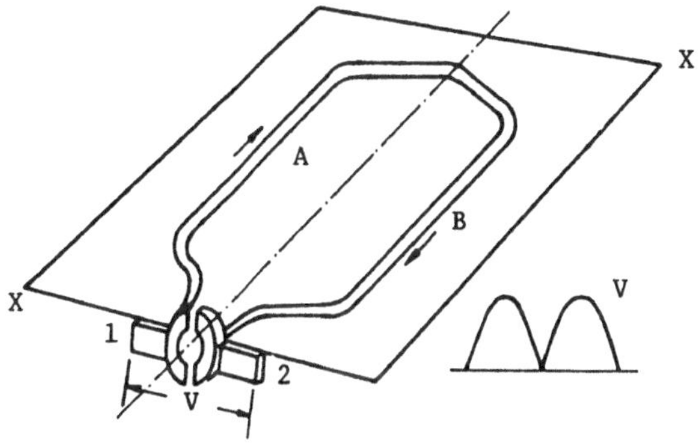

Fig. 4

where ϕ is the shaft angle. This transformation gives the fixed frame 1-forms in terms of the moving frame 1-forms. Computing the first structure equation for these fixed frame 1-forms

$$d\dot{\theta} = -\omega \wedge \theta + \Theta \qquad (8)$$

it is easily verified that the torsion is nonzero. The equations for this fixed frame machine can be obtained from the equations (2) by using the transformation from moving frames to fixed frames. The equations for the fixed frame machine are formally the same where the Riemannian connection is replaced by a metric connection with torsion. Thus the connection is again obtained from the metric g but in this case it has torsion. The torsion can be obtained from the first structure equation (8).

To study the topology of electrical machines it is necessary to impose some natural restrictions on the systems. For the topology of linear systems the McMillan degree is fixed. Reciprocal linear electrical networks have symmetric transfer functions. The number of topological components of the space of symmetric transfer functions with fixed McMillan degree and fixed input and output spaces can be counted by the Cauchy-Maslov index [4]. A homotopy between transfer functions of the same Cauchy-Maslov index can be constructed. By the covering homotopy theorem a homotopy of the base space of transfer functions can be lifted to the principal bundle of state space realizations. For the topology of machines the input and the output spaces will be fixed and isomorphic. In fact these will usually be $\mathbb{R}^1$. Since the equations for a machine depend only on the electrical variables through the currents and not the charges, a machine can be described geometrically as a vector bundle over the circle S^1. For the topology of machines these vector bundles over S^1 will have fixed rank, the linear systems in the fibers of the bundle will be reachable and observable and the one dimensional system in the base manifold will be locally reachable and observable. The linear systems in the fibers will have fixed McMillan degree and fixed Cauchy-Maslov index [4]. For the topology of electrical machines the McMillan degree and the Cauchy-Maslov index can be considered as local invariants because they are properties of the fibers of the vector bundle.

Consider the primitive machine in Fig. 1. Assume that machine A is this primitive machine with slip rings on the rotor that provide moving frames and that machine B is this machine with a commutator on the rotor that provides fixed frames. Consider inputs to the stator and rotor coils that are the same for the two machines and that give reachability and observability of the linear systems in the fibers.

<u>Proposition 2</u>. Machines A and B are not homotopic.

<u>Proof</u>. Note that there is no local obstruction to the homotopy because the linear systems for the two machines have the same McMillan degree and Cauchy-Maslov index. The transformation from the current 1-forms of the rotor for A to the current 1-forms of the rotor for B is given by (7). Since no function relates these 1-forms for A to the 1-forms for B, they are not in the same deRham cohomology class. This fact will verify the result. This property can also be verified more geometrically by recalling the figures that give one loop of wire with either slip rings or commutator. Cutting the circle in half to have trivial vector bundles it is seen that the graphs of the voltages in the latter diagram represents a twist when the two semicircles are put together. This fact will also verify the result. ∎

Let C be the machine with only a field in the direct axis of the stator and the armature is represented by one coil on the rotor. This machine is given by the primitive machine in Fig. 1. Assume that moving frames on the rotor are used so that it has slip rings. A very elementary family of linear systems will be connected to this machine. The input voltage will be applied to the stator inductor. Positive and negative inductances will be connected to the stator and rotor and additional mutual inductance may be used so that the McMillan degree is two and the Cauchy index is zero for the linear systems in the fibers. Positive and negative resistors can also be connected to the machine. This family of systems will be denoted as Ξ.

__Theorem 3__. The family of systems Ξ is not contractible.

__Proof__. This result will be verified by constructing two systems in Ξ that are not homotopic.

A brief description will be given of the technique of the proof. Let S_1^1 be the circle that describes the rotor angle and let S_2^1 be the circle that is obtained by restricting the Laplace transform variable to the real line in the complex plane. By taking the Cayley map of a scalar transfer function, this transfer function defines a map $S_2^1 \longrightarrow S^1$. For a machine these transfer functions are in the fibers of the vector bundle so an element of Ξ defines a continuous map $S_1^1 \times S_2^1 \longrightarrow S^1$ by applying the aforementioned procedure for transfer functions to the linear systems in the fibers. Likewise a homotopy between two elements in Ξ defines a map $S_1^1 \times S_2^1 \times [0,1] \longrightarrow S^1$ which is jointly continuous. The joint continuity arises basically because the McMillan degree and the Cauchy index are fixed. This map induces two maps $S_1^1 \times [0,1] \longrightarrow S^1$ and $S_2^1 \times [0,1] \longrightarrow S^1$ by keeping the other variable in the domain fixed. Now consider two elements of Ξ. These two elements define maps $f_i\colon S_1^1 \times S_2^1 \longrightarrow S^1$ $i=1,2$ by the above procedure. There will be an obstruction to the homotopy between these two systems if the degrees of the maps

$$f_i(s_1, \cdot)\colon S_2^1 \longrightarrow S^1 \qquad i=1,2$$

or

$$f_i\ (\cdot, s_2)\ :\ S_1^1 \longrightarrow S^1 \qquad i=1,2$$

are not equal. For the first maps this is merely the requirement that the Cauchy indices are the same. The following construction will exhibit two elements of Ξ that have different degrees for the second maps.

The inductance of the machine is given by the metric

$$\begin{bmatrix} L_1 & M_{12}\cos\phi \\ M_{12}\cos\phi & L_2 \end{bmatrix}$$

where $\phi \in S^1$ and the resistances of the machine are given by the diagonal form

$$\begin{bmatrix} R_1 & 0 \\ 0 & R_2 \end{bmatrix}$$

The resistor $\tilde{R}_1$, and the inductor $\tilde{L}_1$, will be inserted in series with R_1 and L_1 in the rotor circuit and similarly $\tilde{R}_2$ and $\tilde{L}_2$ will be inserted in series with R_2 and L_2. In addition a resistor R_3 will be in series in the stator circuit along with the voltage input and R_3 will also be connected across the rotor circuit. These quantities will be chosen so that the McMillan degree is two and the Cauchy index is zero for the linear systems in the fibers. The transfer function in the fibers is

$$\frac{I}{E} = \frac{a_1 s_1 + a_0}{b_2 s^2 + b_1 s + b_0}$$

where

$$a_1 = L_1 + \tilde{L}_1$$

$$a_0 = R_1 + \tilde{R}_1 + R_3$$

$$b_2 = -M_{12}^2 \cos^2 \phi + (L_1 + \tilde{L}_1)\ (L_2 + \tilde{L}_2)$$

$$b_1 = -2R_3\ M_{12} \cos \phi + (L_2 + \tilde{L}_2)\ (R_2 + \tilde{R}_2 + R_3) + (L_1 + \tilde{L}_1)\ (R_2 + \tilde{R}_2 + R_3)$$

$$b_3 = (R_2 + \tilde{R}_2 + R_3)\ (R_1 + \tilde{R}_1 + R_3) - R_3^2$$

A description will be given of the specifications of $\tilde{L}_1$, $\tilde{L}_2$, $\tilde{R}_1$, $\tilde{R}_2$, R_3. Choose $\tilde{L}_1$ and $\tilde{L}_2$ so that $(L_1 + \tilde{L}_1)$ and $(L_2 + \tilde{L}_2)$ are small in absolute value and that b_2 is always negative. Choose $\tilde{R}_1$, $\tilde{R}_2$, R_3, $\tilde{L}_1$, and $\tilde{L}_2$ in a neighborhood of $\phi = \frac{\pi}{4}$ and $\phi = \frac{3\pi}{4}$ so that $b_1^2 - 4\, b_0 b_2 < 0$ and in the neighborhood of $\phi = \frac{\pi}{4}$

$$L_1 + \tilde{L}_1 = \delta_1 \cos (\phi + \frac{\pi}{4})$$

$$R_1 + \tilde{R}_1 + R_3 = \delta_2 \sin (\phi + \frac{\pi}{4})$$

where sgn δ_1 = sgn δ_2. Let δ_2 be fixed sign and nonzero only in this neighborhood of $\frac{\pi}{4}$. In the neighborhood of $\frac{3\pi}{4}$

$$L_1 + \tilde{L}_1 = \varepsilon_1 \cos (\phi + \frac{\pi}{4})$$

$$R_1 + \tilde{R}_2 + R_3 = \varepsilon_2 \sin (\phi + \frac{\pi}{4})$$

where sgn ε_1 = sgn δ_1 = -sgn ε_2 and ε_2 has fixed sign and is nonzero only in this neighborhood of $\frac{3\pi}{4}$.

Globalize the choice of these parameters so that the McMillan degree and the Cauchy index remain fixed and that $(L_1 + \tilde{L}_1)$ does not change sign again.

The transfer function as a map $S_1^1 \longrightarrow S^1$ has winding number two. By choosing appropriate constant values for $\tilde{L}_1$, $\tilde{L}_2$, $\tilde{R}_1$, $\tilde{R}_2$, R_3 it is elementary to construct an example where the transfer function as a map $S_1^1 \longrightarrow S^1$ has winding number zero. ∎

It should be noted that these nonlinear systems obtained from electrical machines can be generalized to a family of nonlinear systems that are described geometrically by a vector bundle. The fibers contain linear systems which provide local invariants for the systems while the piecing together of these linear systems provides global invariants.

REFERENCES

1. W. Ambrose, The Cartan structural equations in classical Riemannian geometry, J. Indian Math. Soc. 24 (1960), 23-76.

2. M.F. Atiyah and R. Bott, On the periodicity theorem for complex vector bundles, Acta Math., 112 (1964), 229-247.

3. R. Bott, The stable homotopy of the classical groups, Ann. of Math. 70 (1959), 313-337

4. C.I. Byrnes and T.E. Duncan, On certain topological invariants arising in system theory, to appear in New Directions in Applied Mathematics (P. Hilton and G.S. Young, eds.) Springer-Verlag, 1981.

5. S. Kobayashi and K. Nomizu, Foundations of Differential Geometry I, Interscience, New York, 1963.

6. K. Kondo and Y. Ishizuka, Recapitulation of the geometrical aspects of Gabriel Kron's non-Riemannian electrodynamics, Memoirs of the Unifying Study of the Basic Problems in Engineering Sciences by Means of Geometry, I, 185-239, Tokyo, 1955.

7. G. Kron, Non-Riemannian dynamics for rotating electrical machinery, J. Math. Phys. 13(1934), 103-195.

8. G. Kron, The Application of Tensors to the Analysis of Rotating Electrical Machinery, General Electric Review, Schenectady, NY, 1938.

9. D.P. Sen Gupta and J.W. Lynn, Electrical Machine Dynamics, Macmillan, London, 1980.

FINITE-DIMENSIONAL OBSERVATION-SPACES FOR NON-LINEAR SYSTEMS'

Michel FLIESS

Laboratoire des Signaux et Systèmes
C.N.R.S. - E.S.E.
Plateau du Moulon
91190 Gif-sur-Yvette, France (mailing address)

and

Université Paris VIII
2, rue de la Liberté
93526 Saint-Denis Cédex 02, France.

Sommario. Un sistema non-lineare ha lo stesso comportamento entrata-uscita di un sistema regolare (o bilineare) se, e soltante se, un certo spàzio funzionale, detto spàzio d'osservazione, é di dimensione finita.

Zusammenfassung. Ein nichtlineares System besitzt das selbe Eingang-Ausgang Verhalten wie ein reguläres (oder bilineares) System genau dann, wenn ein gewisser Funktionenraum, gennant Beobachtungsraum, endliche Dimension hat.

Résumé. Un système non linéaire a le même comportement entrée-sortie qu'un système régulier (ou bilinéaire) si, et seulement si, un certain espace fonctionnel, dit espace d'observation, est de dimension finie.

Abstract. A non-linear system has the same input-output behaviour as a regular (or bilinear) system if, and only if, a certain function-space, called the observation-space, is finite-dimensional.

Introduction.

In this note we give a necessary and sufficient condition such that a linear analytic system has the same input-output behaviour as a regular (or bilinear) system. This problem has already been studied by Krener [4], Lo [5] and Hijab [3]. The solution we give is in fact the continuous-time analogue of a result which was already obtained by Sontag [6] for discrete-time systems.

Our proof deals with non-commutative generating power series [1]. In [2] a more general approach is proposed which includes C^∞ non-linear systems.

I. - Main definitions and statement of the result.

Consider the linear analytic system (Σ)

$$\begin{cases} \dot{q}(t) \ (= dq/dt) = A_0(q) + \sum_{i=1}^{n} u_i(t)\, A_i(q) \\ y(t) = h(q) \end{cases}$$

The state q belongs to a connected real analytic manifold Q. The vector fields $A_o, A_1, \dots, A_n$ and the function $h : Q \to \underline{R}$ are analytic. The controls $u_1, \dots, u_n$ are real-valued and piecewise continuous.

Call (Σ_{q_o}) the system (Σ) when it is initialized at an arbitrary state $q_o \in Q$. It is known that the input-output behaviour of (Σ_{q_o}) is characterized by the non-commutative generating power series

$$\underline{\underline{g}}_{q_o} = h|_{q_o} + \sum_{\nu \geq o} \sum_{j_o,\dots,j_\nu = o}^{n} A_{j_o} \dots A_{j_\nu} h|_{q_o} \; x_{j_\nu} \dots x_{j_o}$$

(the bar $|_{q_o}$ indicates the evaluation at q_o).

Take another linear analytic system (Σ') with state-space Q' and the same controls u_i as (Σ) :

$$\begin{cases} \dot{q}'(t) = A'_o(q') + \sum_{i=1}^{n} u_i(t)\, A'_i(q') \\ y'(t) = h'(q') \end{cases}$$

A <u>subordination</u> $\tau : (\Sigma) \to (\Sigma')$ is an analytic map $\tau : Q \to Q'$ such that at the points $q \in Q$ and $q' = \tau(q) \in Q'$ the systems (Σ_q) and $(\Sigma'_{q'})$ have the same generating series.

<u>Remark</u>. The purpose of this definition is to say that systems (Σ_q) and $(\Sigma'_{q'})$ have the same input-output behaviour. By dealing with generating series we have not to worry about possible difficulties with the manifolds or the vector fields (see [2] for an approach which is independant of the generating series).

A regular (or bilinear) system (B) is of the form

$$\begin{cases} \dot{\eta}(t) = (M_o + \sum_{i=1}^{n} u_i(t)\, M_i)\; \eta(t) \\ y(t) = \lambda\, \eta(t) \end{cases}$$

The state η belongs to a d-dimensional $\underline{R}$-vector space E. The maps $M_o, M_1, \dots, M_n$: $E \to E$ and $\lambda : E \to \underline{R}$ are R-linear.

In order to study possible subordinations $(\Sigma) \to (B)$, we have to introduce the R-vector space spanned by the function h and the iterated Lie derivatives $A_{j_o} \dots A_{j_\nu} h$ ($\nu \geq o$; $j_o, \dots, j_\nu = 0, 1, \dots, n$). Following Sontag [6], we will call this vector space, which is in general infinite-dimensional, the observation-space of system (Σ).

Theorem. A linear analytic system is subordinated to a regular (or bilinear) system if, and only if, its observation-space is finite-dimensional.

II. - Proof.

α) Necessity. Suppose there exists a subordination $\tau : (\Sigma) \to (B)$, i.e., a map $\tau : Q \to E$. Then at the points $q \in Q$ and $\eta = \tau(q) \in E$, the generating series of system (Σ_q)

$$h|_q + \sum_{\nu \geq o} \sum_{j_o,\ldots,j_\nu = o}^{n} A_{j_o} \ldots A_{j_\nu} h|_q \; x_{j_\nu} \ldots x_{j_o} ,$$

and the generating series of system (B_η)

$$\lambda\eta + \sum_{\nu \geq o} \sum_{j_o,\ldots j_\nu = o}^{n} \lambda M_{j_\nu} \ldots M_{j_o} \eta \; x_{j_\nu} \ldots x_{j_o} \tag{1}$$

are equal. But the coefficients of the rational series (1) obviously generate a finite-dimensional vector-space. This proves that the observation-space of system (Σ) is also finite-dimensional.

β) Sufficiency. Suppose the observation-space $\mathcal{S}$ of system (Σ) is of finite dimension d.

Let X^* be the free monoid generated by the letters $\{x_o, x_1, \ldots, x_n\}$(1). For any word $w \in X^*$, we can define a right canonical action on $\mathcal{S}$ in the following way :

- if $w = 1$, this action is the identity;
- if $w = x_{k_\mu} \ldots x_{k_o}$, the action on $A_{j_o} \ldots A_{j_\nu} h \in \mathcal{S}$ is given by

$$A_{j_o} \ldots A_{j_\nu} h . \; x_{k_\mu} \ldots x_{k_o} = A_{k_o} \ldots A_{k_\mu} A_{j_o} \ldots A_{j_\nu} h .$$

It yields a left canonical action of X^* on the dual ${}^t\mathcal{S}$ of $\mathcal{S}$, and defines a representation $\alpha : X^* \to \mathrm{End}\,({}^t\mathcal{S})$ of X^* by $\underline{R}$-linear endomorphims of ${}^t\mathcal{S}$.

---

(1)The elements of X^* are sequences $x_{j_\nu} \ldots x_{j_o}$, called words. The product of two words $x_{j_\nu} \ldots x_{j_o}$ and $x_{k_\mu} \ldots x_{k_o}$ is defined by concatenation :

$$(x_{j_\nu} \ldots x_{j_o})(x_{k_\mu} \ldots x_{k_o}) = x_{j_\nu} \ldots x_{j_o} x_{k_\mu} \ldots x_{k_o} .$$

The neutral element, called the empty word, is noted 1.

Introduce now the map $\tau : Q \to {}^t\mathcal{S}$, $q \mapsto \eta_q$, where, for $A_{j_0} \ldots A_{j_\nu} h \in \mathcal{S}$, η_q is defined by

$$\eta_q(A_{j_0} \ldots A_{j_\nu} h) = A_{j_0} \ldots A_{j_\nu} h|_q \quad .$$

Take for λ the canonical image of the function $h \in \mathcal{S}$ in the double dual ${}^{tt}\mathcal{S}$ Then the map τ subordinates system $(\sum)$ to the following regular system with state-space ${}^t\mathcal{S}$

$$\begin{cases} \dot{\eta}(t) = (\alpha(x_0) + \sum_{i=1}^{n} u_i(t)\, \alpha(x_i))\, \eta(t) \\ y(t) = \lambda\, \eta\, (t) \end{cases} \qquad (B_{\eta_q})$$

As a matter of fact, the generating series of system (B_{η_q}) is given by

$$\lambda \eta_q + \sum_{\nu \geq 0} \sum_{j_0, \ldots, j_\nu = 0}^{n} \lambda\, \alpha(x_{j_\nu}) \ldots \alpha(x_{j_0})\, \eta_q\, x_{j_\nu} \ldots x_{j_0}$$

which, by construction, is equal to

$$h|_q + \sum_{\nu \geq 0} \sum_{j_0, \ldots, j_\nu = 0}^{n} A_{j_0} \ldots A_{j_\nu} h|_q \;\; x_{j_\nu} \ldots x_{j_0} .$$

Remark. If we know a basis of the vector space $\mathcal{S}$, this proof gives an explicit construction of the regular system and of the subordination map.

III. - Two simple examples.

(i) Consider the one-dimensional system on the real line

$$\begin{cases} \dot{q}(t) = u_1(t) \\ y(t) = h(q) \end{cases}$$

The observation-space is spanned by $\{\frac{d^\nu h}{dq^\nu} \mid \nu \geq o\}$. It is finite-dimensional iff the function h is an exponential polynomial.

(ii) The following example has its origin in statistical physics (cf. Suzuki, Kaneko and Sasagawa [7]). The state-space is $\underline{R} - \{0\}$:

$$\begin{cases} \dot{q}(t) = aq - bq^\rho + u_1(t)\, q \quad (\rho \geq 2) \\ y(t) = 1/q^{\rho - 1} . \end{cases}$$

Notice that the Lie algebra generated by the corresponding two vector fields

$A_0 = (aq - b\, q^\rho) \frac{d}{dq}$ and $A_1 = q \frac{d}{dq}$ is two-dimensional, since, with the basis given by A_1 and $B = q^\rho \frac{d}{dq}$, we get $A_0 = a\, A_1 - bB$, $[A_0, A_1] = A_1 A_0 - A_0 A_1 = (1-\rho)\, b\, q^\rho \frac{d}{dq} = (1-\rho)\, bB$.

Consider the map $\tau : \underline{R}-\{0\} \to \underline{R}^2$, $q \mapsto \begin{bmatrix} 1 \\ 1/q^{\rho-1} \end{bmatrix}$. We get

$$A_0 \begin{bmatrix} 1 \\ 1/q^{\rho-1} \end{bmatrix} = \begin{bmatrix} 0 \\ b(\rho-1) + a(1-\rho)/q^{\rho-1} \end{bmatrix}$$

$$= \begin{bmatrix} 0 & 0 \\ b(\rho-1) & a(1-\rho) \end{bmatrix} \begin{bmatrix} 1 \\ 1/q^{\rho-1} \end{bmatrix} ,$$

$$A_1 \begin{bmatrix} 1 \\ 1/q^{\rho-1} \end{bmatrix} = \begin{bmatrix} 0 \\ (1-\rho)/q^{\rho-1} \end{bmatrix} = \begin{bmatrix} 0 & 0 \\ 0 & 1-\rho \end{bmatrix} \begin{bmatrix} 1 \\ 1/q^{\rho-1} \end{bmatrix} .$$

We obtain the following regular system to which our original system is subordinated by the map τ :

$$\begin{cases} \dot\eta_1(t) = 0 \\ \dot\eta_2(t) = b(\rho-1)\, \eta_1 + (1-\rho)\, (a + u_1(t))\eta_2 \\ y(t) = \eta_2 \end{cases}$$

References.

[1] M. Fliess, Fonctionnelles causales non linéaires et indéterminées non commutatives, Bull. Soc. Math. France, 109, 1981, pp.3-40.

[2] M. Fliess and I. Kupka, A finiteness criterion for nonlinear input-output differential systems, submitted for publication.

[3] O.B. Hijab, Minimum energy estimation, Ph. D. Thesis Math., University of California, Berkeley, 1980.

[4] A.J. Krener, Bilinear and nonlinear realizations of input-output maps, SIAM J. Contr., 13, 1975, pp.827-834.

[5] J.T. Lo, Global bilinearization of systems with controls appearing linearly, SIAM J. Contr., 13, 1975, pp. 879-885.

[6] E.D. Sontag, Polynomial response maps, Lect. Notes Contr. Inform. Sci. 13, Springer-Verlag, Berlin, 1979.

[7] M. Suzuki, K. Kaneko and F. Sasagawa, Phase transition and slowing down in non-equilibrium stochastic processes, Prog. Theoret. Physics, 65, 1981, pp. 828-849.

POLYNOMIAL MODELS AND ALGEBRAIC STABILITY CRITERIA

Paul A. Fuhrmann
Department of Mathematics
Ben Gurion University of the Negev
Beer Sheva, Israel

1. Introduction

The problem of finding algebraic stability criteria is one of the first problems to be solved in control theory. Its roots go back to the 19th century to the work of Hermite [1856], Routh [1877] and Hurwitz [1895] and with later important contributions by Liapunov [1893], Schur [1918], Cohn [1922], Lienard and Chipart [1914], Fujiwara [1926] and Kalman [1969] to mention some. The method of Liapunov, via the matrix equation named after him, and Hermite's method of quadratic forms are known to be closely related. This relation has been pointed out in various papers e.g. Parks [1962], Kalman [1969], Muller [1977] and Datta [1978] to mention some. A comprehensive survey can be found in Gantmacher [1959]. Most of the papers relating the Liapunov method and the method of quadratic forms do so mainly by complex matrix manipulation wh which do not make for easy reading nor facilitate understanding.

The object of this paper is to apply the method of polynomial models to this circle of ideas with the hope of establishing a better contact with modern algebraic system theory. In so doing it is hoped that the exposition becomes more streamlined and the conceptual basis of some of the results is more clearly emphasized.

2. Polynomial Models

Since this paper's theme is classical stability criterias we restrict ourselves to polynomial models based on scalar polynomials. For the general theory of polynomial models the reader is referred to Fuhrmann [1976,1977,1979,1981], Fuhrmann and Willems [1980], Emre [1980], Emre and Hautus [1980] and Khargonekar and Emre [1981].

In what follows F denotes an arbitrary field, to be identified later with the real number field. By $F[z]$ we denote the ring of polynomials over F, $F((z^{-1}))$ the set of truncated Laurent series in z^{-1} and by $F[[z^{-1}]]$ and $z^{-1}F[[z^{-1}]]$ the set of all formal power series in z^{-1} and the set of those power series with vanishing constant term respectively. Let π_+ and π be the projections of $F((z^{-1}))$ onto $F[z]$ and $z^{-1}F[[z^{-1}]]$ respectively.

Since $F((z^{-1})) = F[z] \oplus z^{-1}F[[z^{-1}]]$ they are complementary projections. Also $z^{-1}F[[z^{-1}]]$ is isomorphic to $F((z^{-1}))/F[z]$ which is an $F[z]$-module with the module action given by

(2.1) $$z\cdot h = S_- h = \pi_- zh\ .$$

Similarly we define

(2.2) $$S_+ f = zf \quad \text{for} \quad f \in F[z]\ .$$

We define projections π_d and π^d in $F[z]$ and $z^{-1}F[[z^{-1}]]$ respectively by

(2.3) $\qquad \pi_d f = d\pi_- d^{-1} f \quad \text{for} \quad f \in X_d$

and

(2.4) $\qquad \pi^d h = \pi_- d^{-1}\pi_+ dh \quad \text{for} \quad h \in z^{-1}F[[z^{-1}]]$

and corresponding subspaces by

(2.5) $\qquad X_d = \text{Range } \pi_d$

(2.6) $\qquad X^d = \text{Range } \pi^d$,

The maps S_d in X_d and S^d are defined by

(2.7) $\qquad S_d f = \pi_d S_+ f \quad \text{for} \quad f \in X_d$

and

(2.8) $\qquad S^d h = S_- h \quad \text{for} \quad h \in X^d$.

A map Z in X_d commutes with S_d if and only if $Z = p(S_d)$ for some polynomial $p \in F[z]$ and $p(S_d)$ is invertible if and only if p and d are coprime.

We define a pairing of elements of $F((z^{-1}))$ by letting, for $f(z) = \Sigma f_j z^j$ and $g(z) = \Sigma g_j z^j$

(2.9) $\qquad [f,g] = \Sigma f_{-j-1} g_j$.

Clearly, since both series are truncated, the sum in (2.9) is well defined. In terms of this pairing we can make the following identification, Fuhrmann [1981]. The dual of $F[z]$ as a linear space is $z^{-1}F[[z^{-1}]]$. Now, given a nonzero polynomial q the module X_q is isomorphic to $F[z]/qF[z]$. If we define, for a subset M of $F((z^{-1}))$, $M^\perp$ by

(2.10) $\qquad M^\perp = \{g \in F((z^{-1})) \mid [f,g] = 0 \quad \text{for all} \quad f \in M\}$

then in particular $F[z]^\perp = F[z]$ and $(qF[z])^\perp = X_q$. Since, in general $(X/M)^* = M^\perp$ we have

$$X_q^* = (F[z]/qF[z])^* = [qF[z]]^\perp = Xq .$$

But in turn we have $X^q = X_q^*$ and so X_q^* can be identified with X_q. This can be made more concrete through the use of the bilinear form

(2.11) $\qquad \langle f,g \rangle = [q^{-1}f,g]$.

Relative to this bilinear form we have the important relation

(2.12) $\qquad S_q^* = S_q$.

For more details and the multivariable case the reader is referred to Fuhrmann [1981].

Let X be a finite dimensional vector space and X^* its dual space. Let $\{e_1,\ldots,e_n\}$ be a basis for X then the set of vectors $\{f_1,\ldots,f_n\}$ in X^* is called the *dual basis* if

(2.13) $\qquad \langle e_i,f_j \rangle = \delta_{ij} \qquad 1 \leq i, j \leq n$.

Given $q \in F[z]$ with $q(z) = z^n + q_{n-1}z^{n-1} + \ldots + q_0$ then the elements of X_q are all polynomials of degree $\leq n-1$. In particular the subset of X_q given by $B_0 = \{f_1,\ldots,f_n\}$ where

(2.14) $$f_i(z) = z^{i-1} \qquad i=1,\dots,n$$

is a basis for X_q. Since

(2.15) $$S_q z^i = \begin{cases} z^{i+1} & 0 \leq i < n-1 \\ -(q_0 + \dots + q_{n-1}z^{n-1}) & \text{if } i = n-1 \end{cases}.$$

the matrix representation of S_q relative to the basis $\mathcal{B}_0$ is

(2.16) $$\begin{pmatrix} 0 & . & . & . & -q_0 \\ 1 & & & & . \\ & . & & & . \\ & & . & & . \\ & & & 1 & -q_{n-1} \end{pmatrix}$$

i.e., it is the *companion matrix* of q.

It is of considerable interest to characterize the elements of the dual basis. Since we have the identification $X_q^* = X_q$ under the pairing (2.9) the dual basis elements are also polynomials of degree $< n$. Given the polynomial q as above, we define

(2.17) $$e_i(z) = \pi_+ z^{-i} q = q_i + q_{i+1}z + \dots + z^{n-i}$$

$i=1,\dots,n.$

Theorem 2.1: The set $\mathcal{B}_c = \{e_1,\dots,e_n\}$ is the dual basis to the basis $\mathcal{B}_0 = \{1,z,\dots,z^{n-1}\}$.

Proof: We have

$$\begin{aligned} < e_i, f_j > &= [q^{-1}e_i, f_j] = [q^{-1}\pi_+ z^{-i}q, z^{j-1}] = [\pi_- q^{-1}\pi_+ q z^{-i}, z^{j-1}] \\ &= [\pi^q z^{-i}, z^{j-1}] = [z^{-i}, \pi_q z^{j-1}] = [z^{-i}, z^{j-1}] = \delta_{ij} \end{aligned}$$

It follows from the general study of duality that the matrix representation of S_q with respect to the basis $\mathcal{B}_c$ is

(2.18) $$[S_q]_{\mathcal{B}_c}^{\mathcal{B}_c} = \begin{pmatrix} 0 & 1 & & & \\ & & . & & \\ & & & . & \\ & & & & 1 \\ -q_0 & . & . & . & -q_{n-1} \end{pmatrix}$$

which is the control form. For this reason $\mathcal{B}_c$ is referred to as the *control basis*. Of course the matrix representation (2.18) can be verified directly by observing that

(2.19) $$S_q e_i = \begin{cases} e_{i-1} - q_{j-1}e_n & \text{if } i=2,\dots,n \\ -q_0 e_n & i = 1 \end{cases}.$$

As it is obvious that

(2.20) $$IS_q = S_q I$$

the next corollary follows trivially,

Corollary 2.2: If $\mathcal{B}_o$, $\mathcal{B}_c$ are the standard and control bases of X_q then

$$[I]_{\mathcal{B}_c}^{\mathcal{B}_o}[S_q]_{\mathcal{B}_c}^{\mathcal{B}_c} = [S_q]_{\mathcal{B}_o}^{\mathcal{B}_o}[I]_{\mathcal{B}_c}^{\mathcal{B}_o} . \tag{2.21}$$

But

$$[I]_{\mathcal{B}_c}^{\mathcal{B}_o} = \begin{pmatrix} q_1 & . & . & . & q_{n-1} & 1 \\ . & & & . & . & \\ . & & . & . & & \\ q_{n-1} & . & & & & \\ 1 & & & & & \end{pmatrix} \tag{2.22}$$

so it follows that this Hankel matrix intertwines the companion matrices (2.16) and (2.18). This result appears in Taussky [197], Barnett [197], Gohberg et al. [1978] and Kailath [1980].

3. Bezoutians

Given two polynomials p and q and two, not necessarily commuting, variables z and w we can write

$$p(z)q(w)-q(z)p(w) = \Sigma b_{ij}z^{i-1}(z-w)w^{j-1} \tag{3.1}$$

holds where (b_{ij}) is a symmetric matrix. The matrix $B(p,q) = (b_{ij})$ is called the *Bezoutian* of the polynomials p and q. A proof of this fact and some interesting related results can be found in Kravitsky [1980]. A multivariable Bezoutian is introduced in Anderson and Jury [1976]. If the variables z and w commute we can rewrite (3.1) as

$$(p(z)q(w)-p(w)q(z))/(z-w) = \Sigma b_{ij}z^{i-1}w^{j-1} \tag{3.2}$$

but we prefer to make (3.1) the starting point of the following discussion.

We assume now that $\deg p \leq \deg q$ and choose a monic $r \in \mathbf{F}[z]$ such that $\deg r = \deg q$ and r and q are coprime. This is possible, clearly so if F is an infinite field. Furthermore let us substitute in (3.1) $z = S_r$ and $w = S_q$. Then, since $q(S_q) = 0$, it follows that

$$-q(S_r)p(S_q) = \Sigma b_{ij}S_r^{i-1}(S_r-S_q)S_q^{j-1} . \tag{3.3}$$

Now S_q and S_r both act in the space of all polynomials of degree $\leq n-1$. Clearly

$$(S_q-S_r)z^i = \begin{cases} 0 & \text{for } i=0,\dots,n-2 \\ r-q & \text{for } i=n-1 . \end{cases} \tag{3.4}$$

Note that, since both q and r are monic, $r-q$ is of degree $< n$. Define

$$b = r-q = -p_r q1 = -q(S_r)1 . \tag{3.5}$$

Given two polynomials $b,c \in X_q$ we define the rank one operator $b \otimes \tilde{c}$ by

(3.6) $(b \otimes \tilde{c})f = \langle f,c \rangle b$.

With this definition we can rewrite (3.4) as

(3.7) $S_q - S_r = b \otimes \tilde{c}$

where $c(z) = e_n(z) = 1$. Equality (3.4) can also be rewritten as

(3.8) $$q(S_r)p(S_q) = \Sigma b_{ij} S_r^{i-1} b \otimes \tilde{c}\, S_q^{j-1} = \Sigma b_{ij} (S_r^{i-1} b) \otimes \widetilde{(S_q^{j-1} c)}$$
$$= (b\ S_r b, .., S_r^{n-1} b) B(p,q) \otimes \widetilde{(c\ S_q c, .., S_q^{n-1} c)}$$

Now, as $b = -q(S_r)1$, it follows that

(3.9) $$S_r^{i} b = -S_r^{i} q(S_r) 1 = -q(S_r) S_r^{i} 1 .$$

By our choice r and q are coprime and this quarantees, by Theorem 4.7 in Fuhrmann [1976], the invertibility of $q(S_r)$. There for (3.8) implies

(3.10) $$p(S_q) = -\Sigma\, b_{ij} (S_r^{i-1} 1) \otimes \widetilde{(S_q^{j-1} c)}$$
$$= -\Sigma\, b_{ij} z^{i-1} \otimes \widetilde{S_q^{j-1} c} = -\Sigma\, b_{ij} f_i \otimes \tilde{g}_j$$

with $f_i(z) = z^{i-1}$ and $g_j = S_q^{j-1} c$. Let us define maps $V: F^n \to X_q$ and $W: X_q \to F^n$ by

(3.11) $$Wf = \begin{pmatrix} \langle f,c \rangle \\ \cdot \\ \cdot \\ \cdot \\ \langle f, S_q^{n-1} c \rangle \end{pmatrix}$$

and

$$V \begin{pmatrix} a_1 \\ \cdot \\ \cdot \\ \cdot \\ a_n \end{pmatrix} = \Sigma\, a_i f_i .$$

We will denote by B the standard basis in F^n. Since $\Sigma\, b_{ij} f_i \otimes \tilde{g}_j = VBW$ it follows that

(3.12) $$[p(S_q)]_{B_c}^{B_o} = -\, [V]_{B}^{B_o} B [W]_{B_c}^{B} .$$

Now it is easy to check that, since the standard basis elements in F^n are mapped by V to the polynomials f_i, that

(3.13) $$[V]_{B}^{B_o} = I .$$

We will show that also $[W]_{B_c}^{B} = I$. To this end we note that $(We_i)_j$, the j-th component of We_i in F^n, satisfies

$$(We_i)_j = \langle e_i, S_q^{\,j-1} 1 \rangle = \langle e_i, S_q^{\,j-1} f_1 \rangle = \langle e_i, f_j \rangle = \delta_{ij} .$$

Summing up we have proved the following theorem.

Theorem 3.1: Let $p,q \in R[z]$ with $\deg p \leq \deg q$. Then the Bezoutian $B = B(p,q)$ of p and q satisfies

$$B(p,q) = -[p(S_q)]_{B_c}^{B_0} . \tag{3.14}$$

As an immediate consequence we get the following result, due to Barnett [1972].

Corollary 3.2: Let p and q be polynomials with $\deg p \leq \deg q$ then

$$B(p,q) = -[I]_{B_c}^{B_o} p([S_q]_{B_c}^{B_c}) = -p([S_q]_{B_o}^{B_o})[I]_{B_c}^{B_o} .$$

Proof: This follows from the equality

$$p(S_q) = Ip(S_q) = p(S_q)I$$

and the fact that

$$[p(S_q)]_B^B = p([S_q]_B^B) .$$

The main advantage of Theorem 3.1 is in the reduction of the analysis of the Bezoutian to that of $p(S_q)$ which is easier to handle. This allows us to derive in an easy manner the following classical result, one proof of which can be found in Householder [1970].

Theorem 3.3: Given two polynomials $p,q \in R[z]$ then

(i) $B(p,q)$ is invertible if and only if p and q are coprime

(ii) $\dim \operatorname{Ker} B(p,q)$ is equal to the degree of the g.c.d. of p and q.

Proof: Part (i) follows from the fact that $p(S_q)$ is invertible if and only if p and q are coprime (this is a very special case of Theorem 4.7 in Fuhrmann [1976]). To prove (ii) we note that $\dim \operatorname{Ker} B = \dim \operatorname{Ker} p(S_q)$. Now if r is the g.c.d. of p and q then $p = rp'$, $q = rq'$ and $\operatorname{Ker} p(S_q) = q'X_r \subset X_q$ and $\dim q'X_r = \dim X_r = \deg r$.

Another corollary of Theorem 3.1 gives a direct link between the polynomials, their Bezoutian and Liapunov's method. This result seems to be due to Datta [1978].

Corollary 3.4: Given polynomials $p,w \in R[z]$ with $\deg p \leq \deg q$ then

$$B(p,q)[S_q]_{B_c}^{B_c} = [S_q]_{B_0}^{B_0} B(p,q) . \tag{3.15}$$

Proof: From the commutativity of S_q and $p(S_q)$ it follows that

$$[p(S_q)]_{B_c}^{B_o} [S_q]_{B_c}^{B_c} = [S_q]_{B_o}^{B_o} [p(S_q)]_{B_c}^{B_o} . \tag{3.16}$$

We note that

(3.17) $$[S_q]_{B_c}^{B_c} = [S_q]_{B_o}^{\sim B_o} .$$

4. Algebraic Stability Criteria

A polynomial q with real coefficients will be called *stable* or a *Hurwitz* polynomial if all its zeroes lie in the open left half plane. The two basic approaches to the characterization of stable polynomials are through Liapunov's equation or through Hermite's method of quadratic forms. In this exposition we use, following Datta [1978], Liapunov's theorem and inertia theorems by Carlson and Schneider [1963] and Wimmer [1974] to obtain the positive definiteness of the Hermite-Fujiwara quadratic form.

We begin with some notation. Given the polynomial

(4.1) $$q(z) = z^n + q_{n-1} z^{n-1} + \dots + q_o$$

with real coefficients we identify X_q with the space of all (real) polynomials of degree $\leq n-1$, and with the module action given by (2.7). We define a map $J:X_q \to X_q$ by

(4.2) $$(Jf)(z) = f(-z)$$

and we define q_* by

(4.3) $$q_*(z) = (Jq)(z) = q(-z) .$$

Let q_+ and q_- be defined by

(4.4) $$q_+(z) = \Sigma\, q_{2j} z^j \qquad q_-(z) = \Sigma\, q_{2j+1} z^j$$

then

(4.5) $$q(z) = q_+(z^2) + zq_-(z^2) \qquad q_*(z) = q_+(z^2) - zq_-(z^2)$$

and

(4.6) $$q(z) + q_*(z) = 2q_+(z^2) \quad \text{if } n \text{ is odd} \qquad q(z) - q_*(z) = 2zq_-(z^2) \quad \text{if } n \text{ is even.}$$

Since $q(S_q) = 0$ we have

(4.7) $$q_*(S_q) = (q_*+q)(S_q) = (q_*-q)(S_q)$$

It follows from (4.6) that

(4.8) $$q_*(S_q) = \begin{cases} 2q_+(S_q^2) & \text{when } n \text{ is odd} \\ -2S_q q_-(S_q^2) & \text{when } n \text{ is even .} \end{cases}$$

The relation between the actions of J defined by (4.2) and S_q is given by the following.

Lemma 4.1: Let J be defined by (4.2) then

(i) if n is odd

$$(4.9)\qquad (JS_q + S_qJ)z^i \quad \begin{cases} 0 & 0 \leq i \leq n-2 \\ -2q_+(z^2) & i=n-1 \end{cases}$$

and

(ii) if n is even

$$(4.10)\qquad (JS_q + S_qJ)z^i = \begin{cases} 0 & 0 \leq i \leq n-2 \\ 2zq_-(z^2) & \end{cases}$$

Given the Bezoutian $B = (b_{ij})$ of q and q_* the corresponding Hermite-Fujiwara matrix is defined by $H(q,q_*) = (h_{ij})$ where

$$(4.11)\qquad h_{ij} = (-1)^i b_{ij} .$$

Theorem 4.2: Assume q is a real polynomial of degree n having no zeroes on the imaginary axis and let A be its companion matrix defined by (2.16). Then the Hermite-Fujiwara matrix H is a solution of a Liapunov equation

$$(4.12)\qquad AH + HA^* = -Q$$

with Q a nonnegative definite quadratic form.

Proof: Clearly, since $q_*(S_q)$ and S_q commute we have

$$(4.13)\qquad Jq_*(S_q)S_q = JS_q q_*(S_q) .$$

Now $JS_q = -S_qJ + (JS_q + S_qJ)$ so

$$(4.14)\qquad Jq_*(S_q)S_q + S_qJq_*(S_q) = (JS_q + S_qJ)q_*(S_q) .$$

We note that if for $f \in X_q$, $[f]^{\mathcal{B}}$ denotes the column vector of the coordinates of f with respect to a basis $\mathcal{B}$ and if $\mathcal{B}^*$ denotes the dual basis then

$$< f,g > = [\tilde{g}]^{\mathcal{B}^*}[f]^{\mathcal{B}} .$$

In particular this implies the following equality

$$< (JS_q + S_qJ)q_*(S_q)f,f > = [\tilde{f}]^{\mathcal{B}_c}[JS_q + S_qJ]^{\mathcal{B}_o}_{\mathcal{B}_o}[q_*(S_q)]^{\mathcal{B}_o}_{\mathcal{B}_c}[f]^{\mathcal{B}_c}$$

so it suffices to show that for $f \neq 0$, $-< (JS_q + S_qJ)q_*(S_q)f,f > \geq 0$.

We will deal separately with the case of odd and even n.

With $\mathcal{B}_c$ and $\mathcal{B}_o$ denoting the control and standard bases respectively of X_q we have the following matrix equality

$$(4.15)\qquad [J]^{\mathcal{B}_o}_{\mathcal{B}_o}[q_*(S_q)]^{\mathcal{B}_o}_{\mathcal{B}_c}[S_q]^{\mathcal{B}_c}_{\mathcal{B}_c} + [S_q]^{\mathcal{B}_o}_{\mathcal{B}_o}[J]^{\mathcal{B}_o}_{\mathcal{B}_o}[q_*(S_q)]^{\mathcal{B}_o}_{\mathcal{B}_c}$$

$$= [JS_q + S_qJ]^{\mathcal{B}_o}_{\mathcal{B}_o}[q_*(S_q)]^{\mathcal{B}_o}_{\mathcal{B}_c}$$

Since $[J]^{\mathcal{B}_o}_{\mathcal{B}_o} = \mathrm{diag}(1,-1,\dots,(-1)^{n-1})$ and $B(q,q_*) = -[q_*(S_q)]$ it follows that with $A = [S_q]_{B_c}^{B_c}$ the companion matrix of q

(4.16) $$HA + A^*H = -[JS_q + S_qJ]_{B_o}^{B_o}\,[q_*(S_q)]_{B_c}^{B_o}$$

and we will show that the matrix on the right corresponds to a nonnegative definite form.

Assume first that n is odd. In this case we use (4.8) and (4.9) and the expansion $f = \Sigma < f,e_j > z^{i-1}$ to obtain

$$- <(JS_q + S_qJ)q_*(S_q)f,f> = - < (JS_q + S_qJ)\Sigma < q_*(S_q)f,e_i > z^{i-1},f >$$

$$= - \Sigma < q_*(S_q)f,e_i > < (JS_q + S_qJ)z^{i-1},f > = < q_*(S_q)f,e_n > < 2q_+(z^2)1,f >$$

$$= < q_*(S_q)f,e_n > < e_n,q_*(S_q)f > = < (q_*(S_q)e_n \times q_*(S_q)e_n)f,f > \geq 0 .$$

Here we used the fact that

$$< 2q_+(z^2)1,f > = < \pi_q(q + q_*)e_n,f > = < e_n,q_*(S_q)f > .$$

Next we assume n is even. In this case we use the same argument but (4.10) to obtain

$$- < (JS_q + S_qJ)q_*(S_q)f,f > = - < q_*(S_q)f,e_n > < (JS_q + S_qJ)z^{n-1},f >$$

$$= - < q_*(q)f,e_n > < 2zq_-(z^2)1,f > = - < q_*(S_q)f,e_n > < e_n,2S_qq_-(S_q^2)f >$$

$$= < q_*(S_q)f,e_n > < e_n,q_*(S_q)f > .$$

The last equality follows using (4.6).

Let us define

(4.17) $$b_+(z) = 2q_+(z^2) \quad \text{and} \quad b_-(z) = 2zq_-(z^2) .$$

We quote next the Carlson-Schneider [1963] and Wimmer [1974] inertia theorems which are extensions of Liapunov's original result. We recall that the *inertia* of a matrix A, $In(A)$, is defined by

(4.18) $$In(A) = (\pi(A), \nu(A), \delta(A))$$

where $\pi(A)$, $\nu(A)$ and $\delta(A)$ denote the number of eigenvalues of A with positive, negative and zero real part respectively.

<u>Theorem 4.3</u>: If the pair (A,C) is reachable, $C \geq 0$ and H a solution of the Liapunov equation

(4.19) $$AH + H\tilde{A} = C$$

then $In(A) = In(H)$.

If H is a nonsingular solution of (4.19) then $In(A) = In(H)$.

<u>Remark</u>. In particular if A is stable, i.e., $\pi(A) = \delta(A) = 0$, then under the assumption of the reachability of the pair (A,C) a solution of (4.19) is unique and necessarily positive definite. Conversely, if H is a positive definite solution of

(4.19) with $C \gtrless 0$ and (A,C) reachable then A is stable.

We can state now the Hermite-Fujiwara theorem.

Theorem 4.4: Let q be a real polynomial and let H be the Hermite-Fujiwara matrix. If H is nonsingular then the number of zeroes of q with positive (negative) real part is equal to the number of negative (positive) eigenvalues of H. The polynomial q is stable if and only if the Hermite-Fujiwara matrix is positive definite.

Proof: By Theorem 4.2 H is a solution of the Liapunov equation with a non positiv definite right hand side. Since H is nonsingular so is the Bezoutian $B(q,q_*)$ and hence, by Theorem 3.3, q and q_* are coprime. Since for even n we have $b_- = q-q_*$ and for odd n we have $b_+ = q+q_*$. It follows that q is coprime with both b_- and b_+. Thus the reachability condition of Theorem 4.3 is satisfied and the result follows from that theorem. If we assume q is stable then q_* has all its zeroes in the right half plane and so q and q_* are coprime. Thus the Bezoutian B is nonsingular and so is the Hermite-Fujiwara matrix. Since H is a solution of the Liapunov equation, which in this case is unique, it is necessarily symmetric and as $-\mathrm{In}(A) = \mathrm{In}(H)$, it follows that H is positive definite.

We compute now the Bezoutian of q and q_* using relations (4.5).

$$\begin{aligned}
B(q,q_*) &= (q(z)q_*(w)-q_*(z)q(w))/(z-w)\\
&= ((q_+(z^2)+zq_-(z^2))(q_+(w^2)-wq_-(w^2))-(q_+(z^2)-zq_-(z^2))(q_+(w^2)+wq_-(w^2)))/(z-w)\\
&= (2zq_-(z^2)q_+(w^2)-2wq_+(z^2)q_-(w^2))/(z-w)\\
&= 2(zq_-(z^2)q_+(w^2)-wq_+(z^2)q_-(w^2))(z+w)/(z^2-w^2)\\
&= 2(z^2q_-(z^2)q_+(w^2)-q_+(z^2)w^2q_-(w^2))/(z^2-w^2)+2zw(q_-(z^2)q_+(w^2)-q_+(z^2)q_-(w^2))/(z^2-w^2)
\end{aligned}$$

Note that the first term contains only even powers of z and w whereas the second term contains only odd powers. Thus we have the following

Theorem 4.5: Given a real polynomial q then we have the following isomorphism.

(4.18) $$B(q,q_*) \cong B(zq_-,q_+) \oplus B(q_-,q_+)$$

whereas the Hermite-Fujiwara form $H(q,q_*)$ satisfies

(4.19) $$H(q,q_*) = 2B(zq_-,q_+) \oplus 2(-1)B(q_-,q_+) = 2B(zq_-,q_+) \oplus 2B(q_+,q_-)$$

Just as reachability and realizability criterias for rational functions can be given in terms of Hankel matrices so can stability criterias. Here we present a simple derivation of a theorem of this type Gantmacher [1959,p.232].

As before let $q(z) = q_+(z^2) + zq_-(z^2)$. Since for $n = 2m$, $\deg q_+ = m$, $\deg q_- = m-1$ and for $n = 2m+1$ $\deg q_+ = \deg q_- = m$ the rational function g defined by

(4.20) $$g(z) = q_-(z)/q_+(z)$$

is proper for odd n and strictly proper for even n. The rational function g can be expanded in a power series in z^{-1}

(4.21) $$g(z) = g_0 + g_1 z^{-1} + \ldots$$

with $g_0 = 0$ for n even.

Theorem 4.6: The real monic polynomial q is stable if and only if the Hankel matrices

$$H^{(m)} = \begin{pmatrix} g_1 & g_2 & \cdot & \cdot & \cdot & g_{m-1} \\ g_2 & g_3 & & & \cdot & \\ \cdot & & & \cdot & & \\ \cdot & & \cdot & & & \\ \cdot & \cdot & & & & \\ g_{m-1} & & & & & g_{2m-3} \end{pmatrix}$$

and

$$(\sigma H)^{(m)} = \begin{pmatrix} g_2 & g_3 & \cdot & \cdot & \cdot & g_m \\ g_3 & g_4 & \cdot & \cdot & \cdot & \\ \cdot & & & \cdot & & \\ \cdot & & \cdot & & & \\ \cdot & \cdot & & & & \\ g_m & & & & & g_{2m-2} \end{pmatrix}$$

are positive definite and $g_0 \geq 0$.

Proof: The functional representation of the Hankel map $H_g: R[z] \to z^{-1}R[[z^{-1}]]$ induced by g is given by

(4.22) $$H_g u = \pi_- gu \quad \text{for} \quad u \in R[z] .$$

To this map correspond the infinite quadratic form H on $R[z]$ given by

(4.23) $$\tilde{x}Hx = [H_g u_x, u_x]$$

where, for $\tilde{x} = (x_0, \ldots, x_{n-1})$ $u_x(z) = \Sigma\, x_i z^i$.

Now for $x \in R^m$

$$\tilde{x}Hx = [H_g u_x, u_x] = [\pi_- g u_x, u_x] = [\pi_- q_+^{-1} \underset{\sim}{q_-} u_x, u_x] = [q_+^{-1} q_+ \pi_- q_+^{-1} q_- u_x, u_x]$$
$$= \langle q_-(S_{q_+}) u_x, u_x \rangle = [u_x]\, B(q_+, q_-)\, [u_x] > 0 .$$

Similarly

$$x(\sigma H)x = [H_{zg} x, x] = [\pi_- zg u_x, u_x] = [\pi_- q_+^{-1} q_- z u_x, u_x] = [q_+^{-1} q_+ \pi_- q_+^{-1} q_- z u_x, u_x]$$
$$= \langle S_{q_+ q_-}(S_{q_+}) u_x, u_x \rangle = [\tilde{u}_x]\, B(zq_-, q_+)\, [u_x]$$

which also implies the positive definiteness of $H^{(m)}$ and $(\sigma H^{(m)})$.

Clearly if n is odd, since q is monic and stable, all q_i are positive and so $g_0 = q_{2m}^{-1} > 0$. Conversely assuming the quadratic forms $H^{(m)}$ and $(\sigma H)^{(m)}$ are positive definite we use the same formulas to deduce the positive definiteness of the Bezoutians $B(q_+, q_-)$ and $B(zq_-, q_+)$. This, by Theorem 4.2, implies the stability of q.

References

[1976] B.D.O. Anderson and E.I. Jury, "Generalized Bezoutian and Sylvester matrices in multivariable control", IEEE Trans. Aut. Control, AC-21, 551-556.

[1972] S. Barnett, "A Note on the Bezoutian matrix", SIAM J. Appl. Math. 22, 84-86.

[1963] D. Carlson and H. Schneider, "Inertia theorems for matrices: the semidefinite case", J. Math. Anal. Appl. 6, 430-446.

[1922] A. Cohn, "Uber die Anzahl der Wurzeln einer Algebraischen Gleichung in einem Kreise", Math. Z. 14, 110-138.

[1978] B.N. Datta, "An elementary proof of the stability criterion of Lienard and Chipart", Linear Algebra and Appl. 22, 89-96.

[1978] B.N. Datta, "On the Routh-Hurwitz-Fujiwara and the Schur-Cohn-Fujiwara theorems for the root-separation problem", Linear Algebra and Appl. 22, 235-246.

[1980] E. Emre, "Nonsingular factors of polynomial matrices and (A,B)-invariant subspaces", SIAM J. Contr. 18, 288-296.

[1980] E. Emre and M.L.J. Hautus, "A polynomial characterization of (A,B)-invariant and reachability subspaces", SIAM J. Contr. Optimiz. 18, 420-436.

[1976] P.A. Fuhrmann, "Algebraic system theory: An analyst's point of view", J. Franklin Inst. 301, 521-540.

[1977] P.A. Fuhrmann, "On strict system equivalence and similarity", Int. J. Contr. 25, 5-10.

[1979] P.A. Fuhrmann, "Linear feedback via polynomial models", Int. J. Contr. 30, 363-377.

[1981] P.A. Fuhrmann, "Duality in polynomial models with some applications to geometric control theory", IEEE Trans. Aut. Control, AC-26, 284-295.

[1980] P.A. Fuhrmann and J.C. Willems, "A study of (A,B)-invariant subspaces via polynomial models", Int. J. Contr. 31, 467-494.

[1926] M. Fujiwara, "Uber die algebraischen Gleichungen, deren Wurzeln in einem Kreise oder in einer Halbebene liegen", Math. Z., 24, 161-169.

[1959] F.B. Gantmacher, *The Theory of Matrices*, Chelsea, New York.

[1978] I. Gohberg, P. Lancaster and L. Rodman, "Representation and divisibility of operator polynomials", Canadian J. Math. 30, 1045-1069.

[1856] C. Hermite, "Sur le nombre des racines d'une equation algebrique comprise entre des limites donnes", J. Reine Angew. Math. 52, 39-51.

[1970] A.S. Householder, "Bezoutians, elimination and localization", SIAM Review, 12, 73-78.

[1895] A. Hurwitz, "Uber die bedingungen, unter welchen eine Gleichung nur Wurzeln mit negativen reelen Teilen besitzt", Math. Annal. 46, 273-284.

[1980] T. Kailath, *Linear Systems*, Prentice Hall.

[1969] R.E. Kalman, "Algebraic characterization of polynomials whose zeros lie in algebraic domains", Proc. Nat. Acad. Sci. 64, 818-823.

[1981] P.O. Khargonekar and E. Emre, "Further results on polynomial characterization of (F,G)-invariant subspaces", to appear.

[1980] N. Kravitsky, "On the discriminant function of two noncommuting nonself-adjoint operators", Integral Eq. and Operator Theory, 3, 97-124.

[1893] A.M. Liapunov, "Probleme general de la stabilite du mouvement", Ann. Fac. Sci. Toulouse 9 (1907), 203-474. (French translation of the Russian paper published in Comm. Soc. Math. Kharkow).

[1914] A. Lienard and M. Chipart, "Sur le signe de la partie reelle des racines d'une equation algebrique", J. de Math. 10, 291-346.

[1977] P.C. Muller, *Stabilität und Matrizen*, Springer Verlag Berlin.

[1962] P.C. Parks, "A new proof of the Routh-Hurwitz stability criterion using the second method of Lyapunov", Proc. Cambridge Philos. Soc. 58, 694-702.

[1877] E.J. Routh, *A Treatise on the Stability of a Given State of Motion*, Macmillan, London.

[1918] I. Schur, "Uber Potenzreihen die im Innern des Einheitskreises beschrankt sind", J. fur Math. 148, 122-145.

[1972] O. Taussky, "The role of symmetric matrices", Linear Algebra and Appl. 5, 147-154.

[1974] H.K. Wimmer, "Inertia theorems for matrices, controllability and linear vibrations", Linear Algebra and Appl. 8, 337-344.

LINEAR SYSTEM FACTORIZATION

J. Hammer† and M. Heymann††

1. Introduction

In HAUTUS and HEYMANN [1978], an investigation was initiated of the algebraic structure of discrete time, time invariant, finite dimensional linear systems (or, simply, linear systems) with particular emphasis on static state feedback. This investigation was extended to the study of dynamic as well as static output feedback in HAMMER and HEYMANN [1981]. Pivotal in the extended theory was the problem of causal factorization, i.e., the problem of factoring two system maps over each other through a causal factor. The theory was further extended in HAMMER and HEYMANN [1980] where the structural invariants of precompensation orbits and the concept of strict observability were studied in detail. Algebraically, the theory of strict observability hinges on the problem of polynomial factorization, i.e., the problem of factoring two system maps over each other through a polynomial factor.

It has since become increasingly clear, that the theory of linear systems can be formulated in a very general algebraic setup in which the central concepts of causality (and hence of feedback), of stability and of realization are investigated in a unified framework. In the present paper we present some of the essentials of this theory with particular emphasis on the issue of system stability. Proofs of theorems are omitted because of space limitations and will appear in a future expanded paper HAMMER and HEYMANN [1982].

2. The Mathematical Setup

We assume that the reader has basic familiarity with the setup and terminology of HAUTUS and HEYMANN [1978], HAMMER and HEYMANN [1981] as well as HAMMER and HEYMANN [1980]. We review the principal aspects of this setup very briefly.

For a field K and a K-linear space S, we denote by ΛS the set of all formal Laurent series in z^{-1} with coefficients in S, i.e., series of the form

$$(2.1)\qquad s = \sum_{t=t_0}^{\infty} s_t z^{-t} \quad ; \quad s_t \in S .$$

---

† Center for Mathematical System Theory, University of Florida, Gainesville, Florida, 32611. The research of this author was supporated in part by US Army Research Grant DAAG 29-80-G0050 and US Air Force Grant AFOSR76-3034D through the Center for Mathematical System Theory, University of Florida.

†† Department of Electrical Engineering, Technion, Haifa, Israel. Supported in part by the Technion Fund for Promotion of Research.

In ΛS, the set of polynomial elements of the form $\sum_{t\leq o} s_t z^{-t}$, is denoted by Ω^+S, and the set of causal elements, that is, the set of power series of the form $\sum_{t\geq o} s_t z^{-t}$, is denoted by Ω^-S.

The set ΛK is a field under coefficientwise addition and convolutional multiplication and, under similar operations, the set ΛS becomes a ΛK-linear space. The polynomial subset Ω^+K of ΛK and the set of causal elements Ω^-K are subrings (principal ideal domains) of ΛK. The field ΛK is then an Ω^+K-module and an Ω^-K-module as well.

The Ω^-K-order of an element $s = \sum s_t z^{-t} \in \Lambda S$ is defined by

$$\text{ord}_{\Omega^-K}\, s := \begin{cases} \min t\in Z \mid s_t\neq 0 & \text{if } s\neq 0 \\ \infty & \text{if } s=0 \end{cases} \tag{2.2}$$

where Z denotes the integers.

Let the K-linear spaces U and Y be given. A ΛK-linear map $\bar{f}:\Lambda U\to\Lambda Y$ represents a linear time invariant system, having U as the input value space and Y as the output value space. It is assumed throughout the paper that all underlying K-linear (value) spaces, and, in particular, U and Y are finite dimensional. The Ω^-K-order (or, simply, order) of a ΛK-linear map $\bar{f}:\Lambda U\to\Lambda Y$ is defined as

$$\text{ord}\, \bar{f} := \inf\{\text{ord}\, \bar{f}(u) - \text{ord}\, u \mid 0\neq u\in\Lambda U\}. \tag{2.3}$$

The map $\bar{f}$ is said to be of finite order if $\text{ord}\, \bar{f} > -\infty$.

If $\bar{f}$ is a ΛK-linear map of finite order t_o, we associate with it its transfer function, i.e., an element

$$T = \sum_{t=t_o}^{\infty} T_t z^{-t} \in \Lambda L ,$$

where L is the K-linear space of K-linear maps $U\to Y$ as follows. We define the K-linear maps p_t and i_u by

$$\begin{cases} i_u : U\to\Lambda U : u \mapsto u \text{ (canonical injection)} \\ p_k : \Lambda Y\to Y : \Sigma y_t z^{-t} \mapsto y_k \end{cases} \tag{2.4}$$

and then for all integers $t\geq t_o$ we let $T_t := T_t(\bar{f}) := p_t \cdot \bar{f} \cdot i_u$. Conversely, with each element $T = \sum T_t z^{-t} \in \Lambda L$ we associate a ΛK-linear map $\bar{f}=\bar{f}_T$ of finite order whose action on elements $u = \sum u_t z^{-t} \in \Lambda U$ is defined through the convolution formula

$$\bar{f}_T \cdot u := \sum_t \Big(\sum_k T_k u_{t-k}\Big) z^{-t} .$$

For a map $\bar{f}: \Lambda U\to\Lambda Y$ and a subset $A\subset\Lambda U$, we denote by $\bar{f}[A]$ the image of A under $\bar{f}$, i.e., $\bar{f}[A] = \{\bar{f}(u) \mid u\in A\}$. A ΛK-linear map $\bar{f}: \Lambda U\to\Lambda Y$ is called causal if $\text{ord}\, \bar{f}\geq 0$ or, equivalently, if $\bar{f}[\Omega^-U]\subset\Omega^-Y$. Similarly, $\bar{f}$ is called strictly causal if $\text{ord}\, \bar{f}\geq 1$ or, equivalently, if $\bar{f}[\Omega^-U]\subset z^{-1}\Omega^-Y$. We have the following

(2.5) DEFINITION. A ΛK-linear map $\bar{f}: \Lambda U \to \Lambda Y$ is called a linear input/output (or i/o) map if it is strictly causal and of finite order.

Associated with a linear i/o map $\bar{f}: \Lambda U \to \Lambda Y$ are two further maps as follows. First, we restrict the inputs to the Ω^+K-module Ω^+U, and consider the projection of the corresponding outputs on the quotient Ω^+K-module $\Gamma^+Y := \Lambda Y/\Omega^+Y$. Then we obtain the restricted linear i/o map $\tilde{f}: \Omega^+U \to \Gamma^+Y$ associated with $\bar{f}$ through

$$\tilde{f} = \pi^+ \cdot \bar{f} \cdot j^+ \quad ,$$

where $j^+: \Omega^+U \to \Lambda U$ is the canonical injection and $\pi^+: \Lambda Y \to \Gamma^+Y$ is the canonical projection. It is readily seen that $\tilde{f}$ is an Ω^+K-homomorphism. Next, we associate with $\bar{f}$ the output response map $f: \Omega^+U \to Y$ given by $f := p_1 \cdot \bar{f} \cdot j^+$ or, more explicitly,

$$f: \Omega^+U \to Y: u \mapsto f(u) := p_1 \bar{f}(u) \quad .$$

Since the map p_1 is K-linear, so is also the output response map f. The case in which f is an Ω^+K-homomorphism as well, is of particular importance and we have

(2.6) DEFINITION. A linear i/o map $\bar{f}: \Lambda U \to \Lambda Y$ is called an input/state (or i/s) map if there exists an Ω^+K-module structure on Y, compatible with its K-linear structure, such that the output response map $f = p_1 \cdot \bar{f} \cdot j^+$ is an Ω^+K-homomorphism.

3. Rationality and Stability: General Considerations

An element $s \in \Lambda S$ is called Ω^+K-rational (or sometimes simply rational) if there exists a nonzero polynomial $\psi \in \Omega^+K$ such that $\psi s \in \Omega^+S$.† The set of Ω^+K-rationals in ΛS is denoted $Q_{\Omega^+K}S$. For an element $s \in Q_{\Omega^+K}S$, the set of polynomials $\psi \in \Omega^+K$ for which $\psi s \in \Omega^+S$ is easily seen to be an ideal in Ω^+K. Since Ω^+K is a principal ideal domain, this ideal is generated by a monic polynomial ψ_s, which we call the least denominator of s. The zeros of ψ_s are called the poles of s. (In case $K=\mathbb{R}$, the field of real numbers, it is customary to consider not only poles in $\mathbb{R}$ but also in $\mathbb{C}$, the field of complex numbers). The definition of Ω^+K-rationality applies, in particular, also to transfer functions of ΛK-linear maps and we call a ΛK-linear map $\bar{f}: \Lambda U \to \Lambda Y$ Ω^+K-rational (or, simply, rational) if so is its transfer function.

We turn now to the concept of stability. If $\mathcal{D}$ is a set of polynomials, we say that an Ω^+K-rational map is $\mathcal{D}$-stable if its least denominator is in $\mathcal{D}$. We impose a number of restrictions on the set $\mathcal{D}$ of stable denominators (see MORSE [1976]) as follows :

(3.1) DEFINITION. A set $\mathcal{D}$ of (monic) polynomials over K is called a denominator set if it satisfies the following conditions :

(i) $\mathcal{D}$ is multiplicatively closed, i.e., $p \in \mathcal{D}$, $q \in \mathcal{D}$ imply $p \cdot q \in \mathcal{D}$.

(ii) The unit polynomial 1 belongs to $\mathcal{D}$ but the zero polynomial does not belong to $\mathcal{D}$.

---

† Throughout the paper S denotes a finite dimensional K-linear space.

(iii) $\mathcal{D}$ contains at least one polynomial of degree one, i.e., there exists $\alpha \in K$ such that $z-\alpha \in \mathcal{D}$.

(iv) $\mathcal{D}$ is saturated, i.e., if $p \in \mathcal{D}$ and q is a monic divisor of p, then $q \in \mathcal{D}$.

Conditions (i) and (ii) say that $\mathcal{D}$ is a multiplicative set so that one can define the set $\Omega_{\mathcal{D}}K$ as the set of fractions p/q, where $p \in \Omega^+K$ and $q \in \mathcal{D}$. Conditions (iii) and (iv) are motivated by considerations that are discussed shortly. We now introduce the following

(3.2) DEFINITION. Let $\mathcal{D}$ be a denominator set. Then an element $s \in Q_{\Omega^+K}S$ is called stable (or, explicitly, $\mathcal{D}$-stable) if there exists $\psi \in \mathcal{D}$ such that $\psi s \in \Omega^+S$, or, equivalently, if the least denominator $\psi_s \in \mathcal{D}$. The set of stable elements in $Q_{\Omega^+K}S$ is denoted by $\Omega_{\mathcal{D}}S$. The set of stable and causal elements is denoted by $\Omega_{\mathcal{D}}^-S$, i.e.,

(3.3)
$$\Omega_{\mathcal{D}}^-S = \Omega_{\mathcal{D}}S \cap \Omega^-S \quad .$$

The above definition of stability is easily seen to be a generalization to arbitrary fields of the usual concept of stability in system theory defined in an algebraic framework.

Definition 3.2 applies, in particular, to the case $S = L$, the space of all linear maps $U \mapsto Y$ and we have a definition of stable transfer functions and stable ΛK-linear maps. In particular, we have the following

(3.4) PROPOSITION. The map $\bar{f} \in \Omega_{\mathcal{D}}L$ if and only if $\bar{f}[\Omega_{\mathcal{D}}U] \subset \Omega_{\mathcal{D}}Y$.

The set $\Omega_{\mathcal{D}}K$ is easily seen by direct computation to be a subring (with identity) of the rational field Q_{Ω^+K} ($=Q_{\Omega^+K}K$), and is actually a principal ideal domain. In fact, we have even more :

(3.5) PROPOSITION. The ring $\Omega_{\mathcal{D}}K$ is a Euclidean domain.

Since we are interested in causal systems, we shall be interested in the ring $\Omega_{\mathcal{D}}^-K$ which, as was proved in MORSE [1976] is also a principal ideal domain and, in fact, just as $\Omega_{\mathcal{D}}K$, is also a Euclidean domain. We generalize now our framework of consideration so as to include the preceding examples as special cases. In particular, since we encountered as substructures of ΛK the rings Ω^+K, Ω^-K, $\Omega_{\mathcal{D}}K$ and $\Omega_{\mathcal{D}}^-K$ all of which are Euclidean domains or, more generally, principal ideal domains, we consider now a more general framework as follows :

Let ΩK be a principal ideal domain (P.I.D.) properly contained as a subring in ΛK. The ΛK-linear space ΛS is then also an ΩK-module. Define ΩS to be the ΩK-submodule of ΛS generated by S, i.e., if $s_1,\dots,s_n$ is a basis for S then

(3.6)
$$\Omega S := \{s \in \Lambda S \mid s = \sum_{i=1}^{n} \alpha_i s_i \ , \ \alpha_i \in \Omega K, \quad i = 1,\dots,n\} \ .$$

We now extend some basic concepts and terminology to the P.I.D. ΩK. An element $s \in \Lambda S$ is called ΩK-rational if there exists a nonzero element $\psi \in \Omega K$ such that $\psi s \in \Omega S$. The set of ΩK-rationals in ΛS is denoted $Q_{\Omega K}S$. Just as in the case Ω^+K, the definition of ΩK-rationality also applies to transfer functions of ΛK-linear maps and we call a ΛK-linear map ΩK-rational if so is its transfer function. It is readily seen that $\bar{f}: \Lambda U \to \Lambda Y$ is an ΩK-rational map if and only if $\bar{f}[Q_{\Omega K}U] \subset Q_{\Omega K}Y$. (The sufficiency of this condition depends on the finite dimensionality of U). An element $s \in \Lambda S$ is called an ΩK-element if $s \in \Omega S$. Thus, a ΛK-liner map $\bar{f}: \Lambda U \to \Lambda Y$ is an ΩK-map in case its transfer function is an ΩK-element of ΛL. $\bar{f}$ is called ΩK-unimodular if it is an invertible ΩK-map and its inverse is also an ΩK-map.

We shall make use of the following notation :

(3.7)
$$\begin{cases} j_{\Omega K}: \Omega S \to \Lambda S: s \mapsto s & \text{(natural injection)} \\ \pi_{\Omega K}: \Lambda S \to \Lambda S/_{\Omega S} =: \Gamma_{\Omega K}S & \text{(canonical projection)} \end{cases}$$

We can write the following

(3.8) THEOREM. Let $\bar{f}: \Lambda U \to \Lambda Y$ be a ΛK-linear map. Then $\bar{f}$ is an ΩK-map if and only if $\bar{f}[\Omega U] \subset \Omega Y$ (or, equivalently, if and only if $\Omega U \subset \ker \pi_{\Omega K}\bar{f}$).

The following corollary to Theorem 3.8 is very useful

(3.9) COROLLARY. A ΛK-linear map $\bar{\ell}: \Lambda U \to \Lambda U$ is ΩK-unimodular if and only if $\bar{\ell}[\Omega U] = \Omega U$ (equivalently, $\ker \pi_{\Omega K} \bar{\ell} = \Omega U$).

4. The Order and Adapted Bases

Our main objective in this section is to obtain finitary characterizations of ΩK-submodules of ΛK-linear spaces and of related properties of ΛK-linear maps. As before, we let ΩK be a principal ideal domain properly contained as a subring in ΛK and let $Q_{\Omega K}$ $(=Q_{\Omega K}K)$ denote the field of quotients generated by ΩK.

For an element $s \in \Lambda S$ we define the order of s, denoted $\text{ord}_{\Omega K}s$ (or, simply, ord s when the underlying ring is clear) as the set of all elements $\alpha \in Q_{\Omega K}$ for which $\alpha s \in \Omega S$. When $s=0$ we obviously have that ord s = $Q_{\Omega K}$, i.e., the whole quotient field generated by ΩK. In general, it is an easy exercise to verify that ord s is an ΩK-module (submodule of $Q_{\Omega K}$). In fact, we have the following :

(4.1) THEOREM. If $s \in \Lambda S$ is nonzero, then ord s is a cyclic ΩK-module.

Let $0 \neq s \in \Lambda S$ be any element and let $\alpha \in Q_{\Omega K}$ be any generator of ord s (possibly zero). If $\alpha' \in Q_{\Omega K}$ is another generator of ord s, then it is clearly an associate of α with respect to ΩK, i.e. $\alpha' = \mu\alpha$ where $\mu \in \Omega K$ is a unit (i.e., an invertible). It follows that α is uniquely defined modulo units in ΩK, and it will sometimes be convenient to identify ord s with one of its generators.

Before we proceed with our discussion, let us consider some examples of special interest.

First, let ΩK be the ring $\Omega^- K$ of causal elements. It is easily seen that $Q_{\Omega^- K} = \Lambda K$ since for every $\alpha \in \Lambda K$, either α or α^{-1} is in ΩK (or both). Further, for every element

$0\neq\alpha\in\Lambda K$ there is a unique integer k such that $\alpha=\mu z^{-k}$ for some unit $\mu\in\Omega^- K$. Thus, for each $0\neq s\in\Lambda S$, there exists a unique integer k such that $\text{ord}_{\Omega^-K}s = (z^{-k})_{\Omega^-K}$ and we may identify $\text{ord}_{\Omega^-K}s$ with the integer k associated with it. This definition of order of an element as an integer is precisely the (standard) definition of order as given in (2.2) above. (See also HAUTUS and HEYMANN [1978] and HAMMER and HEYMANN [1980],[1981]).

As the second example let ΩK be the ring Ω^+K of polynomials. In this case Q_{Ω^+K} is the usual field of rationals. For an element $s\in\Lambda S$, $\text{ord}_{\Omega^+K}s\neq 0$ if and only if $s\in Q_{\Omega^+K}S$, i.e., if and only if s is rational (in the classical sense). Let $0\neq s\in Q_{\Omega^+K}S$ be given as $s = (s_1,\dots,s_m)$ with $s_i = \frac{p_i}{q_i}$, $p_i,q_i\in\Omega^+K$ being coprime for all $i=1,\dots,m$. Then $\text{ord}_{\Omega^+K}s$ is generated by the rational element q/p where q and p are the monic polynomials $q = \text{l.c.m.}(q_1,\dots,q_m)$ and $p = \text{g.c.d.}(p_1,\dots,p_m)$ (l.c.m. and g.c.d. denoting, respectively, the <u>least common multiple</u> and the <u>greatest common divisor</u>). To see this, write $p_i = p\bar{p}_i$ and $q = q_i\bar{q}_i$ for polynomials $\bar{p}_i$, $\bar{q}_i$, $i = 1,\dots,m$. Then $\frac{q}{p}\cdot s = (\frac{q}{p}s_1,\dots,\frac{q}{p}s_m) = (\bar{q}_1\bar{p}_1,\dots,\bar{q}_m\bar{p}_m) \in \Omega^+S$ so that $(\frac{q}{p})_{\Omega^+K}\subset\text{ord}_{\Omega^+K}s$. Conversely, let $\frac{r}{t}$ be any element in $\text{ord}_{\Omega^+K}s$ where r and t are coprime polynomials. Then for each $i=1,\dots,m$, $\frac{r}{t}\frac{p_i}{q_i}\in\Omega^+K$. Thus, q_i is a divisor of r for each i, and since q is the l.c.m. of the q_i's it follows that q is a divisor of r as well, that is, $r = q\bar{r}$ for some $\bar{r}\in\Omega^+K$. Similarly, t is a divisor of each of the p_i's and hence also of p, so that $p = t\bar{p}$ for some $\bar{p}\in\Omega^+K$. Thus, $\frac{r}{t} = \frac{q\bar{r}}{t} = \frac{q\bar{r}\bar{p}}{t\bar{p}} = \frac{q}{p}(\bar{r}\bar{p})$ and it follows that $\frac{r}{t}\in(\frac{q}{p})_{\Omega^+K}$, and combining with our previous observations, we have that $\text{ord}_{\Omega^+K}s=(\frac{q}{p})_{\Omega^+K}$.

Finally, we consider the case when ΩK is the ring $\Omega^-_{\mathcal{D}}K$ of causal and stable elements. The quotient field $Q_{\Omega^-_{\mathcal{D}}K}$ again coincides with the usual field of rationals Q_{Ω^+K} and an element $s\in\Lambda S$ has nonzero $\Omega^-_{\mathcal{D}}K$-order if and only if $s\in Q_{\Omega^+K}S$. Let $s = (s_1,\dots,s_m)\in Q_{\Omega^+K}S$ be a nonzero element and write each entry s_i, $i = 1,\dots,m$ as $s_i = \frac{p_i r_i}{q_i}$ where $r_i,q_i\in\mathcal{D}$ are coprime (with respect to Ω^+K) and where $(0\neq)p_i\in\Omega^+K$ is coprime with every element of $\mathcal{D}$. Then it can be verified by direct computation that $\text{ord}_{\Omega^-_{\mathcal{D}}K}s$ is generated by an element $\frac{q}{rp}\in Q_{\Omega^+K}$ as follows: $p = \text{g.c.d.}(p_1,\dots,p_m)$ and q and r are any coprime elements of $\mathcal{D}$ such that $\text{ord}_{\Omega^-K}(\frac{q}{pr}) = -\text{ord}_{\Omega^-K}s$.

We proceed now with the discussion of some general properties of the order.

(4.2) THEOREM. <u>Let</u> $s\in\Lambda S$ <u>be any element. Then</u> ord $s\neq 0$ <u>if and only if</u> $s\in Q_{\Omega K}S$.

Next, we have the following simple characterization of elements in ΩS.

(4.3) PROPOSITION. <u>Let</u> $s\in\Lambda S$ <u>be any element. Then</u> $s\in\Omega S$ <u>if and only if</u> $\Omega K\subset\text{ord } s$.

Let $s_1,\dots,s_m\in Q_{\Omega K}S$ be a set of elements with orders $\text{ord } s_i = (\gamma_i)_{\Omega K}$, $i=1,\dots,m$. Then the intersection $\text{ord } s_1\cap\dots\cap\text{ord } s_m$ is also a cyclic ΩK-module, and hence there is a generator $\gamma\in Q_{\Omega K}$ such that $\text{ord } s_1\cap\dots\cap\text{ord } s_m = (\gamma)_{\Omega K}$. It is easily seen that γ is a least common multiple over ΩK of $\gamma_1,\dots,\gamma_m$, (i.e., γ divides every element $\gamma'\in Q_{\Omega K}$ satisfying the condition that there exists for each i an element $\bar{\gamma}_i\in\Omega K$ such that $\gamma' = \gamma_i\bar{\gamma}_i$). If $s\in Q_{\Omega K}S$ and $\alpha\in Q_{\Omega K}$ are any elements, then $\text{ord }\alpha s = \alpha^{-1}\text{ ord } s$ so that if ord $s = (\gamma)_{\Omega K}$, then ord $\alpha s = (\alpha^{-1}\gamma)_{\Omega K}$. In particular if $\alpha \in \Omega K$, then ord $s \subset$ ord αs. Furthermore, if $s_1,\dots,s_m \in Q_{\Omega K}S$ is any set of elements, then

(4.4) $$\text{ord } s_1 \cap \dots \cap \text{ord } s_m \subset \text{ord } (s_1+\dots+s_m) .$$

Finally, we shall say that a set of elements $s_1,\dots,s_m \in \Lambda S$ is ΩK-<u>ordered</u> (or simply <u>ordered</u>) if ord $s_1 \subset \dots \subset$ ord s_m.

We turn now to characterization of when a ΛK-linear map $\bar{f}: \Lambda U \to \Lambda Y$ is an ΩK-map. Recall that $\bar{f}$ is an ΩK-map if $\bar{f}[\Omega U] \subset \Omega Y$ and let $0 \neq u \in Q_{\Omega K}U$ be any element. Then ord $u = (\gamma)_{\Omega K}$ for some $\gamma \in Q_{\Omega K}$ and $\gamma u \in \Omega U$. If $\bar{f}$ is an ΩK-map, then $\bar{f}(\gamma u) = \gamma \bar{f}(u) \in \Omega Y$ so that $\Omega K \subset$ ord $\bar{f}(\gamma u)$ (see Proposition 4.3), or, equivalently, $\Omega K \subset$ ord $\gamma \bar{f}(u) = \gamma^{-1}$ ord $\bar{f}(u)$. Thus we conclude that $(\gamma)_{\Omega K} \subset$ ord $\bar{f}(u)$, and a necessary condition for $\bar{f}$ to be an ΩK-map is that ord $u \subset$ ord $\bar{f}(u)$. This condition is actually also sufficient and we have the following

(4.5) THEOREM. <u>Let</u> $\bar{f}: \Lambda U \to \Lambda Y$ <u>be a</u> ΛK-<u>linear map. Then</u> $\bar{f}$ <u>is an</u> ΩK-<u>map if and only if</u> ord $u \subset$ ord $\bar{f}(u)$ <u>for each</u> $u \in Q_{\Omega K}U$.

The condition of Theorem 4.5 is, of course, not easily tested directly and we would like to find a finite "test set" of elements in $Q_{\Omega K}U$ which is sufficient for verification that a ΛK-linear map is an ΩK-map. That a basis for $Q_{\Omega K}U$ may not be appropriate for this purpose is seen in the following simple example.

(4.6) EXAMPLE. Let $\Omega K = \Omega^- K$ and let $Y = U = K^2$. Take as basis for $Q_{\Omega K}K^2$ the elements $u_1 = \binom{z^{-1}}{1}$ and $u_2 = \binom{z^{-2}}{1}$ and define $\bar{f}: \Lambda K^2 \to \Lambda K^2$

$$\bar{f}(u_1) = u_1 + u_2$$
$$\bar{f}(u_2) = u_2 \quad .$$

Obviously, $\Omega^- K = \text{ord}_{\Omega^- K}\, u_1 = \text{ord}_{\Omega^- K}\, \bar{f}(u_1) = \text{ord}_{\Omega^- K}\, u_2 = \text{ord}_{\Omega^- K}\, \bar{f}(u_2)$. Thus, $\bar{f}$ satisfies the condition of Theorem 4.5 for the basis u_1, u_2, yet it is not an $\Omega^- K$-map (that is, not causal). Indeed, since $\bar{f}(u_1 - u_2) = u_1$ and since $u_1 - u_2 = \binom{z^{-1} - z^{-2}}{0}$, we have

$$\text{ord}_{\Omega^- K}(u_1 - u_2) = z\Omega^- K \not\subset \text{ord}_{\Omega^- K}\, u_1 = \Omega^- K .$$

Let us explore now the cause of difficulty encountered in the above example. If $s_1,\dots,s_m \in Q_{\Omega K}S$ is a given set of elements and $\alpha_1,\dots,\alpha_m \in Q_{\Omega K}$ is any set of scalars, then by formula (4.4),

$$\bigcap_{i=1}^{m} \text{ord } \alpha_i s_i \subset \text{ord } \sum_{i=1}^{m} \alpha_i s_i \quad .$$

But, the above inclusion, in general, need not hold with equality (even when the s_i are $Q_{\Omega K}$-linearly independent). This order "deficiency" also occurs in the example and therefore the basis selected there failed as a test set for causality. Indeed, we have there

$$\bigcap_{i=1}^{2} \text{ord}_{\Omega^- K}\, u_i = \Omega^- K \neq \text{ord}_{\Omega^- K}(u_1 - u_2) = z\Omega^- K \quad .$$

Thus, we are motivated to introduce the following

(4.7) DEFINITION. A set of nonzero elements $s_1,\dots,s_m \in Q_{\Omega K}S$ is called ΩK-adapted if for every set of scalars $\alpha_1,\dots,\alpha_m \in Q_{\Omega K}$ the condition

(4.8) $$\bigcap_{i=1}^{m} \operatorname{ord} \alpha_i s_i = \operatorname{ord} \sum_{i=1}^{m} \alpha_i s_i$$

holds. A basis of ΩK-adapted elements $s_1,\dots,s_n$ of $Q_{\Omega K}S$ is called an ΩK-adapted basis.

It is easily verified that in Definition 4.7 we could replace $Q_{\Omega K}$ by ΩK, i.e., $s_1,\dots,s_m$ is ΩK-adapted if and only if (4.8) holds for every set $\alpha_1,\dots,\alpha_m \in \Omega K$.

In the case when $\Omega K=\Omega^- K$, it can be seen that $\Omega^- K$-adapted sets coincide with properly independent sets (see HAMMER and HEYMANN [1981]) and minimal bases (see FORNEY [1975]) which have found many applications in system theory (see also WOLOVICH [1974], HAUTUS and HEYMANN [1978] and KAILATH [1980]).

Next we have the following theorem

(4.9) THEOREM. An ΩK-adapted set of nonzero elements $s_1,\dots,s_m \in Q_{\Omega K}S$ is ΛK-linearly independent.

Let $s_1,\dots,s_m \in \Lambda S$ be a set of elements and let $\Lambda[s_1,\dots,s_m]$ denote the ΛK-linear space spanned by $s_1,\dots,s_m$. We then have the following characterization of ΩK-adapted sets.

(4.10) THEOREM. Consider a set of nonzero elements $s_1,\dots,s_m \in Q_{\Omega K}S$ with $\operatorname{ord} s_i = (\gamma_i)_{\Omega K}$, $i=1,\dots,m$. Then $\{s_1,\dots,s_m\}$ is an ΩK-adapted set if and only if $\{\gamma_1 s_1,\dots,\gamma_m s_m\}$ forms a basis for the ΩK-module $\Lambda[s_1,\dots,s_m] \cap \Omega S$.

As an immediate consequence of the above theorem we have the following characterization of ΩK-adapted bases.

(4.11) COROLLARY. Assume the set $s_1,\dots,s_n \in Q_{\Omega K}S$ is a basis for ΛS with $\operatorname{ord} s_i = (\gamma_i)_{\Omega K}$, $i=1,\dots,n$. Then the set $\{s_1,\dots,s_n\}$ is ΩK-adapted if and only if $\{\gamma_1 s_1,\dots,\gamma_n s_n\}$ generates ΩS.

(4.12) EXAMPLE. Corollary 4.11 provides a particularly simple way for determining whether a basis $s_1,\dots,s_n$ of a ΛK-linear space ΛS is ΩK-adapted. Indeed, the main clause of the Corollary can be restated to read: The basis $s_1,\dots,s_n$ of ΛS is ΩK-adapted if and only if $\det[s_1,\dots,s_n] = \gamma_1^{-1}\cdot\gamma_2^{-1}\cdot\dots\cdot\gamma_n^{-1}$. Using this simple criterion, we show that the columns

$$s_1 = \begin{bmatrix} z \\ z^3 \\ z^4 \end{bmatrix}, \quad s_2 = \begin{bmatrix} z^2+1 \\ (z^2+1)^2 \\ z^4(z^2+1) \end{bmatrix}, \quad s_3 = \begin{bmatrix} 0 \\ 0 \\ z^3+1 \end{bmatrix}$$

form an (unordered) Ω^+K-adapted basis of ΛK^3. Indeed, we have $\operatorname{ord}_{\Omega^+K} s_1 = (z^{-1})_{\Omega^+K}$, $\operatorname{ord}_{\Omega^+K} s_2 = ((z^2+1)^{-1})_{\Omega^+K}$ and $\operatorname{ord}_{\Omega^+K} s_3 = ((z^3+1)^{-1})_{\Omega^+K}$, whence $\gamma_1^{-1}\cdot\gamma_2^{-1}\cdot\gamma_3^{-1} = z(z^2+1)(z^3+1)$

which is equal to $\det[s_1,s_2,s_3]$. If however, s_1, say, is replaced by $s_1'=(2z,z^3,z^4)^T$, the resulting set will no longer be Ω^+K-adapted since
$\det[s_1',s_2,s_3] = (z^3+1)(z^2+1)(z^3+2z)$.
□

We turn now to the characterization of ΩK-maps with the aid of ΩK-adapted bases. As a further consequence of Theorem 4.10 we have the following

(4.13) PROPOSITION. Let $\bar{f}: \Lambda U \to \Lambda Y$ be a ΛK-linear map and assume that $u_1,\dots,u_n$ is an ΩK-adapted basis for ΛU. Then $\bar{f}$ is an ΩK-map if and only if ord $u_i \subset$ ord $\bar{f}(u_i)$ for all $i=1,\dots,n$.

(4.14) DEFINITION. A ΛK-linear map $\bar{f}: \Lambda U \to \Lambda Y$ is called ΩK-order preserving (or, simply, order preserving) if for each $u \in Q_{\Omega K}U$, ord u = ord $\bar{f}(u)$.

(4.15) THEOREM. Let $\bar{f}: \Lambda U \to \Lambda Y$ be a ΛK-linear map and let $u_1,\dots,u_n \in Q_{\Omega K}U$ be an ΩK-adapted basis for ΛU. Then $\bar{f}$ is ΩK-order preserving if and only if (i) $\bar{f}(u_1),\dots,\bar{f}(u_n)$ is ΩK-adapted and (ii) for all $i=1,\dots,n$, ord u_i = ord $\bar{f}(u_i)$.

(4.16) THEOREM. Let $\bar{f}: \Lambda U \to \Lambda U$ be a surjective ΛK-linear map. Then $\bar{f}$ is ΩK-unimodular if and only if it is ΩK-order preserving.

5. Bounded ΩK-Modules

Let $\Delta \subset \Lambda S$ be an ΩK-module. We say that Δ is ΩK-bounded (or simply bounded) if there exists a nonzero element $\gamma \in Q_{\Omega K}$ such that $\gamma \in$ ord s for all $s \in \Delta$ (i.e., $\gamma s \in \Omega S$ for all $s \in \Delta$). It is clear that if Δ is a bounded ΩK-submodule of ΛS, it consists only of ΩK-rational elements. An ΩK-module consisting of ΩK-rational elements is called rational. If $\Delta \subset \Lambda S$ is bounded ΩK-submodule, we define the order of Δ, denoted ord Δ, as the class of all elements $\gamma \in Q_{\Omega K}$ such that $\gamma \in$ ord s for all $s \in \Delta$. It is easily seen that ord $\Delta = \bigcap_{s\in\Delta}$ ord s whence if $\Delta \neq 0$, ord Δ is a cyclic ΩK-module and is generated by an element $\psi \in Q_{\Omega K}$. Explicitly, ψ is a least common ΩK-multiple of all order generators $\gamma = \gamma(s)$ of elements $s \in \Delta$.

Next, we have the following :

(5.1) LEMMA. Let $\Delta \subset \Lambda S$ be a rational ΩK-submodule. Then Δ is bounded if and only if Δ has finite rank (i.e., is finitely generated) in which case rank $\Delta \leq$ dim S.

Below we make use of the Smith canonical form theorem for matrices over a principal ideal domain (see e.g. MACDUFFEE [1934] and NEWMAN [1972]). We shall identify ΛK-linear maps with their transfer function matrices. In particular, we shall speak of an ΩK-matrix if its entries are in ΩK and of an ΩK-unimodular matrix if both it and its inverse are ΩK-matrices. Smith's theorem is stated as follows:

(5.2) THEOREM. Let T be an mxn ΩK-matrix. Then there are ΩK-unimodular matrices M_L and M_R of dimensions mxm and nxn, respectively, and elements $\delta_1,\dots,\delta_r \in \Omega K$, uniquely defined up to multiples of units of ΩK, with $r \leq \min(m,n)$ and $\delta_{i+1} | \delta_i$, $i=1,\dots,r-1$, such that

(5.3) $$T = M_L\, D\, M_R$$

where D is the mxn matrix given by $D = \text{diag}(\delta_1,\ldots,\delta_r, 0,\ldots,0)$.

The elements $\delta_1,\ldots,\delta_r$ in Theorem 5.2 are called the invariant factors of T and the theorem itself is sometimes called the invariant factor theorem.

Assume now that $\Delta\subset\Lambda S$ is a nonzero and bounded ΩK-module with ord $\Delta = (\psi)_{\Omega K}$ and (in view of Lemma 5.1) let $d_1,\ldots,d_r\in\Delta$ be a basis for Δ. Then $\psi d_1,\ldots,\psi d_r\in\Omega S$ and the mxr matrix $\psi T: = [\psi d_1,\ldots,\psi d_r]$ (where ψd_i is viewed as a column vector) has Smith representation

(5.4) $$\psi T = M_L\, D\, M_R$$

where

$$D = \left[\begin{array}{ccc} \delta_1 & & 0 \\ & \ddots & \\ 0 & & \delta_r \\ \hline & 0 & \end{array}\right],$$

and the $\delta_i\in\Omega K$ (with $\delta_{i+1}|\delta_i$) are the invariant factors of ψT. We note that, by assumption, $\delta_1,\ldots,\delta_r$ are nonzero. Dividing both sides of (5.4) by ψ yields

(5.5) $$T = M_L\, D_o\, M_R$$

where D_o is the Mcmillan form of D, and is given by

$$D_o = \left[\begin{array}{ccc} \delta_1/\psi & & 0 \\ & \ddots & \\ 0 & & \delta_r/\psi \\ \hline & 0 & \end{array}\right]$$

Let d_{oi} denote the ith column of D_o. It is easily observed that the columns $d_{o1},\ldots,d_{or}\in Q_{\Omega K}S$ constitute an ΩK-adapted set. Indeed, for every set $\alpha_1,\ldots,\alpha_r\in Q_{\Omega K}$ we have that

$$d: = \sum_{i=1}^{r}\alpha_i d_{oi} = \begin{bmatrix} \alpha_1\frac{\delta_1}{\psi} \\ \vdots \\ \alpha_r\frac{\delta_r}{\psi} \\ 0 \\ \vdots \\ 0 \end{bmatrix}$$

and clearly ord $d = (\frac{\alpha_1\delta_1}{\psi})_{\Omega K}\cap\ldots\cap(\frac{\alpha_r\delta_r}{\psi})_{\Omega K}$.

Furthermore, we have

$$\Delta = T[\Omega S] = M_L D_o M_R[\Omega S] = M_L D_o[\Omega S] ,$$

the last equality following since M_R is ΩK-unimodular (see Corollary 3.9). Now, M_L is ΩK-unimodular, so that by Theorem 4.15 the columns of $M_L D_o$, given by $\frac{\delta_1}{\psi}M_{L1},\ldots,\frac{\delta_r}{\psi}M_{Lr}$ (where M_{Li} is the ith column of M_L) are also ΩK-adapted. Further, since M_L is ΩK-unimodular, it also follows that ord $M_{Li} = \Omega K$, whence, ord $\frac{\delta_i}{\psi}M_{Li} = \frac{\psi}{\delta_i}\Omega K = \left(\frac{\psi}{\delta_i}\right)_{\Omega K}$. We see immediately that the set $\frac{\delta_1}{\psi}M_{L1},\ldots,\frac{\delta_r}{\psi}M_{Lr}$ constitutes an ordered ΩK-adapted basis for Δ. We make the following further observation. Since $\delta_{i+1}|\delta_i$, it follows that

$$\left(\frac{\psi}{\delta_r}\right)_{\Omega K} \subset \left(\frac{\psi}{\delta_{r-1}}\right)_{\Omega K} \subset\ldots\subset\left(\frac{\psi}{\delta_1}\right)_{\Omega K}$$

so that

$$\text{ord } \Delta = (\psi)_{\Omega K} = \left(\frac{\psi}{\delta_r}\right)_{\Omega K}$$

and we conclude that δ_r is a unit in ΩK which, in particular, can always be chosen as $\delta_r = 1$.

We summarize the foregoing discussion with the following important theorem

(5.6) THEOREM. Let $\Delta\subset\Lambda S$ be a nonzero bounded ΩK-module. Then

- (i) Δ has an ordered ΩK-adapted basis $d_1,\ldots,d_r$.
- (ii) If $d_1',\ldots,d_r'$ is any other ordered ΩK-adapted basis of Δ then ord d_i' = ord d_i, $i = 1,\ldots,r$.

If $\Delta\subset\Lambda S$ is a bounded ΩK-module with ordered ΩK-adapted basis $d_1,\ldots,d_r$, then the set of ΩK-modules ord $d_i = \left(\frac{\delta_i}{\psi}\right)_{\Omega K}$, $i = 1,\ldots,r$ constitutes an important invariant of Δ and we call it the order trace of Δ.

Let $\Delta\subset\Lambda S$ be a bounded ΩK-module of rank r and let $d_1,\ldots,d_r$ be a basis of Δ. We can form the matrix $D: = [d_1,\ldots,d_r]$ and view Δ as the image of an ΩK-homomorphism $\Omega K^r\to\Lambda S$ defined by $e_i \mapsto De_i = d_i$. With this convention we then write Δ as $\Delta = D\Omega K^r$. We say that Δ is full (in ΛS) if rank Δ = dim S, i.e., if $\Delta = D\Omega S$ and D is nonsingular.

(5.7) THEOREM. Let Δ_1, $\Delta_2 \subset \Lambda S$ be bounded ΩK-submodules given by $\Delta_1 = D_1\Omega S$ and $\Delta_2 = D_2\Omega S$, respectively. Then $\Delta_2 \subset \Delta_1$ if and only if there exists an ΩK-matrix R (i.e., with entries in ΩK) such that $D_2 = D_1R$.

(5.8) COROLLARY. Let Δ_1, $\Delta_2 \subset \Lambda S$ be bounded ΩK-submodules given by $\Delta_1 = D_1\Omega S$ and $\Delta_2 = D_2\Omega S$. Assume Δ_1 is full and define $R: = D_1^{-1}D_2$. Then $\Delta_2 \subseteq \Delta_1$ if and only if R is an ΩK-matrix, with equality holding if and only if R is ΩK-unimodular.

We turn now to the existence of ΩK-adapted bases for ΛK-linear spaces. A ΛK-linear subspace $R\subset\Lambda S$ is called ΩK-rational if it has a basis $s_1,\ldots,s_k$ consisting of ΩK-rational vectors.

(5.9) THEOREM. Let dim S = n and let $R \subset \Lambda S$ be a nonzero ΩK-rational ΛK-linear subspace. Then (i) R has an ΩK-adapted basis, and (ii) every ΩK-adapted subset $s_1,\ldots,s_\ell \in R$ can be extended to an ΩK-adapted basis for R.

Next, we give the following characterization of the order trace.

(5.10) PROPOSITION. Let $\Delta, \Delta' \subset \Lambda S$ be nonzero and bounded ΩK-modules of equal rank m. Then there exists an ΩK-unimodular map M: $\Lambda S \to \Lambda S$ such that $M[\Delta] = \Delta'$ if and only if Δ and Δ' have the same order traces.

Related to the notion of ΩK-adapted bases is also the following

(5.11) DEFINITION. Let $R_1,\ldots,R_k \subset \Lambda S$ be ΩK-rational ΛK-linear subspaces. Then $R_1,\ldots,R_k$ are called ΩK-adapted if for every set of elements $s_1,\ldots,s_k$ where $s_i \in R_i$, $i = 1,\ldots,k$,

$$\text{ord }(s_1 +\ldots+ s_k) = \bigcap_{i=1}^{k} \text{ord } s_i \quad .$$

It follows readily from the above definition that the concept of ΩK-adapted subspaces is equivalent to the following : Let $R_1,\ldots,R_k \in \Lambda S$ be ΩK-rational ΛK-linear subspaces and let $d_{i1},\ldots,d_{i\ell_i}$ be a basis for R_i, $i = 1,\ldots,k$. Then the subspaces $R_1,\ldots,R_k$ are ΩK-adapted if and only if $d_{11},\ldots,d_{1\ell_1},\ldots,d_{k1},\ldots,d_{k\ell_k}$ is an ΩK-adapted basis for $R_1 +\ldots+ R_k$. Naturally, ΩK-adapted spaces are ΛK-linearly independent so that the above sum of subspaces is, in fact, a direct sum. Accordingly, we speak of ΩK-adapted direct sums of ΛK-linear spaces.

The concept of ΩK-adapted subspaces is of course a generalization to arbitrary P.I.D.'s of the concept of properly independent and stably independent spaces as defined in HAMMER and HEYMANN [1981], and in HAUTUS and HEYMANN [1980a], [1980b].

Theorem 5.9 leads to the following useful result.

(5.12) COROLLARY. Let $R_1 \subset R_2$ $(\subset \Lambda S)$ be ΩK-rational ΛK-linear subspaces. Then R_1 has an ΩK-adapted direct summand in R_2.

6. ΩK-Factorization and Invertibility

Consider two ΛK-linear maps $\bar{f}_1$: $\Lambda U \to \Lambda Y$ and $\bar{f}_2$: $\Lambda U \to \Lambda W$ and assume there exists an ΩK-map h: $\Lambda Y \to \Lambda W$ such that $\bar{f}_2 = \bar{h}\cdot\bar{f}_1$. Let $u \in \Lambda U$ satisfy the condition that $\bar{f}_1(u) \in \Omega Y$, or, in the notation of (3.7), that $u \in \ker\pi_{\Omega K}\bar{f}$. Then, obviously, $\bar{f}_2(u) = \bar{h}\cdot\bar{f}_1(u) \in \Omega W$ so that $u \in \ker\pi_{\Omega K}\bar{f}_2$, and the existence of the ΩK-map $\bar{h}$ such that $\bar{f}_2 = \bar{h}\cdot\bar{f}_1$, implies that $\ker\pi_{\Omega K}\bar{f}_1 \subset \ker\pi_{\Omega K}\bar{f}_2$. In case the maps $\bar{f}_1$ and $\bar{f}_2$ are ΩK-rational, the converse of the above statement is also true and we have the following central

(6.1) THEOREM. Let $\bar{f}_1$: $\Lambda U \to \Lambda Y$ and $\bar{f}_2$: $\Lambda U \to \Lambda W$ be ΩK-rational ΛK-linear maps. There exists an ΩK-map $\bar{h}$: $\Lambda Y \to \Lambda W$ such that $\bar{f}_2 = \bar{h}\cdot\bar{f}_1$ if and only if $\ker\pi_{\Omega K}\bar{f}_1 \subset \ker\pi_{\Omega K}\bar{f}_2$.

Theorem 6.1 depends on the following lemmas.

(6.2) LEMMA. <u>Let</u> $\bar{f}$: $\Lambda U \to \Lambda Y$ <u>be an</u> ΩK-<u>rational</u> ΛK-<u>linear map. Let</u> $r := \dim_{\Lambda K} \mathrm{Im}\bar{f}$ <u>and let</u> $Y_o \subset Y$ <u>be any</u> r-<u>dimensional subspace. Then there exists an</u> ΩK-<u>unimodular map</u> M: $\Lambda Y \to \Lambda Y$ <u>such that</u> $\mathrm{Im} M \cdot \bar{f} = \Lambda Y_o$.

(6.3) LEMMA. <u>Let</u> $\bar{f}$: $\Lambda U \to \Lambda Y$ <u>be a</u> ΛK-<u>linear map. If</u> $\mathcal{R} \subset \ker \pi_{\Omega K} \bar{f}$ <u>is a</u> ΛK-<u>linear subspace, then</u> $\mathcal{R} \subset \ker \bar{f}$.

Theorem 6.1 admits the following

(6.4) COROLLARY. <u>Let</u> $\bar{f}_1, \bar{f}_2$: $\Lambda U \to \Lambda Y$ <u>be</u> ΩK-<u>rational</u> ΛK-<u>linear maps. There exists an</u> ΩK-<u>unimodular map</u> M: $\Lambda Y \to \Lambda Y$ <u>such that</u> $\bar{f}_2 = M \cdot \bar{f}_1$, <u>if and only if</u> $\ker \pi_{\Omega K} \bar{f}_1 = \ker \pi_{\Omega K} \bar{f}_2$.

We call a ΛK-linear map $\bar{f}$: $\Lambda U \to \Lambda Y$ ΩK-<u>left invertible</u> if it has an ΩK-map as a left inverse. The following further corollary to Theorem 6.1 is also useful.

(6.5) COROLLARY. <u>An</u> ΩK-<u>rational</u> ΛK-<u>linear map</u> $\bar{f}$: $\Lambda U \to \Lambda Y$ <u>is</u> ΩK-<u>left invertible if and only if</u> $\ker \pi_{\Omega K} \bar{f} \subset \Omega U$.

Before concluding the section, we wish to express in an explicit form the main quantities that appeared in our discussion. Let $\bar{f}$: $\Lambda U \to \Lambda Y$ be an ΩK-rational ΛK-linear map. We start with an explicit representation of the ΩK-module $\ker \pi_{\Omega K} \bar{f}$. We shall identify the map $\bar{f}$ with its transfer matrix, and shall denote $r := \dim_{\Lambda K} \mathrm{Im}\bar{f}$. Let M_L: $\Lambda Y \to \Lambda Y$ and M_R: $\Lambda U \to \Lambda U$ be ΩK-unimodular maps such that $\bar{f} = M_L \cdot D \cdot M_R$, where the matrix D: $\Lambda U \to \Lambda Y$ is of the form $D = \begin{pmatrix} D_o & 0 \\ 0 & 0 \end{pmatrix}$, with D_o: $\Lambda K^r \to \Lambda K^r$ (square) nonsingular. One possible choice of D is, of course, the McMillan canonical form of $\bar{f}$. Also, we let $U_o \oplus U_1 = U$ be a direct sum decomposition, where $\Lambda U_o = \ker D$ and ΛU_1 is the domain of D_o.

Now, $\ker \pi_{\Omega K} \bar{f} = \ker \pi_{\Omega K} M_L D M_R = M_R^{-1}[\ker \pi_{\Omega K} M_L D]$, and, applying corollary 6.4, we obtain that $\ker \pi_{\Omega K} \bar{f} = M_R^{-1}[\ker \pi_{\Omega K} D]$. Further, it is readily seen that $\ker \pi_{\Omega K} D = D_o^{-1}[\Omega U_1] \oplus \Lambda U_o$, and, consequently, we have

(6.6) $$\ker \pi_{\Omega K} \bar{f} = M_R^{-1}[D_o^{-1}[\Omega U_1] \oplus \Lambda U_o], \qquad \text{and}$$

(6.7) $$\ker \bar{f} = M_R^{-1}[\Lambda U_o].$$

Defining now the map

$$\bar{f}_* := M_R^{-1} \begin{pmatrix} D_o^{-1} \\ 0 \end{pmatrix} : \Lambda U_1 \to \Lambda U ,$$

we have that

(6.8) $$\ker \pi_{\Omega K} \bar{f} = \bar{f}_*[\Omega U_1] + \ker \bar{f} ,$$

so that $\bar{f}_*$ generates the "bounded part" of $\ker \pi_{\Omega K} \bar{f}$.

Next, let $\bar{f}'$: $\Lambda U \to \Lambda Y'$ be a linear i/o map. We express now the condition of theorem 6.1 in more explicit form. The condition $\ker \pi_{\Omega K} \bar{f} \subset \ker \pi_{\Omega K} \bar{f}'$ is clearly equivalent to $\bar{f}'[\ker \pi_{\Omega K} \bar{f}] \subset \Omega Y'$. Substituting now (6.8), and noting that $\ker \bar{f}$ is a ΛK-linear subspace, the latter condition can be split into the two conditions: (i) $\bar{f}' \bar{f}_*[\Omega U_1] \subset \Omega Y'$, and (ii) $\bar{f}'[\ker \bar{f}] = 0$. These conditions are then equivalent to simply

(i_a) $\bar{f}\bar{f}_*$ *is an* ΩK-*map, and* (ii_a) $\ker\bar{f}\subset\ker\bar{f}'$, respectively.

Returning now to theorem 6.1, we can summarize as follows: *There exists an* ΩK-*map* $\bar{h}: \Lambda Y \to \Lambda Y$ *such that* $\bar{f}' = \bar{h}\cdot\bar{f}$ *if and only if* $\bar{f}'\bar{f}_*$ *is an* ΩK-*map, and* $\ker\bar{f}\subset\ker\bar{f}'$. Moreover, through a direct computation, one can show that, if $\bar{h}$ exists, then it is necessarily of the form

$$\bar{h} = (\bar{f}'\bar{f}_*,\ y_1,\ldots,y_{p-r})M_L^{-1} ,$$

where $p := \dim_K Y$, and $y_1,\ldots,y_{p-r}$ are (arbitrary) elements in $\Omega Y'$. Thus, the map $\bar{f}_*$, which generates the "bounded part" of $\ker\pi_{\Omega K}\bar{f}$, plays a central role in factorization theory.

7. Precompensation and Stable Output Feedback

We turn now to a brief discussion of some applications of the above factorization theory to stable (and causal) output feedback. We assume throughout the section that ΩK is either the ring $\Omega_{\mathcal{D}}K$ or the ring $\Omega_{\mathcal{D}}^{-}K$.

Let $\bar{f}: \Lambda U \to \Lambda Y$ be a linear i/o map and let $\bar{\ell}: \Lambda U \to \Lambda U$ be a bicausal ΛK-linear map (i.e., $\Omega^{-}K$-unimodular) which we regard as a precompensator for $\bar{f}$. We can express $\bar{\ell}^{-1}$ as

$$\bar{\ell}^{-1} = L^{-1}(I + \bar{h}) \tag{7.1}$$

where L is static (see HAUTUS and HEYMANN [1978]) and where $\bar{h}$ is strictly causal. If, additionally, we can express $\bar{h}$ as $\bar{h} = \bar{g}\bar{f}$ for some causal map $\bar{g}: \Lambda Y \to \Lambda U$ then we can give $\bar{\ell}$ an output feedback interpretation through the formula

$$\bar{f}\bar{\ell} = \bar{f}(I + \bar{g}\bar{f})^{-1}L ,$$

which is the i/o map of the composite system

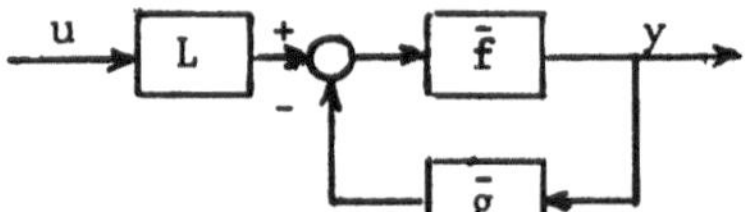

The map $\bar{g}$ is then clearly a causal (dynamic) output feedback compensator and L is a coordinate transformation map in the input value space. We may require additionally that the feedback compensator $\bar{g}$ be *stable*, i.e., an $\Omega_{\mathcal{D}}K$-map. We are then faced with the question of when can $\bar{h}$ of (7.1) be factored over $\bar{f}$ through an $\Omega_{\mathcal{D}}^{-}K$-map $\bar{g}$. The answer is provided by Theorem 6.1 and we have the following

(7.2) THEOREM. *Let* $\bar{f}: \Lambda U \to \Lambda Y$ *be an* $\Omega_{\mathcal{D}}K$-*rational linear i/o map, let* $\bar{\ell}: \Lambda U \to \Lambda U$ *be an* $\Omega_{\mathcal{D}}K$-*rational bicausal precompensator for* $\bar{f}$ *and express* $\bar{\ell}$ *as in* (7.1). *There exists a causal and stable output feedback representation for* $\bar{\ell}$ *if and only if* $\ker\pi_{\Omega_{\mathcal{D}}^{-}K}\bar{f}\subset\ker\pi_{\Omega_{\mathcal{D}}^{-}K}\bar{h}$.

We say that a linear i/o map $\bar{f}: \Lambda U \to \Lambda Y$ is $\Omega_{\mathcal{D}}K$-*minimum phase* (or, simply, *minimum phase*)

if it is an $\Omega_{\mathcal{D}}K$-map (i.e., stable) and is $\Omega_{\mathcal{D}}K$-left invertible. Thus $\bar{f}$ is $\Omega_{\mathcal{D}}K$-minimum phase precisely whenever

$$\ker \pi_{\Omega_{\mathcal{D}}K} \bar{f} = \Omega_{\mathcal{D}} U \ . \tag{7.3}$$

We recall further (see HAMMER and HEYMANN [1981]) that a linear i/o map $\bar{f}$ is called <u>nonlatent</u> if and only if

$$\ker \pi_{\Omega^- K} \bar{f} = z\Omega^- U \ ,$$

i.e., if and only if $z\bar{f}$ has a causal left inverse. Now, if $\bar{f}$ is $\Omega_{\mathcal{D}}K$-left invertible, so is also $(z+\alpha)\bar{f}$ where $(z+\alpha)\in\mathcal{D}$. In case $\bar{f}$ is nonlatent as well, then $(z+\alpha)\bar{f}$ also has a causal left inverse. Thus, one can readilly see that an i/o map $\bar{f}$ is <u>nonlatent and minimum phase</u> if and only if

$$\ker \pi_{\Omega^-_{\mathcal{D}} K} \bar{f} = (z+\alpha)\Omega^-_{\mathcal{D}} U \ . \tag{7.4}$$

We now have the following Theorem which is an analog to Corollary 5.4 in HAMMER and HEYMANN [1981]

(7.5) THEOREM. <u>Assume that for some</u> $\alpha,\beta\in K$, <u>both</u> $(z+\alpha)$ <u>and</u> $(z+\beta)$ <u>are in</u> $\mathcal{D}$, <u>and let</u> $\bar{f}: \Lambda U \to \Lambda Y$ <u>be an</u> $\Omega_{\mathcal{D}}K$<u>-rational and stable linear i/o map. Then</u> $\bar{f}$ <u>is nonlatent and minimum phase if and only if every</u> $\Omega^-_{\mathcal{D}}K$<u>-unimodular</u> ΛK<u>-linear precompensator</u> $\bar{\ell}: \Lambda U \to \Lambda U$ <u>has a causal and stable feedback representation</u> $(L,\bar{g})$, <u>i.e., there exists a pair</u> $(L,\bar{g})$ <u>with</u> L <u>static and</u> $\bar{g}$ <u>causal and</u> $\mathcal{D}$<u>-stable such that</u> $\bar{\ell} = (I+\bar{g}\bar{f})^{-1}L$.

The interest in Theorem 7.5 derives from the fact that stable injective linear i/s maps are always nonlatent and minimum phase. This fact is seen as follows. It was shown in HAMMER and HEYMANN [1980] that if $\bar{f}: \Lambda U \to \Lambda Y$ is an injective linear i/s map, it is <u>strictly observable</u>, i.e., $\ker \pi_{\Omega^+ K}\bar{f} \subset \Omega^+ U$. Let D be an Ω^+K-adapted basis matrix for $\ker \pi_{\Omega^+ K}\bar{f}$, that is, $D\Omega^+ U = \ker \pi_{\Omega^+ K}\bar{f}$. It is easily verified that we then also have that $D\Omega_{\mathcal{D}} U = \ker \pi_{\Omega_{\mathcal{D}} K}\bar{f}$. Now, the strict observability of $\bar{f}$ implies that D is a polynomial matrix and thus $D\Omega_{\mathcal{D}} U \subset \Omega_{\mathcal{D}} U$ (since $\Omega^+ K \subset \Omega_{\mathcal{D}} K$). We conclude that $\ker \pi_{\Omega_{\mathcal{D}} K}\bar{f} \subset \Omega_{\mathcal{D}} U$, and if the i/s map $\bar{f}$ is also stable, the minimum phase property (see (7.3)) follows. That injective linear i/s maps are nonlatent was proved in HAMMER and HEYMANN [1981] (Theorem 5.5). We summarize the above in the following.

(7.6) PROPOSITION. <u>If</u> $\bar{f}: \Lambda U \to \Lambda Y$ <u>is a stable injective linear i/s map, then it is nonlatent and minimum phase</u>.

We can now combine Theorem 7.5 with Proposition 7.6 to obtain the following result.

(7.7) COROLLARY. <u>Let</u> $\bar{f}: \Lambda U \to \Lambda Y$ <u>be a stable, injective linear i/o map and let</u> $\bar{\ell}: \Lambda U \to \Lambda U$ <u>be an</u> $\Omega^-_{\mathcal{D}}K$<u>-unimodular precompensator for</u> $\bar{f}$. <u>Then</u> $\bar{\ell}$ <u>has a stable causal (dynamic) state feedback representation in every stable realization of</u> $\bar{f}$.

ACKNOWLEDGEMENT

Helpful discussions with M.L.J. Hautus are gratefully acknowledged.

REFERENCES

G.D. FORNEY, Jr.
[1975] "Minimal bases of rational vector spaces with applications to multivariable linear systems", SIAM J. Control, 13, pp. 493-520.

J. HAMMER and M. HEYMANN
[1980] "Strictly observable rational linear systems", Preprint.

[1981] "Causal factorization and linear feedback", SIAM J. Control and Optimization, 19, pp. 445-468.

[1982] "Linear system factorization: feedback and stability", to appear.

M.L.J. HAUTUS and M. HEYMANN
[1978] "Linear feedback - an algebraic approach", SIAM J. Control and Optimization, 16, pp. 83-105.

[1980a] "New results on linear feedback decoupling" in Analysis and Optimization of Systems, A. Bensoussan and J.L. Lions, Eds., pp. 562-577, Lecture Notes in Control and Information Sciences, Vol. 28, Springer Verlag, New York.

[1980b] "Linear feedback decoupling: transfer function analysis", Preprint.

T. KAILATH
[1980] Linear Systems, Prentice Hall.

C.C. MACDUFFEE
[1934] The Theory of Matrices, Chelsea Publishing Co., New York.

A.S. MORSE
[1976] "System invariants under feedback and cascade control" in Mathematical System Theory, Udine 1975, pp. 61-74, Lecture Notes in Economics and Mathematical Systems, Vol. 131, Springer Verlag, New York.

M. NEWMAN
[1972] Integral Matrices, Academic Press, New York.

W.A. WOLOVICH
[1974] Linear Multivariable Systems, Springer Verlag, New York.

CONTROLLED INVARIANCE IN SYSTEMS OVER RINGS

by M.L.J. Hautus[†]

Abstract. The definition of controlled invariant (i.e. (A,B)-invariant) subspaces of a linear system is extended to systems over rings. It is observed that in this more general setting, the equivalence of the geometric and the feedback characterization is no longer true. Particular attention is paid to the weakly unobservable space V^*, and conditions are given for this space to satisfy the feedback characterization. These conditions have the form of the existence of a factorization of the transfer function. An application to the disturbance rejection problem is given.

[†]Dept. of Mathematics, University of Technology, Eindhoven

1. Introduction

The concept of controlled invariant subspace (abbreviated C.I.S.) (see [2]) (or (A,B)-invariant subspace, see [16]) has played a significant role in the development of linear system theory.

In view of the great potentiality of the theory of systems over rings (see, e.g. [15]), it is tempting to generalize the concept of controlled invariance to systems over rings. However, efforts in this direction are met by a serious obstacle. There are various equivalent characterizations for a C.I.S., the most well-known being the geometric characterization "$A\mathcal{V} \subseteq \mathcal{V} + \text{im } B$" and the feedback characterization: "there exists F such that $(A + BF)\mathcal{V} \subseteq \mathcal{V}$" (see [2,16]). These properties are no longer equivalent in the ring case! It is easily seen that the feedback characterization implies the geometric condition, but the converse is not true. The main reason of this difficulty is that for spaces over rings (i.e. modules), subspaces are not necessarily direct summands, so that the map F can be defined on $\mathcal{V}$ (supposing that $\mathcal{V}$ is free) but it cannot be extended to a map defined on the whole state space $\mathcal{X}$. As a consequence of this state of affairs, we introduce in addition to a C.I.S (i.e. a space $\mathcal{V}$ satisfying $A\mathcal{V} \subseteq \mathcal{V} + \text{im } B$) another type of subspace, viz. a C.I.S of the feedback type, abbreviated C.I.S.F., i.e. a space for which there exists F such that $(A + BF)\mathcal{V} \subseteq \mathcal{V}$. A C.I.S. is more manageable than a C.I.S.F. and it behaves like in the field case. For example, the sum of two C.I.S.'s is again a C.I.S. and if $\mathcal{K}$ is an arbitrary subspace, there exists a largest C.I.S. contained in $\mathcal{K}$. Neither of these statements is true for C.I.S.F.'s! This is very inconvenient, because a C.I.S.F. is the type of space we need in applications.

We will spend most of our attention to a particular C.I.S., the space $\mathcal{V}^*$ of weakly unobservable states (compare [14]), which in the case of a strictly causal system reduces to the largest C.I.S. contained in ker C (for details on notation see section 2), and we will investigate the question of when $\mathcal{V}^*$ has the feedback property. A necessary and sufficient condition for this to be the case will be given in the form of a factorization condition on the transfer function, assuming that the system is reachable and injective. Under these assumptions, it will follow that for a single input system, $\mathcal{V}^*$ has always the feedback property. Also, it follows from the factorization condition that is does notdepend on the realization whether $\mathcal{V}^*$ has the feedback property or not, as long as the realization is reachable. For a similar situation we refer to [5].

In section 5 a result by G. Conte and A. Perdon is given, which states that in the case when $\mathcal{R}$ is a principal ideal domain, $\mathcal{V}^*$ has the feedback property if and only if it is a direct summand.

Finally, in section 6 an application is given to the disturbance rejection problem.

2. Controlled invariance and the feedback property

In this section, R denotes an integral domain with unit element and A,B,C,D are matrices over R of dimensions $n \times n$, $n \times m$, $r \times n$, $r \times m$, respectively. The matrix quadruple (A,B,C,D) will be called a (free) system and denoted by Σ. We have in mind particularly the discrete time interpretation of Σ:

(2.1) $$x_{t+1} = Ax_t + Bu_t \ , \ y_t = Cx_t + Du_t \ .$$

The quantities u_t, x_t and y_t are called input, state and output, respectively, and they are elements of $U := R^m$, $X := R^n$ and $Y := R^r$, respectively. For a given input sequence $\underline{u} = (u_t)_{t=0}^{\infty}$ and $x_0 \in X$ we denote by $x_t(x_0,\underline{u})$ the state at time t resulting via (2.1) from initial value x_0 and input $\underline{u}$. The corresponding output $Cx_t(x_0,\underline{u}) + Du_t$ is denoted as $y_t(x_0,\underline{u})$.

Σ is called reachable if for every $\tilde{x} \in X$ a number $\tau > 0$ exists and an input $\underline{u}$ such that $x_\tau(0,\underline{u}) = \tilde{x}$. Necessary and sufficient for Σ to be reachable is that the $n \times nm$ matrix $[B,AB,\ldots,A^{n-1}B]$ be right invertible.

A subspace $V \subseteq X$ is called a controlled invariant subspace (=: C.I.S.) if for each $x_0 \in V$ there exists an input sequence $\underline{u}$ such that $x_t(x_0,\underline{u}) \in V$ for $t = 0,1,\ldots$. The following criterion is immediate:

(2.2) PROPOSITION. V is a C.I.S. iff $AV \subseteq V + \operatorname{im} B$.

A subspace $V \subseteq X$ is called a controlled invariant subspace of the feedback type (=: C.I.S.F.) if there exists $F \in R^{m\times n}$ such that $(A + BF)V \subseteq V$. A C.I.S.F. is easily seen to be a C.I.S. but the converse is not true.

(2.3) EXAMPLE. Let $R := \mathbb{R}[\sigma]$, $X := R^2$,

$$A := \begin{bmatrix} 1 & 1 \\ 0 & 1 \end{bmatrix} , \ B := \begin{bmatrix} \sigma & 0 \\ 0 & 1 \end{bmatrix} , \ V = \operatorname{im} \begin{bmatrix} 0 \\ \sigma \end{bmatrix} .$$

We have

$$A\begin{bmatrix} 0 \\ \sigma \end{bmatrix} = \begin{bmatrix} \sigma \\ \sigma \end{bmatrix} = \begin{bmatrix} 0 \\ \sigma \end{bmatrix} + \begin{bmatrix} \sigma & 0 \\ 0 & 1 \end{bmatrix} \begin{bmatrix} 1 \\ 0 \end{bmatrix} .$$

Hence $AV \subseteq V + \operatorname{im} B$. Now suppose that for

$$F = \begin{bmatrix} f_{11} & f_{12} \\ f_{21} & f_{22} \end{bmatrix}$$

we have $(A + BF)V \subseteq V$, i.e.

$$\begin{bmatrix} 1 + \sigma f_{11} & 1 + \sigma f_{12} \\ f_{21} & 1 + f_{22} \end{bmatrix} \begin{bmatrix} 0 \\ \sigma \end{bmatrix} = p(\sigma) \begin{bmatrix} 0 \\ \sigma \end{bmatrix}$$

for some polynomial $p(\sigma)$. The first equation reads $\sigma + \sigma^2 f_{12} = 0$, i.e. $f_{12} = -1/\sigma$, so that $f_{12} \notin R$. Notice that the system in this example is reachable. □

A simple way of finding counterexamples to various conjectures about C.I.S.'s and C.I.S.F.'s is given by G. Conte and A.M. Perdon; see section 5.

We will concentrate on a special C.I.S.

(2.4) DEFINITION. Given Σ, a state x_0 is called <u>weakly unobservable</u> if there exists an input $\underline{u}$ such that $y_t(x_0,\underline{u}) = 0$ for $t = 0,1,\ldots$. The set of weakly unobservable states is called the <u>weakly unobservable space</u> and is denoted by $V^*(\Sigma)$, or simply by V^*, if the underlying system is clear.

The following result is easily shown:

(2.5) PROPOSITION. V^* <u>is a</u> C.I.S. If $D = 0$, <u>then</u> V^* <u>is the largest</u> C.I.S. <u>contained</u> in ker C.

PROOF. That V^* is a subspace (i.e. a submodule of X) is immediately obvious. To show that V^* is a C.I.S. assume that $x_0 \in V$ and $\underline{u} = (u_0,u_1,\ldots)$ are such that $y_t(x_0,\underline{u}) = 0$ for $t = 0,1,2,\ldots$. Then $x_1 = Ax_0 + Bu_0$ is also in V^* since $y_t(x_1,\tilde{\underline{u}}) = 0$ for $t = 0,1,\ldots$, where $\tilde{\underline{u}} := (u_1,u_2,\ldots)$. Hence $AV^* \subseteq V^* + \mathrm{im}\, B$. The second statement follows from the definition. □

We denote by $N(\Sigma)$ (or by N) the subspace of <u>unobservable states</u>, i.e. the set of initial states x_0 for which $y_t(x_0,\underline{0}) = 0$ for $t = 0,1,\ldots$, where $\underline{0}$ stands for the input sequence $(0,0,\ldots)$. Obviously,

$$N \subseteq V^*. \tag{2.6}$$

Furthermore, N is well known and easily seen to be A-invariant (i.e. $AN \subseteq N$).

A feedback transformation has the form

$$u_t = Fx_t + v_t, \tag{2.7}$$

where $F \in R^{m\times n}$ and where v_t is considered a new input variable. Such a transformation yields a new system

$$\Sigma_F := (A + BF, B, C + DF, D),$$

in discrete time interpretation:

$$\begin{aligned} x_{t+1} &= (A + BF)x_t + Bv_t, \\ y_t &= (C + DF)x_t + Dv_t. \end{aligned} \tag{2.8}$$

It is easily seen that the set of state trajectories (i.e. state sequences resulting from some input) for a given initial state is invariant under a feedback transformation. In particular, for a given x_0, if there exists an input $\underline{u}$ for Σ such that $y_t(x_0,\underline{u}) = 0$ for all t, then there exists an input $\underline{v}$ for system Σ_F such that the output is identically zero. Consequently,

(2.9)PROPOSITION. V^* is feedback invariant, i.e. $V^*(\Sigma_F) = V^*(\Sigma)$ for all F.

Combining this result with (2.6) we find that

$$N(\Sigma_F) \subseteq V^*$$

for all F.

By definition, for every $x_0 \in V^*$ there exists $\underline{u}$ such that $y_t(x_0,\underline{u}) = 0$ for all t. We say that V^* has the feedback property if there exists a feedback $F \in R^{m\times n}$ such that for each $x_0 \in V^*$, the feedback input $\underline{u}$ defined by $u_t = Fx_t$ for $t = 0,1,\ldots$ yields zero output. For systems over a field, V^* has always the feedback property (see [14]), but for rings this is not the case (for an example see Example 5.6). We have the following characterizations:

(2.10). PROPOSITION. The following statements are equivalent

i) V^* has the feedback property with feedback F,
ii) $V^* = N(\Sigma_F)$,
iii) $(A + BF)V^* \subseteq V^*$ and $V^* \subseteq \ker(C + DF)$.

PROOF. i) ⇔ ii): V has the feedback property with feedback F iff the transformed system (2.8) yields zero output for zero input v_t, for every $x_0 \in V^*$. This will be the case iff $V^* \subseteq N(\Sigma_F)$. The converse inclusion is always satisfied.
ii) ⇔ iii): Since $N(\Sigma_F)$ is the largest $(A + BF)$-invariant subspace contained in $\ker(C + DF)$, iii) is equivalent to ii). □

It follows in particular that, if V^* has the feedback property, it is a C.I.S.F. If $D = 0$, the converse is true, since in this case $V^* \subseteq \ker C$.
Finally we mention

(2.11) COROLLARY. If V is a subspace for which there exists a feedback F such that

(2.12) $\quad (A + BF)V \subseteq V \;,\; V \subseteq \ker(C + DF),$

then $V \subseteq V^*$.

PROOF. Condition (2.12) implies $V \subseteq N(\Sigma_F)$. □

3. Input-output conditions for the feedback property

In this section we want to formulate conditions for V^* to have the feedback property in terms of the input-output behavior of Σ, specifically, in terms of the transfer function of Σ. For this it is convenient to identify input or output sequences with formal power series. We want to take a slightly more general point of view than in the previous section in the sense that we allow input sequences which start at an arbitrary, possibly negative, time. That is, an input sequence will be a doubly infinite sequence $(u_t)_{t\in\mathbb{Z}}$ with the property that a number t_0 exists such that $u_t = 0$ for $t < t_0$. Such a sequence will be identified with the formal Laurent series $\Sigma u_t z^{-t}$. Similarly we proceed for output sequences.

A rational function $\varphi = n/d$ ($\in R(z)$) is called <u>expandable</u> is there exists a formal Laurent series ψ such that $d\psi = n$. In this case we identify φ with ψ. Using long division one can show that φ is expandable if d is <u>monic</u>, i.e. has leading coefficient equal to 1.

(3.1) LEMMA. <u>If</u> R <u>is Noetherian then any expandable rational function has a representation n/d with monic</u> d.

For a proof see [15], or Corollary A.4.

A rational function ϕ is <u>causal</u>, if it is expandable and if its formal Laurent series is causal, i.e. has $u_t = 0$ for $t < 0$. It is easily seen that for an expandable n/d to be causal it is necessary and sufficient that $\deg n \leq \deg d$. Finally, φ is called bicausal if φ and $1/\varphi$ are causal. Similar terminology is used for rational matrices. In particular, a matrix L is bicausal if L and L^{-1} are causal.

Let us return to the system Σ given by (2.1). To Σ is associated its <u>transfer function</u>

(3.2) $$T(z) := C(zI - A)^{-1}B + D.$$

If $x_t = 0$ for sufficiently negative t, we have the relation

$$y(z) = T(z)u(z).$$

The matrix T(z) has the representation $T(z) = N(z)/d(z)$ where N(z) is a polynomial matrix and $d(z) := \det(zI - A)$ is monic. It follows that T(z) is causal (see (3.2)).

(3.3) DEFINITION. Let T(z) be a rational matrix. Then T(z) is called <u>injective</u> if $T(z)u(z) = 0$ implies $u(z) = 0$ for every formal Laurent series u(z). Further, T(z) is called <u>strongly injective</u> if every formal Laurent series u(z) for which T(z)u(z) is polynomial, is itself a polynomial.

If T(z) is strongly injective, it is also injective, for, if $T(z)u(z) = 0$ then $T(z)(z^{-k}u(z)) = 0$ for all k. Hence $z^{-k}u(z)$ is polynomial for all k, which is only possible if $u(z) = 0$. In the definition of injectivity we could have used polynomial, or rational, or causal, or expandable rational functions u(z) instead of formal Laurent series. This would have resulted in an equivalent concept of injectivity. The concept of strong injectivity is more sensitive, however. Instead of formal Laurent series we could have used expandable rational functions, as follows easily from

Lemma A1, but if we would have used polynomial or rational u(z)'s a different concept of strong injectivity would have resulted. The definition uses formal Laurent series in order that for transfer functions T(z) the system theoretic meaning be immediately obvious. In fact, we introduce:

(3.4) DEFINITION. System Σ of (2.1) is called injective if for any pair of inputs $\underline{u}$ and $\underline{v}$ and any $x_0 \in X$ we have that $y_t(x_0,\underline{u}) = y_t(x_0,\underline{v})$ for all t implies that $\underline{u} = \underline{v}$. System Σ is called strongly injective if for any pair of inputs $\underline{u}$ and $\underline{v}$, any $x_0 \in X$ and any $t_1 \geq 0$ we have that $y_t(x_0,\underline{u}) = y_t(x_0,\underline{v})$ for $t \geq t_1$ implies that $u_t = v_t$ for $t \geq t_1$.

It is straightforward that Σ is (strongly) injective iff T_Σ is (strongly) injective.

(3.5) REMARK. The concept of strong injectivity for systems over a field has appeared in literature under various names: A strongly invertible system is called strictly observable in [7], irreducible in [12] and feedback irreducible in [11]. For injective systems strong injectivity is equivalent to the absence of zeros and it is closely related to the concept of strong observability as discussed in [14]. □

The following result connects strong injectivity with the concepts of the previous section:

(3.6) THEOREM. Suppose that Σ is reachable. Then Σ is strongly injective iff Σ is injective and $N(\Sigma) = V^*(\Sigma)$.

PROOF. "if": Let $y_t(0,\underline{u}) = 0$ for $t \geq t_1$. Then $\tilde{x}_0 := x_{t_1} \in V^*$. Consequently, $x_{t_1} \in N(\Sigma)$ and hence $y_t(\tilde{x}_0,0) = 0$. On the other hand, $y_t(\tilde{x}_0,\tilde{u}) = y_{t+t_1}(0,\underline{u}) = 0$ for $t \geq t_1$, where $\tilde{u} := (u_{t_1}, u_{t_1+1},\ldots)$. Injectivity implies $\tilde{u} = 0$.
"only if": Let $x_1 \in V^*$. Since Σ is reachable there exists $\underline{\tilde{u}}$ and $t_1 \geq 0$ such that $x_1 = x_{t_1}(0,\underline{\tilde{u}})$. In addition, there exists $\underline{\hat{u}}$ such that $y_t(x_1,\underline{\hat{u}}) = 0$ $(t \geq 0)$. Concatenation of $\underline{\tilde{u}}$ and $\underline{\hat{u}}$ at t_1 yields the input sequence $\underline{u} := (\tilde{u}_0,\tilde{u}_1,\ldots,\tilde{u}_{t-1},\hat{u}_0,\hat{u}_1,\ldots)$, which has the property that $y_t(0,\underline{u}) = 0$ for $t \geq t_1$. By the strong injectivity of Σ this implies that $u_t = 0$ for $t \geq t_1$, i.e. $\underline{\hat{u}} = \underline{0}$. Hence $y(x_1,\underline{0}) = 0$. We see that $x_1 \in N(\Sigma)$. □

Now we are in the position to formulate a criterion in terms of the transfer function for V^* to have the feedback property.

(3.7) THEOREM. Let Σ be injective and reachable and let $T := T_\Sigma$. Then V^* has the feedback property iff there exists a bicausal L such that TL is strongly injective.

PROOF. "only if": If $(A + BF)V^* \subseteq V^*$, $V^* \subseteq \ker (C + DF)$, then Proposition 2.10 and Theorem 3.6 imply that Σ_F is strongly injective. Since $T_{\Sigma_F} = T_\Sigma L$, where

$$L := L_F := (I - FT_s)^{-1}$$

and $T_s(z) := (zI - A)^{-1}B$, and L is bicausal, the condition of the theorem follows. "if": Let L be bicausal and TL = S be strongly injective. By the extension to systems over rings of [9, Thm. 5.7] (see also [5]) we know that L can be realized by feedback (i.e. there exists F such that $L = L_F$) iff for any polynomial u we have: If $T_s u$ and u are polynomial then $L^{-1}u$ is polynomial. If $T_s u$ and u are polynomial then $Tu = CT_s u + Du$ is polynomial and hence $SL^{-1}u$ is polynomial. Since S is strongly injective, if follows that $L^{-1}u$ is polynomial. Hence there exists F such that $L = L_F$ and $TL = T_{\Sigma_F}$. Because Σ_F is strongly injective it follows that $N(\Sigma_F) = V^*(\Sigma_F) = V^*$. Hence V^* has the feedback property. □

As a consequence of this theorem, it does not depend on the realization whether or not $V(\Sigma)$ has the feedback property, as long as the realization is reachable.

A further conclusion can be drawn from Theorem 3.7. By definition, V^* has the feedback property if there exists a feedback control $u_t = Fx_t$ such that the output will be identically zero for every $x_0 \in V^*$. Now suppose we want to relax this condition by allowing dynamic state feedback, i.e. a system Φ with input x and output u given by the relation $u = F(z)x + v$ where $F(z) = T_\Phi(z)$. This yields a combined system with transfer function S := TL where $L(z) := (I - F(z)T_s(z))^{-1}$. We claim that the resulting system Σ_F is strongly injective. In fact, the compensator is chosen in such a way that the input $\underline{v} = \underline{0}$ yields $\underline{y} = \underline{0}$ for every $x_0 \in V^*$, so that $V^* = N(\Sigma_F)$. Since TL = S is strongly injective and L bicausal, it follows from Theorem 3.7 that V^* has the feedback property, so that invariance could have been obtained by static state feedback. Nothing was gained by allowing dynamic feedback (compare [5]).

The following is a modified version of Theorem 3.7. The condition of Theorem 3.7 can be interpreted as the possibility to factorize the transfer function into $T = SL^{-1}$, where S is strongly injective and L^{-1} is bicausal. Now we give a characterization in which less stringent conditions are imposed on the factorization.

(3.8) THEOREM. Let Σ be injective and reachable. Then V^* has the feedback property iff $T := T_\Sigma$ can be factorized as T = PR where P is (not necessarily causal) strongly injective and R is causal and left invertible with an expandable (but not necessarily causal) left inverse S.

Necessity is obvious since the factorization $T = SL^{-1}$, mentioned before, satisfies the conditions. For sufficiency we decompose S as $S = S_+ + S_-$, where S_+ is the polynomial part and S_- is strictly causal. Then $S_+R = I - S_-R$ is rational and bicausal. Let $L := (I - S_-R)^{-1}$. We have $S_+RL = I$. It follows that RL is strongly injective. The result follows from Theorem 3.7. □

4. Systems over Noetherian unique factorization domain

In this section we assume that R is a N.U.F.D. (:= Noetherian unique factorization domain. (see [1, Ch. 4], [13])). Then $R[z]$ is also N.U.F.D. For this type of ring it is possible to give conditions for a rational matrix to be strongly injective.

A prime element p of $R[z]$ is either essentially monic, i.e. of the form $p = \alpha p^-$ where α is a unit and p^- is monic, or p has a noninvertible leading coefficient. When multiplication by units is allowed we will always assume that prime factors have been chosen monic whenever possible. Any element $r \in R[z]$ can be factored as $r = r^+r^-$, where r^- is the product of the monic prime factors of r and hence monic, and r^+ is the product of the nonmonic prime factors of r. We call r^- the monic part of r and r^+ the nonmonic part (see [5] for somewhat more general concepts). We say that p is completely nonmonic if $r^- = 1$. It is easily seen that $p|q$ (p divides q) implies $p^-|q^-$.

(4.2) THEOREM. Let $P = N/d$ be a rational $r \times m$ matrix with monic denominator d and injective numerator matrix N. Then, P is strongly injective if the monic part χ^- of the G.C.D. χ of the $m \times m$ minors of N divides d.

PROOF. Let u be a formal Laurent series and $Pu =: v$ be a polynomial. Since N is injective, it contains a nonzero $m \times m$ minor χ_1. The equality $Nu = dv$ implies that $\chi_1 u$ is a polynomial, hence that u is rational. But, since u is formal Laurent series, it must be expandable. Hence (see Lemma (3.1)), u has a representation of the form $u = w/\psi$ where ψ is monic. The equality $Nw = \psi dv$ implies that for every $m \times m$ submatrix N_i of N we have $N_i w = \psi d v_i$ for some polynomial vector v_i. Multiplying by the adjoint matrix adj N_i we find that $\psi d|\chi_i w$, where χ_i denotes det N_i. Since this is true for all $m \times m$ submatrices it follows that $\psi d|\chi w$. Taking monic parts we obtain $\psi d|\chi^- w^-$. Since, by assumption, $\chi^-|d$ we have $\psi|w^-$ and a fortiori $\psi|w$. Hence $u = w/\psi$ is a polynomial. □

The converse of this theorem is not true, not even when R is a field. However, if $d = 1$, i.e. if $P = N$ is a polynomial matrix it can be shown that the condition $\chi^- = 1$ is necessary. In fact, let α be a (monic) prime factor of χ^- and let $\bar{P}$ denote the matrix with entries which are the residues modulo α in $R/(\alpha)$, which is an integral domain. Since all $m \times m$ determinants of $\bar{P}$ are zero there exists a nonzero m-vector $\bar{u}$ over $R/(\alpha)$ such that $\bar{P}\bar{u} = 0$. If u is a representative of $\bar{u}$, then $\bar{u} \neq 0$ implies $\alpha \nmid u$. We have $Pu = \alpha v$ for some polynomial v. Hence $P(u/\alpha)$ is a polynomial, u/α is expandable but not a polynomial.

We give some applications of the above result:

(4.2) COROLLARY. If Σ is injective and reachable, and $m = 1$ then V^* has the feedback property.

Proof. Let the ith entry of T be n_i/d_i, where d is monic. We have the factorization

$$T = \mathrm{diag}(n_i^+)\mathrm{col}(n_i^-/d_i),$$

where $\mathrm{diag}(\alpha_i)$ and $\mathrm{col}(\alpha_i)$ denote the diagonal matrix and column, respectively, with entries α_i. The polynomial matrix $P := \mathrm{diag}(n_i^+)$ is strongly injective because of Theorem 4.1 since det P is completely nonmonic. Also, the matrix $R := \mathrm{col}(n_i^-/d_i)$ has an expandable left inverse. In fact, choose any $n_k^- \neq 0$ and $S := [0,\dots,0,d_k/n_k^-,0,\dots,0]$ will do. □

More generally, we have

(4.3) COROLLARY. If Σ is reachable and injective, $T_\Sigma = N/d$ and χ^- (as defined in Theorem 4.1) satisfies

$$\deg \chi^- \leq \deg d,$$

then V^* has the feedback property.

PROOF. We can factorize as follows:

$$T = (N/\chi^-)(\chi^-/d).$$ □

(4.4) EXAMPLE. Let

$$T(z) := z^{-6}\begin{bmatrix} z^5 - 1 & \sigma z^5 - 2z - \sigma \\ z^3 - z^2 & \sigma z^3 - 1 \end{bmatrix}$$

be the transfer matrix of a reachable system Σ over $R := \mathbb{R}[\sigma]$. Then the determinant of the numerator equals

$$\chi(z) = \sigma z^7 - z^5 + 2z^4 - 2z^3 - \sigma z^2 + 1.$$

This polynomial is nonmonic, so that it contains a nonmonic part of degree at least one. Hence, $\deg \chi^- \leq 6$ so that Corollary 4.3 implies that V^* has the feedback property. Actually, it can easily be seen that χ does not have a nonmonic factor of degree 1, so that $\deg \chi^- \leq 5$. Consequently, even if the denominator is z^{-5}, V^* has the feedback property. □

One might be tempted to conjecture that V^* always has the feedback property. This is not the case, as can been seen from Example 5.6.

Contrary to the theorems of the previous section, the results of this section are completely constructive, provided we have a constructive way of computing prime factors of polynomials over R. Not only conditions for V^* to have the feedback property, but also explicit constructions of V^* and the desired feedback can be derived from the results of this and the previous sections. In [5] and [10] it is indicated how a feedback F can explicitly be constructed for a given bicausal L in Theorem 3.7. Furthermore, the space V^* is computed as the unobservable space of Σ_F.

5. Systems over Principal Ideal Domains

The results of this section are mainly due to G. Conte and A.M. Perdon ([4]). We recall the following definition (see [3, Def. 1.9]).

(5.1) DEFINITION. Given a subspace (i.e. an R-submodule) V of R^n, the closure $\bar{V}$ of V is defined as the set of all $x \in R^n$ such that $\alpha x \in V$ for some $\alpha \in R$. V is said to be closed if $\bar{V} = V$.

We assume throughout this section that R is a principal ideal domain. Then we have:

(5.2) PROPOSITION. A subspace $V \subseteq R^n$ is closed iff it is a direct summand (see [3, Prop. 1.10, iv)]).

The following simple observation is crucial:

(5.3) PROPOSITION. If V is a C.I.S.F. of Σ (defined in (2.1)) then so is $\bar{V}$.

PROOF. If $(A + BF)V \subseteq V$, then it is easily seen that $(A + BF)\bar{V} \subseteq \bar{V}$. □

A similar result for C.I.S.'s is not true. This gives us the possibility of verifying that a given C.I.S. is not a C.I.S.F. Let us reconsider Example 2.3. The space V, which is shown to be a C.I.S. has a closure $\bar{V} = \text{im}[0,1]'$ which is not a C.I.S., since $A\bar{V} = \text{im}[1,1]' \not\subseteq \bar{V} + \text{im}\,B = \text{im}\,B$. It follows again that V is not a C.I.S.F. In a similar way counterexamples may be given to various conjectures one might have. For instance, it is possible to find two C.I.S.F.'s the sum of which is not a C.I.S.F. Also, one might think that V^* is closed if $\text{im}\,B$ is, but an example can be given showing that this is not the case.

The main result of this section is

(5.4) THEOREM. V^* has the feedback property iff it is closed.

PROOF. If V^* has the feedback property then so does $\bar{V}^*$, since $\ker(C + DF)$ is closed. Consequently, $\bar{V}^* \subseteq V^*$ (see Corollary 2.11) and hence $\bar{V}^* = V^*$.

Conversely, if V^* is closed it is a direct summand. Definition 2.4 implies that for each $x_0 \in V^*$ there exists $u \in U$ such that

$$(5.5) \qquad Ax_0 + Bu_0 \in V^* \ , \ Cx_0 + Du_0 = 0.$$

Since V^* is free (being a submodule of free module over a P.I.D., see [6, Thm. 7.8]), it has a basis, say $x_1,\dots,x_k$. Define $F_1 : V^* \to U$ by $Fx_i = u_i$, where u_i is chosen according to (5.5). Since V^* is a direct summand, F_1 can be extended to a map $F : X \to U$. Because of (5.5) we have $(A + BF)V^* \subseteq V^*$ and $(C + DF)V^* = 0$. □

We conclude this section with an example of a reachable injective Σ for which V^* does not have the feedback property.

(5.6) Let $R = \mathbb{R}[\sigma]$

$$A := \begin{bmatrix} 1 & 1 & 0 \\ 0 & 1 & 0 \\ 0 & 0 & 2 \end{bmatrix}, \quad B := \begin{bmatrix} \sigma & 0 \\ 0 & 1 \\ 0 & 1 \end{bmatrix}, \quad C := \begin{bmatrix} 1 & 0 & 0 \\ 0 & 0 & 1 \end{bmatrix}, \quad D := 0.$$

It is easily seen that Σ is injective and $V^* = \text{im}[0,\sigma,0]'$. But $\bar{V}^* = \text{im}[0,1,0]' \neq V^*$ hence V^* is not a C.I.S.F.

6. Disturbance rejection

In the system Σ_1:

(6.1) $$x_{t+1} = Ax_t + Bu_t + Eq_t, y_t = Cx_t + Du_t,$$

where q_t is a disturbance input, we try to find a feedback control $u_t = Fx_t$ such that in the resulting system, y becomes independent of q. If we have found such an F we say that we have solved the disturbance rejection problem and that we have obtained disturbance rejection by state feedback.

The following is a straightforward generalization of a well-known result for systems over fields.

(6.2) PROPOSITION. *Disturbance rejection by a state feedback* F *is achieved iff there exists a* $(A + BF)$*-invariant subspace* V *such that*

$$\text{im}\, E \subseteq V \subseteq \ker(C + DF).$$

The proof is straightforward and omitted. One can make this criterion for the solvability of the disturbance rejection problem more constructive if there exists a largest subspace V for which there exists F such that $(A + BF)V \subseteq V \subseteq \ker(C + DF)$. In general such a subspace does not exist (contrary to the field case). However we have the following result:

(6.3) THEOREM. Let $\Sigma := (A,B,C,D)$ *be such that* $V^*(\Sigma)$ *has the feedback property. Then disturbance rejection by state feedback is possible iff*

$$\text{im}\, E \subseteq V^*.$$

In fact, if V^* has the feedback property, it is the largest space satisfying (2.12), see Corollary (2.11).

Next we give a frequency domain characterization for V^* analogous to a characterization given in [8]. Introducing the formal power series

$$\omega(z) := \sum_0^\infty u_t z^{-t-1} \;, \quad \xi(z) := \sum_0^\infty x_t z^{-t-1} \;,$$

$\eta(z) := \Sigma y_t z^{-t}$, the equations

$$x_{t+1} = Ax_t + Bu_t \;, \quad y_t = Cx_t + Du_t \;, \qquad t \geq 0$$

with initial state x_0 can be written as

$$x_0 = (zI - A)\xi(z) - B\omega(z) \;, \quad \eta(z) = C\xi(z) + D\omega(z).$$

Hence we can write: $x_0 \in V^*$ iff there exist strictly causal formal power series $\xi(z)$ and $\omega(z)$ such that

$$(6.4) \qquad \begin{aligned} (zI - A)\xi(z) - B\omega(z) &= x_0, \\ C\xi(z) + D\omega(z) &= 0 \;. \end{aligned}$$

Because of Lemma A.1, we see that ξ and ω satisfying (6.4) can be chosen rational causal. Hence

(6.5) THEOREM. $x_0 \in V^*$ *iff there exist strictly causal rational functions* ξ *and* ω *satisfying* (6.4). *Equivalently*, $x_0 \in V^*$ *iff there exists a strictly causal rational* ω *such that*

$$(6.5) \qquad T(z)\omega(z) = -C(zI - A)^{-1}x_0$$

where $T := T_\Sigma$.

The second statement of this theorem can be obtained by eliminating ξ from (6.4). If, in addition to T we introduce $T_1(z) := C(zI - A)^{-1}E$, the disturbance to output transfer function, we can rewrite the condition $\operatorname{im} E \subseteq V^*$ as: There exists a strictly causal rational $Q(z)$ such that

$$(6.6) \qquad T_1(z) = T(z)Q(z).$$

This can be seen applying (6.5) to each column of E. Combining this with Theorem 6.3 we have

(6.7) THEOREM. *Let* Σ *be such that* $V^*(\Sigma)$ *has the feedback property. Then, disturbance rejection is possible iff* (6.6) *has a strictly causal solution.*

This result has a system theoretic interpretation. Suppose that instead of the state of Σ_1, the disturbance q is available for measurement. Then one may attempt to achieve disturbance rejection by a strictly causal feedforward compensator Π.

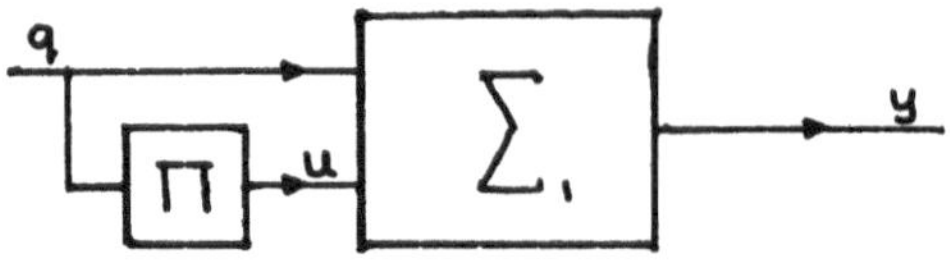

For the problem of disturbance rejection it is no loss of generality to assume that the initial state of Σ_1 is zero. Then (6.1) yields

(6.8) $\quad y(z) = T(z)u + T_1(z)q$.

Suppose that the transfer function of the compensator Π is $R(z)$. Then, assuming (without loss of generality) that Π also has initial state equal to zero, we have $u = R(z)q$. Substitution of this into (6.8) yields

$$y(z) = (T_1(z) + T(z)R(z))q(z) \ .$$

Disturbance rejection will be achieved iff $T_1 + TR = 0$. Hence, the disturbance rejection problem by a feedforward compensator is solvable iff (6.6) has a strictly causal solution. Thus we obtain:

(6.9) COROLLARY. Let Σ be such that $V^*(\Sigma)$ has the feedback property. Then disturbance rejection by state feedback is possible iff disturbance rejection by a strictly causal feedforward compensator is possible.

Appendix

A result is given about the solvability of a linear equation over $\mathcal{R}(z)$.

(A.1) LEMMA. Let $\mathcal{R}$ be a Noetherian domain and let $A(z) \in \mathcal{R}^{m\times n}(z)$, $b(z) \in \mathcal{R}^m(z)$. Consider the linear equation

(A.2) $\quad A(z)x(z) = b(z)$.

Then we have

i) If (A.2) has a formal Laurent series solution then it has a rational solution with monic denominator.

ii) If (A.2) has a causal formal series solution then it has a causal rational solution with monic denominator.

PROOF. i) is an easy consequence of ii). So, we restrict ourselves to the proof of ii). We denote by M the ring of causal rational functions with monic denominator. Without loss of generality we may assume that $A \in M^{m\times n}$ and $b \in M^{m}$, since we may multiply (A.2) with any rational function. Let $x(z) = \Sigma x_t z^{-t}$ be a causal formal solution of (A.2) and define $\xi_k(z) := \sum_{t\leq k-1} x_t z^{-t}$. Then

$$\text{(A.3)} \qquad b - A\xi_k = A(x - \xi_k) \in z^{-k}M^{m} .$$

If

$$N := M^{m}/AM^{n}$$

and $\bar{b}$ is the residue class of b in N, we have to show that $\bar{b} = 0$, because this is equivalent to $b \in AM^{m}$. Relation (A.3) implies that $\bar{b} \in z^{-k}N$, since $\xi_k \in M^{n}$. This holds for every k. Hence

$$\bar{b} \in \bigcap_{k=1}^{\infty} z^{-k}N.$$

Krull's intersection theorem (see [1, Thm 6.21]) implies $\bar{b} = 0$. □

(A.4) COROLLARY. <u>Let in Lemma</u> A.1 <u>the matrix</u> A(z) <u>be nonsingular</u> (<u>i.e</u>. A(z) <u>is invertible over the quotient field of</u> R(z)). <u>If</u> $A^{-1}(z)b(z)$ <u>is expandable, then it has the representation</u> p(z)/q(z), <u>where</u> p(z) <u>is a polynomial vector and</u> q(z) <u>is a monic (scalar) polynomial</u>.

PROOF. If $A^{-1}b$ is expandable, there exists a formal power series x(z) such that (A.2) holds. By Lemma A.1 i) equation A.2 has a solution which is expressible as p(z)/q(z). But since this solution is unique, it follows that $A^{-1}b = p/q$. □

Specializing this result to the scalar case one obtains Lemma 3.1. More generally, the well-known result that a system over a Noetherian domain R is realizable over R if it is realizable over the quotient field of R (see [15, §3B]) is an immediate consequence of the foregoing. Finally, the fact that the existence of formal causal power series ξ and ω satisfying (6.4) implies the existence of a rational causal solution is a consequence of Lemma A.1.

<u>Acknowledgement</u>: The author is indebted to G. Conte and A.M. Perdon for the material of section 5.

References

[1] Barshay, J., Topics in ring theory, W.A. Benjamin, New York, 1969.

[2] Basile, G. & Marro, G., "Controlled and conditioned invariant subspaces in linear system theory", J. Opt. Th. & Appl. 3, 1969, pp. 306-315.

[3] Conte, G. & Perdon, A.M., "Systems over principal ideal domains. A polynomial model approach", to appear in SIAM J. on Cont. and Opt.

[4] Conte, G. & Perdon, A.M., Personal Communication.

[5] Datta, K.B. & Hautus, M.L.J., "Decoupling of multivariable control systems over unique factorization domains" to appear.

[6] Hartley, R. & Hawkes, T.O., Rings, modules and linear algebra, Chapman and Hall Ltd., London, 1970.

[7] Hammer, J. & Heymann, M., "Strictly observable rational linear systems", preprint.

[8] Hautus, M.L.J. "(A,B)-invariant and stabilizability subspaces, a frequency domain description", Automatica, 16, 1980, pp. 703-707.

[9] Hautus, M.L.J. & Heymann, M., "Linear feedback-an algebraic approach", SIAM J. Contr. and Opt,. 16, 1978, pp. 83-105.

[10] Hautus, M.L.J. & Heymann, M., "Linear feedback decoupling-transfer function analysis", submitted for publication.

[11] Heymann, M., Structure and realization problems in the theory of dynamical systems, Springer Verlag, New York, 1975.

[12] Morse, A.S., "System invariants under feedback and cascade control", in Proc. of Int. Symp. on Mathematical system theory, Udine, 1976, Lecture Notes in Econ. and Math. Systems, 131, Springer, New York.

[13] Samuel, P., Anneaux factoriels, Sociedade de Matemática de São Paulo, 1963.

[14] Silverman, L., "Discrete Riccati Equations: Alternative algorithms, asymptotic properties and system theory interpretations", Control and dynamic systems, Vol. 12, 1976, pp. 313-385.

[15] Sontag, E.D., "Linear systems over commutative rings: a survey", Ricerche di Automatica, 7, 1976, pp. 1-34.

[16] Wonham, W.M., Linear Multivariable Control: A Geometric Approach, Springer Verlag, New York, 1979.

CONTROL AND FILTERING OF A CLASS OF NONLINEAR BUT "HOMOGENEOUS" SYSTEMS

Michiel Hazewinkel
Dept. Math., Erasmus Univ. Rotterdam
P.O. Box 1738,
3000 DR ROTTERDAM.
The Netherlands

ABSTRACT. One striking aspect of the class of linear systems is that the controls enter in a way which is independent of the state; that is they are homogeneous, w.r.t. the underlying vectorspace (additive Lie group) structure as far as the controls are concerned, and the autonomous term enjoys reminiscent but not identical "homogeneity properties". Another class of systems which enjoys such properties is the class of systems on Lie groups and coset spaces (E.g. $\dot{g} = (A+\Sigma u_i B_i)g$, $g \in \underline{\underline{GL}}_n$, A, $B_i \in g\ell_n$) studied by Brockett, Jurdjevic-Sussmann, Hirschhorn and others. However, in the case the Lie group G is the additive group this class does not specify to the familiar class of linear systems (but to $\dot{x} = a + \Sigma u_i b_i$, a, $b_i \in \underline{\underline{R}}^n$). Yet the analysis of these two classes of control systems suggests certain "family" characteristics.

In this paper I discuss several aspects of classes of systems, which in one-way or another - there are several different choices one can make - generalize both the familiar linear systems and the class on Lie groups mentioned above.

1. INTRODUCTION.

This paper, or more precisely the research program which this paper tries to describe, resulted from the following two considerations: (i) nonlinear systems theory in general is, at the moment, too difficult and - as a research area - not well enough structured: we have relatively little feeling for the right problems and questions to ask and perhaps little intuition for the phenomena (pathologies) which can occur, and (ii) if in

LQG one changes either L, Q or G things get unstuck immediatelv and rather severely; the three interact rather closely and it seems to follow that to find interesting generalizations all three at once must be adjusted (changed) simultaneously and in a compatible manner.

The lines above are of course the personal opinion of the present author; they may not, as far as I know, reflect the consensus, if such an unlikely thing exists, of the systems theory community.

A situation as described in (i) above is not unusual in mathematics. It has occurred before, e.g. in the theory of Riemannian manifolds. In this particular instance the theory of symmetric spaces came to the rescue. To quote from [Helgason, 1962] (or the revised 1978 edition):

"By their definition, symmetric spaces form a special topic in Riemannian geometry; their theory, however, has merged with the theory of semi-simple Lie groups. This is the source of very detailed and exhaustive information about these spaces. They can therefore often serve as examples on the basis of which general conjectures in differential geometry can be *made* and *tested* ".

At the same time symmetric spaces are general enough to serve as a real testing ground.

It seems to me that nonlinear systems and control theory could do with a class of examples like that. And the classes of "homogeneous", but nonlinear systems described below are mainly intended (by me) as a possible testing ground for ideas, conjectures and concepts in general nonlinear system theory. Special cases, though, do occur naturally in science and engineering, cf. e.g. [Brockett,1972] in connection with theorem 3.14 below.

Consideration (ii) above also points naturally to Lie groups and homogeneous spaces (and some kind of "homogeneous" system on them) as a natural possible class of candidates for generalized

LQG. Especially in view of the theory of "Gaussian processes" on general Lie groups based on Bochner's theorem and a definition of positive definite function which makes sense on any Lie group.

The main philosophy behind what is described below is to study linear systems on $\mathbb{R}^n$ and to formulate their characteristic properties either in terms of the additive Lie group $\mathbb{R}^n$ or in terms of the natural connection on $\mathbb{R}^n$. Not surprisingly these two possible characterization give rise to different possible generalizations when these characteristic properties are formulated for general Lie groups (and homogeneous spaces), even when we restrict attention to (left-) invariant connections on Lie groups.

Two classes of systems arise this way: "Group linear systems" and "connection linear systems". In addition there is a small section on a third class of systems: "fibre linear systems". The "connection linear systems" discussed below are in the torsion-free, zero-curvature case precisely the systems discussed by Brockett in this volume.

What follows below is an outline of a research program rather than a full grown paper. In particular, also to avoid excessive length, I concentrate on ideas and concepts, and proofs are only sketched. A more complete (and longer) account will, hopefully, appear in the future.

All manifolds in the following will be C^∞ and so will all functions and vectorfields defined on them. If M is a C^∞-manifold F(M) denotes the ring of R-valued C^∞-functions (i.e. infinitely often differentiable functions) on M and V(M) denotes the Lie-algebra of all C^∞-vectorfields on M.

2. WHAT MAKES A LINEAR SYSTEM LINEAR

The reason we are asking this question is that we are interested in formulating the conditions for linearity of a system in such a way that natural generalizations on (noncommutative) Lie groups suggest themselves. Let us consider the familiar class of linear systems on $\mathbb{R}^n$

$$(2.1) \qquad \dot{x} = Ax + Bu, \ y = Cx$$

and see whether we can capture its characteristic properties in some "coordinate free way". If $\phi : \mathbb{R}^n \to \mathbb{R}^n$ is any diffeomorphism, then the nonlinear state space transformation $z = \phi(x)$ transforms (2.1) into a set of highly nonlinear looking equations, viz.

$$(2.2) \qquad \dot{z} = (J\phi)(\phi^{-1}(z))(A\phi^{-1}(z) + Bu), \ y = C\phi^{-1}(z)$$

where $(J\phi)(x')$ is the Jacobian matrix of ϕ at z'. These equations still have the form

$$(2.3) \qquad \dot{x} = \alpha(x) + \sum_{i=1}^{m} \beta_i(x)u_i, \ y = \gamma(x)$$

where α, β_i, $i = 1,\dots,m$, are vectorfields on $\mathbb{R}^n$ and γ is a nonlinear function $\mathbb{R}^n \to \mathbb{R}^m$ but beyond that there is little at first sight which might tip one of that we are really dealing with a linear system written down in the wrong coordinates. Up to nonlinear state space equivalence and nonlinear feedback the question of when a system like (2.3) is linear has been considered and solved by [Brockett 1978], and an answer to the question whether a system (2.3) is locally like (2.1) is given by [Krener 1973] in terms of the Lie-algebras generated by the vectorfields $\alpha(x), \beta_i(x)$ (locally around 0).

As a very small simple example consider the example with $A = \begin{pmatrix} 12 \\ 30 \end{pmatrix}$, $B = \begin{pmatrix} 1 \\ 1 \end{pmatrix}$, $C = (2,0)$ in (2.1) and $z = \phi(x)$ given by the diffeomorphism

$$\begin{pmatrix} x_1 \\ x_2 \end{pmatrix} \to z = \begin{pmatrix} 1+x_2^2+2x_2x_1^2+x_1+x_1^4 \\ x_2+x_1^2 \end{pmatrix}$$

which gives us the system

$$(2.5) \qquad \begin{aligned} \dot{z}_1 &= 2z_2+(4+6z_2+8z_2^2)(z_1-1-z_2^2)+(4z_2-2)(z_1-1-z_2^2)^2 \\ &\quad -8z_2(z_1-1-z_2^2)^3 + \{(2+2z_2) + 4z_2(z_1-1-z_2^2)\}u \\ \dot{z}_2 &= (3+4z_2)(z_1-1-z_2^2) + 2(z_1-1-z_2^2)^2 \\ &\quad -4(z_1-1-z_2^2)^3 + (2z_1-1-2z_2^2)u \end{aligned}$$

Returning to our original system (2.1), viewing it as a special case of systems of the form (2.3), and concentrating for the moment on the input part the following "homogeneity properties" could be noticed

(2.6) The input vectorfields $\beta_i(x)$ are invariant with respect to the group structure.

This means the following. Let M be a C^∞-manifold, F(M) the ring of C^∞-functions on M. Then a vectorfield on M is a derivation $X: F(M) \to F(M)$, i.e. an $\mathbb{R}$-linear map with the property $X(fg) = X(f)g + fX(g)$. Let Φ be a diffeomorphism $M \to M$, then the translated vectorfield X^Φ is defined by $(X^\Phi)(f) = (Xf^{\Phi^{-1}})^\Phi$ where $f^{\Phi^{-1}} = f \circ \Phi^{-1}$. If G is a Lie group then X is said to be left invariant if $X^{L_\sigma} = X$ for all $\sigma \in G$ where L_σ stands for the diffeomorphism $g \to \sigma g$, $g \in G$.

Indeed a vectorfield on $\mathbb{R}^n$ can be written as

$$X = \sum f_i(x)\frac{\partial}{\partial x_i} \tag{2.7}$$

Then the requirement that $X^{L_\sigma} = X$ for all $\sigma \in \mathbb{R}^n$ becomes

$$\sum f_i(x-\sigma) \frac{\partial f}{\partial x_i}(x) = \sum f_i(x) \frac{\partial f}{\partial x_i}(x) \tag{2.8}$$

for all functions f (and for all $\sigma \in \mathbb{R}^n$). This means that the $f_i(x)$ in (2.7) must be constants so that the left invariant vectorfields in $\mathbb{R}^n$ are precisely the vectorfields $\sum b_i \frac{\partial}{\partial x_i}$, $b_i \in \mathbb{R}$ which are the vectorfields multiplying the controls in (2.1).

The "vectorfield Ax", or more precisely the vectorfield

$$\alpha(x) = \sum_i (\sum_j a_{ij}x_j) \frac{\partial}{\partial x_i} \tag{2.9}$$

does not have an equally obvious invariance property. But it does have the property

(2.10)   Let $\mathfrak{g}$ be the Lie algebra of left invariant vectorfields
on $\underline{R}^n$, then $[\alpha,X] \in \mathfrak{g}$ for all $X \in \mathfrak{g}$.

The obvious generalization of properties (2.6) and (2.10)
will define the class of what I like to call "group linear
systems". They will be discussed in some more detail below in
section 3. At the moment they are my favourite class of
"nonlinear but homogeneous systems".

A totally different way of saying that the vectorfields
$\beta_i(x)$ in (2.1) are as they are is to remark that the
coefficients b_i in

(2.11) $\sum b_{ij} \dfrac{\partial}{\partial x_j} = \beta_i(x)$

do not vary with x, i.e. that" $\dfrac{\partial}{\partial x_j} b_{ik}$" $= 0$ all k,j. This
concept,however, is not defined on general manifolds but requires
a "manifold with connection" to be properly defined. This will
lead to "connection linear systems" a second class of nonlinear
but homogeneous systems which will probably repay detailed study.
Connection linear systems and their relation with group linear
systems are the topic of section 4 below.

3. GROUP LINEAR SYSTEMS.

3.1. <u>Definition of Group Linear Systems</u>. Let G be a Lie
group, finite dimensional and X a homogeneous space for G, i.e. X
= G/H where H is a closed subgroup of G. Let $\mathfrak{m}$ be the Lie algebra
of G invariant vectorfields on X. (This is a Lie algebra because
$[V_1^\phi, V_2^\phi] = [V_1, V_2]^\phi$ for any two vectorfields V_1, V_2 on a manifold
M and any diffeomorphism ϕ: M $\to$ M). A group linear   system
on X now looks like

(3.2) $\dot{x} = \alpha(x) + \sum \beta_i(x)u_i, \quad y = \gamma(x)$

where

(3.3) $\beta_i(x) \in \mathfrak{m}$ for all i,

(3.4) $[\alpha,\beta] \in \mathfrak{m}$ for all $\beta \in \mathfrak{m}$

(3.5) γ is a collection of quotient maps $X \to G/K_j$
        where K_j is a closed subgroup of G containing H.

   3.6. **Example. Translation Invariant Systems**. An example is
afforded by the systems on Lie groups and spheres studied by
[Brockett 1972,1973], [Jurdjevic-Sussmann, 1972], [Hirschhorn
1977]. Let G be a closed subgroup of $GL_n(\underline{R})$ and $\mathfrak{g}$ the Lie algebra
of G, viewed as a subalgebra of $g\ell_n(\underline{R})$. Consider systems of the
form

$$\dot{g} = g(A + \sum B_i u_i), \quad y = \gamma(g) = Kg$$

The invariant vectorfields on G are the vectorfields gC, $C \in \mathfrak{g}$,
or more explicitly the vectorfields $\sum_{i,j,k} g_{ij} c_{jk} \dfrac{\partial}{\partial g_{ik}}$

(restricted to G) in the coordinates $g_{11},\ldots,g_{nn}$ for $\underline{GL}_n(\underline{R})$
More precisely translation invariant systems are of the form

(3.7) $\dot{g} = \alpha(g) + \sum \beta_i(g)u_i, \quad y = \gamma(g) = gK,$

where α, β_i are left invariant vectorfields, and K is a closed
subgroup of G.

   3.8. **Example. Bilinear systems**. Let $X = \underline{R}^n$ {0} and view X
as a coset space for $GL_n(\underline{R})$ by letting $GL_n(\underline{R})$ act on $\underline{R}^n$ in the
usual manner, i.e. $X = GL_n(\underline{R})/H$ where H is e.g. the stabilizer of
e_1; that is H is the subgroup

$$H = \{ (\begin{smallmatrix} 1 & 0 \\ x & y \end{smallmatrix}): x \in \mathbb{R}^{n-1}, \ y \in GL_{n-1}(\mathbb{R}) \}.$$

Then the vectorfields Ax, $B_i x$ are
right invariant under $GL_n(\underline{R})$, so that (modulo right invariance
versus left invariance) the familiar bilinear systems

(3.9) $\dot{x} = Ax + \sum (B_i x)u_i, \quad y = Cx$

are examples of group linear systems. This also makes it probable that the complete study of group linear systems will not be a totally trivial matter. Note that the equilibrium point $x = 0$ has been removed in the above set up. Results pertaining to this approach to bilinear systems can be found in [Hirschhorn 1977].

3.10. Remark. Consider $\mathbb{R}^n$ as a (vector) Lie group, and consider the systems of type (3.7) on it. E.g. embed $\mathbb{R}^n$ by

$x \mapsto \begin{pmatrix} I_n & x \\ 0 & 1 \end{pmatrix} \in GL_{n+1}(\mathbb{R})$. This gives us systems of the form

$$\dot{x} = a + \sum b_i u_i, \quad a, b_i \in \mathbb{R}^n, \quad y = Cx \tag{3.11}$$

i.e. not the class of systems $\dot{x} = Ax + Bu$, $y = Cx$. This accounts to some extent for the lesser elegance of the results in the inhomogeneous case ($A \neq 0$) with respect to the homogeneous case ($A=0$) in the controllability/reachability results of [Brockett 1972, Jurdjevic-Sussmann 1972].

3.12. Proposition. Consider $\mathbb{R}^n$ as a Lie group. Then the group linear systems (according to definition 3.1) on $\mathbb{R}^n$ are the systems of the form

$$\begin{aligned} &\dot{x} = a + Ax + Bu, \quad y = Cx, \\ &a \in \mathbb{R}^n, \ A \in g\ell_n(\mathbb{R}), \ B \in \mathbb{R}^{n\times m}, \ C \in \mathbb{R}^{p\times n} \end{aligned} \tag{3.13}$$

Proof. Easy exercise. Indeed let $\alpha(x) = \sum f_i(x) \frac{\partial}{\partial x_i}$. Then $[\alpha(x), \frac{\partial}{\partial x_j}]$ left invariant, i.e. constant, means

$(\frac{\partial}{\partial x_j} f_i)(x) = 0$ for all i, j and the result follows.

3.14. Theorem. Let G be a semi-simple or compact Lie group. Then every group linear system over G is of the form (3.7).
Proof. Let G be semisimple and let (Σ) be a system of type (3.2). Let $\mathfrak{g}$ be the Lie algebra of G viewed as a subalgebra of $V(G)$ the Lie algebra of all vectorfields on G. The vectorfield α has the property $[\alpha, \mathfrak{g}] \subset \mathfrak{g}$ and hence defines a derivation of $\mathfrak{g}$. Because $\mathfrak{g}$ is semi-simple every derivation of $\mathfrak{g}$ is inner so that

The following example shows that there are nontrivial intermediate cases.

3.18. Example. The Heisenberg group. Let H be the following subgroup of $GL_3(\mathbb{R})$, the socalled Heisenberg group

$$(3.19) \qquad H = \left\{ \begin{pmatrix} 1 & x & z \\ 0 & 1 & y \\ 0 & 0 & 1 \end{pmatrix} : x,y,z \in \mathbb{R} \right\}$$

Using the global coordinates given by this embedding one finds that all the left invariant vectorfields are linear combinations of

$$(3.20) \qquad b_1 = \frac{\partial}{\partial x},\ b_2 = \frac{\partial}{\partial y} + x\frac{\partial}{\partial z},\ b_3 = \frac{\partial}{\partial z}$$

and that the vectorfields a which have the property that for all $i = 1,2,3$, $[a,b_i] \in \mathfrak{h}$, the Lie algebra spanned by b_1, b_2, b_3 are linear combinations of b_1, b_2, b_3 and the six further vectorfields

$$(3.21) \qquad \begin{array}{l} x\frac{\partial}{\partial x} - y\frac{\partial}{\partial y},\ x\frac{\partial}{\partial y} + \frac{1}{2}x^2\frac{\partial}{\partial z},\ x\frac{\partial}{\partial z} \\ y\frac{\partial}{\partial x} + \frac{1}{2}y^2\frac{\partial}{\partial z},\ y\frac{\partial}{\partial z},\ z\frac{\partial}{\partial z} + y\frac{\partial}{\partial y} \end{array}$$

3.22. A slight generalization. Complete vectorfields and a theorem of Palais. Let M be a differentiable manifold such that there is a finite dimensional Lie algebra of vectorfields $\mathfrak{m}$ such that the vectors $V(x)$, $V \in \mathfrak{m}$ span the tangent space T_xM for all $x \in M$. If $\dim \mathfrak{m} = \dim M$ this makes M parallellizable of course. Now consider systems of the type

$$(3.23) \qquad \dot{x} = \alpha(x) + \sum u_i \beta_i(x)$$

with α such that $[\alpha, \mathfrak{m}] \subset \mathfrak{m}$, $\beta_i \in \mathfrak{m}$. Suppose that the vectorfields α, β_i are all complete. Then the Lie algebra generated by α and the β_i is finite dimensional (it is contained in $\mathfrak{m} + \mathbb{R}\alpha$) and it follows from a theorem of [Palais, 1957] (as was pointed out to me by Roger Brockett) that there will be no finite escape time phenomena for (3.23) (for bounded inputs

$u_i(t))$.

3.24. Reachability Conditions. Both for group linear systems and the slight generalization mentioned just above one expects to find pleasing conditions for reachability/controllability, (and observability, invertability) guided and stimulated by the results of[Brockett 1972], [Jurdjevic-Sussmann 1972], [Hirschhorn 1977] and of course the results of the linear theory. The most natural, coordinate invariant object to consider with respect to controllability is probably the Lie-sub-algebra of $\mathfrak{g}$ generated by the $ad^i\alpha(\beta_j)$, $j = 1, \ldots, m; i=0,1,2,\ldots$.
Here $ad^0\alpha(\beta) = \beta$, $ad^i\alpha(\beta) = [\alpha, ad^{i-1}\alpha(\beta)]$, $i = 1,2,\ldots$. One has e.g.

3.25. Proposition. Let $\dot{x} = \alpha(x) + \Sigma\, u_i\beta_i(x)$ be a group linear control system on the Lie group G with Lie algebra, and suppose that $\alpha(e) = 0$. Then the system is weakly locally reachable around e iff the Lie algebra generated by the $ad^i\alpha(\beta_j)$, $j = 1, \ldots, m$; $i = 0,1,2,\ldots$ is equal to $\mathfrak{g}$.Here locally reachable around e means that for every open neighbourhood U of e the set of points reachable from e such that the trajectory does not leave U contains e in its interior. The sufficiency of the condition for weak local reachability at e is wellknown, cf. e.g. [Hermann-Krener 1977]. Here "weak" means that one is allowed to travel backwards along the vectorfield α (negative time). The example $\alpha = \frac{1}{2}x^2\frac{\partial}{\partial z} + x\frac{\partial}{\partial y}$, $\beta = \frac{\partial}{\partial x}$ on the Heisenberg group (cf. 3.18 above) shows that "weakly" cannot be removed from the statement of the proposition. If all β's are in the centre of $\mathfrak{g}$ (cf. (4.27) below) then weakly can be removed by a result of Hirschhorn.

The proof of the necessity of the condition is most easily done via connections and a sketch is postponed till we have discussed these. That proof in fact yields the stronger result that all trajectories remain in the connected subgroup H of G corresponding to the Lie algebra generated by the $ad^i\alpha(\beta_j)$, so that being able to move far away does not improve the reachability, precisely as in the case of linear systems.

4. CONNECTION LINEAR SYSTEMS.

To be able to say how a vectorfield $\Sigma f_i(x) \frac{\partial}{\partial x_i}$ changes as x varies on a general manifold we need the idea of a connection (or covariant differentiation).

4.1. Connections. Let M be a C^∞-manifold; V(M) the Lie algebra of C^∞-vectorfields on M; F(M) the algebra of C^∞-functions on M. A *linear connection* on M by definition assigns to each $X \in V(M)$ a derivation ∇_X: $V(M) \to V(M)$, of V(M) as a F(M) module; i.e. a map ∇_X which satisfies

$$(4.2) \qquad \nabla_X(fV) = X(f)V + f\nabla_X(V), \quad f \in F(M), \ V \in V(M)$$

Moreover the assignment $X \to \nabla_X$ must satisfy

$$(4.3) \qquad \nabla_{fX+gY} = f\nabla_X + g\nabla_Y, \quad f,g \in F(M); \ Y \in V(M)$$

4.4. Example. Canonical connection on $\mathbb{R}^n$. Assign to $\frac{\partial}{\partial x_i} \in V(\mathbb{R}^n)$ the derivation

$$(4.5) \qquad \Sigma f_j(x) \frac{\partial}{\partial x_j} \to \Sigma \frac{\partial f_j}{\partial x_i}(x) \frac{\partial}{\partial x_j}$$

4.6. Torsion and Curvature. Given a connection ∇ on M its torsion and curvature tensors are defined by

$$(4.7) \qquad T(X,Y) = \nabla_X(Y) - \nabla_Y(X) - [X,Y]$$

$$(4.8) \qquad R(X,Y) = \nabla_X\nabla_Y - \nabla_Y\nabla_X - \nabla_{[X,Y]}$$

The manifold with connection (M,∇) is said to be torsionfree if $T(X,Y) = 0$ and flat if $R(X,Y) = 0$ (in some texts the terminology "flat" is supposed to imply also torsion free). The canonical connection on $\mathbb{R}^n$ is both flat and torsionfree.

4.9. Geodesics and Completeness. Let γ: $(a,b) \to M$ be a curve in M. It is called a geodesic if $\nabla_X(X) = 0$ along γ where X is the vectorfield $\dot{\gamma}(t)$, i.e. $d\gamma(\frac{\partial}{\partial t})$ along $\gamma(a,b) \subset M$.

Given $m \in M$, $v \in T_m M$ there is a unique (local) geodesic $\gamma: (a,b) \to M$, $0 \in (a,b)$ such that $\gamma(0) = m$, $\dot{\gamma}(0) = v$. The manifold with connection (M,∇) is called complete if every geodesic can be extended indefinitely.

4.10. Flat, torsion free manifolds. Let (M,∇) be a flat, torsion-free manifold with connection. The universal covering space $\tilde{M}$ of a manifold with connection carries a natural connection $\tilde{\nabla}$ (cf. e.g. [Wolf, 1976]) and if (M,∇) is flat torsion free then $(\tilde{M},\tilde{\nabla})$ is diffeomorphic to $(\mathbb{R}^n,\nabla_0)$ where ∇_0 is the canonical connection on $\mathbb{R}^n$ described above in example 4.4.

More precisely let E_n be the Lie group of affine motions of $\mathbb{R}^n$, i.e. $E(n) = \mathbb{R}^n \times GL_n(\mathbb{R})$ as a space acting on $\mathbb{R}^n$ by $(x,g)(v) = x + g(v)$, which also defines the group action on E_n. Then every flat, torsion free, connected manifold M with connection is diffeomorphic to $\mathbb{R}^n/\Gamma$ where Γ is a discrete subgroup of E_n acting properly discontinuously, so that M is a product of a torus and an $\mathbb{R}^m$

In particular if (M,∇) is flat, torsion free, connected and simply connected then $M = \mathbb{R}^n$ with the canonical connection (up to connection preserving diffeomorphism) and this gives a not very practical answer to the question of what makes a system (2.3) linear up to diffeomorphism (neglecting outputs). This will be the case if and only if there is a flat, torsion free connection ∇ such that $\nabla\beta_i = 0$ for all i and all vectorfields V (such vectorfields are called constant) and $\nabla_X\alpha$ is constant for all constant vectorfields X and finally there is an equilibrium point for zero controls.

4.11. Connection Linear Systems. This brings us quite naturally to the definition of a connection linear system. A control system

$$\dot{x} = \alpha(x) + \sum \beta_i(x)u_i \tag{4.12}$$

on a manifold with connection (M,∇) will be called *connection*

there exists a vectorfield $V \in \mathfrak{g}$ such that $[\alpha,\beta] = [V,\beta]$ for all $\beta \in \mathfrak{g}$. Now the vectorfields β for every $g \in G$ span a basis for the tangent space
T_gG at g and it follows by the easy lemma below that $\alpha = V$ proving the theorem in this case.

If G is compact consider the translated vectorfields α^{L_σ} for all $\sigma \in G$. Let $d\mu$ be unit mass left invariant Haar measure on G, and define $V = \int \alpha^{L_\sigma} d\mu$. Then V is left invariant and the remaining bit of the proof is as before.

3.15. Lemma. Let $V_1, \ldots, V_n$ be a set of vectorfields on the connected manifold M such that $V_1(x), \ldots, V_n(x)$ is a basis for the tangent space T_xM for all $x \in M$. Let V,W be two more vectorfields on M and suppose that $[V_i,V] = [V_i,W]$, $i = 1, \ldots, n$ and $V(x_0) = W(x_0)$ for some $x_0 \in M$. Then $V = W$.
Proof. This is an immediate consequence of standard uniqueness results for solutions of differential equations.

Another pleasing consequence of lemma 3.15 is that the dimension of the space of all group linear systems on a Lie group G is finite, exactly as in the case of linear systems. This is a property of the space of all linear systems (of a given dimension, with a given number of outputs and inputs) which is important in identification problems.

3.16. Proposition. Let G be an n-dimensional Lie group. Then the space of all systems $\dot{x} = \alpha(x) + \sum_{i=1}^{m} u_i\beta_i(x)$ satisfying (3.3), (3.4) is of dimension $\leq n^2 + n + mn$.

Indeed, the control vectorfields β_i, $i = 1, \ldots, m$ account for mn dimensions. The vectorfield α induces an endomorphism of the n-dimensional vectorspace $\mathfrak{g}$, the Lie algebra of G and is uniquely determined by this endomorphism and its value $\alpha(e)$ (by lemma 3.15). Note that if $G = \mathbb{R}^n$ then the upper bound $n^2 + n + mn$ is reached. It is maybe also worth noticing that the control systems (3.2) satisfying (3.3) - (3.5) are automatically analytic.

3.17. Remarks. Thus the familiar linear systems $\dot{x} = Ax + Bu$ and the systems (3.7) are the extreme examples of the class of group equivariant systems, corresponding respectively to the abelian and semi-simple cases. Then theory though exhibits considerable similarity which gives reasonable grounds for optimism for the whole class.

linear if

(4.13) $$\nabla_V \beta_i = 0 \text{ all } V \in V(M)$$

so that the β_i are constant vectorfields, and

(4.14) $$\nabla_X \alpha = \text{constant for all constant vectorfields } X.$$

It would I think perhaps be even more interesting to consider the class of control systems (4.12) which satisfy (4.13) and

(4.15) $$[\alpha, V] = \text{constant for all constant } V.$$

<u>Warning</u>. On an arbitrary manifold with connection (M,∇) there may very well be no constant vectorfields other than the zero vectorfield.

A last interesting class of connection defined systems, more or less analogous to 3.22 above, consists of systems (4.12) such that the β_i belong to a finite dimensional Lie algebra $\mathfrak{m}$ such that the $\mathfrak{m}(x)$ form a basis (or span) T_xM for all $x \in M$ and which satisfy

(4.16) $$\nabla_X \alpha \in \mathfrak{m} \quad \text{for all } X \in \mathfrak{m}$$

In the case of a connected, simply connected, flat torsion free manifold both (4.13) + (4.14) and (4.13) + (4.15) lead to control systems $\dot{x} = a + Ax + Bu$. If the manifold with connection (M,∇) is connected, flat, torsion free (but not simply connected) then these conditions result in the class of systems described by Roger Brockett in these proceedings (and some of these naturally occur in engineering, loc. cit.).

4.17. <u>Intermezzo on foliations and distributions and the distributions defined by a control system</u>. A *foliation* of an n-dimensional manifold M by q-dimensional submanifolds is a collection of q-dimensional submanifolds (called the leaves) such that through every $x \in M$ there passes exactly one leaf and such that locally around every point the partitioning of M by the leaves looks like $\mathbb{R}^n$ partioned by the

$a + \mathbb{R}^q$, $a \in \{x \in \mathbb{R}^n: x_1 = \ldots = x_q = 0\}$,

$\mathbb{R}^q = \{x \in \mathbb{R}^n: x_{q+1} = \ldots = x_n = 0\}$.

A *distribution* of dimension q on M assigns to every x M a q-dimensional subspace $D(x) \subset T_xM$ of the tangent space of M at x such that D(x) varies differentiably with x.

Obviously a q-dimensional foliation defines a distribution, viz. $x \mapsto T_xF_x$ where F_x is the unique leaf of the foliation passing through x. Such distributions are called *integrable* .They have the following property (obviously): if X,Y are two vectorfields on M such that $X(x), Y(x) \in D(x)$ for all x then also $[X,Y](x) \in D(x)$. Such distributions are called *involutive* .It is a theorem of Frobenius that such distributions are integrable, i.,e., come from foliations.

Now consider a control system (2.3). For each $x \in M$ define a nested series of subspaces of the tangent space T_xM

(4.18) $$B_i(x) = \text{subspace spanned by } ad^j\alpha(\beta_k)(x), \quad j = 0,\ldots,i;\ k = 1,\ \ldots,\ m$$

If the system (2.3) is linear the B_i form a nested system of integrable distributions. And inversely [Brockett 1979] for a control system (2.3) on $\mathbb{R}^n$, if dim $B_i(x)$ is constant as a function of x (so that the B_i are distributions) and these distributions are all integrable then the control system is linear up to nonlinear feedback (and nonlinear base change in input and state space).

There is a version of the results described in 4.10 above relative to a foliation [Blumenthal, 1980] (in which the conditions are stated in terms of a connection "adapted to" the foliation, a socalled basic connection) which - it seems to me - will be worth considering in this connection (e.g. to obtain similar results on more general spaces like the $\mathbb{R}^n/\Gamma$, Γ a discrete subgroup of $\mathbb{R}^n \times GL_n(\mathbb{R})$).

4.19. <u>Parallel displacement</u>. Let (M,∇) be a manifold with connection. Let $X \in V(M)$ and $\gamma: [a,b] \to M$ an integral curve of X, i.e. $d\gamma(\frac{\partial}{\partial t}) = X(\gamma(t))$ for all $t \in [a,b]$. Let Y be another vectorfield. The vectorfield Y is called *parallel along* γ if $\nabla_X(Y)(\gamma(t)) = 0$ for all t. This definition does not depend of course on the vectorfield X but only on γ. This notion can be used to identify the tangent spaces T_xM for $x \in \gamma[a,b]$ (parallel displacement along γ) with $v \in T_xM$ corresponding to $v' \in T_{x'}M$ iff there is a parallel vectorfield Y along γ with $v = Y(x)$, $v' = Y(x')$.

4.20. Intermezzo on Riemannian manifolds and the Levi-Civita connection. A pseudo-Riemannian (resp. Riemannian) manifold is a manifold equipped with a nondegenerate (resp. positive definite) symmetric bilinear form on each tangent space T_xM which varies differentiably with x. Given a pseudo-Riemannian manifold there exists a unique torsion-free connection which preserves the bilinear form (inner product) under parallel displacements along geodesics. This connection is called the Levi-Civita connection. It will perhaps be advantageous to analyse connection linear systems first for connections of this type.

4.21. Group-linear versus connection linear systems. Now let G be a Lie group. More generally similar things can be discussed for homogeneous spaces. There are at least three rather special connections on G which stand out and seem to deserve special attentión. All three are left-invariant where a connection ∇ on G is called left invariant if for all $X,Y \in V(M)$ we have

$$\nabla_X(Y) = \nabla_{X^\sigma}(Y^\sigma)^{\sigma^{-1}} \tag{4.22}$$

where I have simply written σ for the left translation $L_\sigma: G \to G$, $g \to \sigma g$.

Left-invariant connections on G correspond biuniquely to bilinear forms $\alpha: \mathfrak{g} \times \mathfrak{g} \to \mathfrak{g}$, where $\mathfrak{g}$ is the Lie algebra of G. Here α is simply equal to $\alpha(X,Y) = \nabla_{\tilde{X}}(\tilde{Y})(e)$, where $\tilde{X},\tilde{Y}$ are the left-invariant vectorfields whose tangent vectors at $e \in G$ are equal to $X,Y \in \mathfrak{g}$ respectively. Cf. e.g. [Helgason 1978] for this.

Let $\nabla^1,\nabla^2,\nabla^3$ be the three connections on G defined by the bilinear forms

(4.23) $\alpha^1(X,Y) = 0$ (the zero-connection)

(4.24) $\alpha^2(X,Y) = [X,Y]$ (the + connection)

(4.25) $\alpha^3(X,Y) = \frac{1}{2}[X,Y]$ (the - connection)

Under ∇^1 the constant vectorfields are precisely the left-invariant ones. So that using ∇^1 conditions (4.12) and (4.14) together precisely define what we called a group linear system in section 3 above.

∇^3 is the only torsion free connection among these 3 and seems to be by far the most natural torsion free connection on G. It is perhaps worth remarking here that there exist no left-invariant torsion free flat connections on reductive homogeneous spaces ([Doi 1979], cf. also [Matsushima-Okamoto 1979] for the case of real semisimple Lie groups. This very nicely distinguishes $\mathbb{R}^n$ from the reductive homogeneous spaces (such as $\mathbb{R}^n \setminus \{0\}$, the natural state space of bilinear systems).

Finally ∇^2 is such that $\nabla^2_X(V)$ is left-invariant for all left invariant X if and only if $[X,V]$ is left-invariant for all left-invariant X so that under ∇^3 conditions (4.16) and (4.4) are equivalent, cf. also 4.15.

Indeed any vectorfield Y on G can be written as $\sum f_i(x)X_i$ where $X_1,\ldots, X_n$ is a basis for $\mathfrak{g}$. So that for $X \in \mathfrak{g}$

$$\nabla^2_X(Y) = \sum_i X(f_i)X_i + \sum_i f_i[X,X_i]$$

On the other hand $[X,Y](\phi) = \Sigma X((f_iX_i)(\phi)) - \Sigma f_iX_i(X(\phi)) =$
$= \Sigma X(f_i)X_i(\phi) + \Sigma f_i X(X_i(\phi)) - \Sigma f_iX_i(X(\phi))$ So that for the + connection,

(4.26) $\nabla^2_X(Y) = [X,Y],\ X \in \mathfrak{g},\ Y \in V(M).$

However, under ∇^2 the left-invariant vectorfields are no longer the constant ones, so that if G is noncommutative "connection linear" systems and "group linear" systems are different objects.

But of course the vectorfields in the centre of $\mathfrak{g}$ are constant. This defines a special class of systems

$$\dot{x} = \alpha(x) + \Sigma\, u_i \beta_i(x) \tag{4.27}$$

with $\beta_i \in Z(\mathfrak{g})$, the centre of $\mathfrak{g}$ and $[\alpha, \mathfrak{g}] \subset \mathfrak{g}$. This class is intermediate between linear systems and group linear (and bilinear) systems and certainly will repay detailed further investigation. I would also not be surprised if this class yielded further examples of finite dimensional estimation algebras (cf. section 6 below for this notion).

4.28. <u>On the necessity of the controllability condition of proposition 3.25.</u>
Consider a group linear control system on the Lie group G. Let H be the connected Lie subgroup of G corresponding to the sub Lie algebra $\mathfrak{h}$ of $\mathfrak{g}$ generated by the $\mathrm{ad}^i\alpha(\beta_j) \in \mathfrak{g}$. We show that any trajectory starting in $e \in G$ remains in H. To see this consider the + connection on G. First notice that this connection restricts to a connection on H so that parallel displacements of vectors tangent to H at e along a curve γ in H results in vectors in $T_{\gamma(t)}G$ which are tangent to H. Now let $h \in H$ and γ a curve from e to $h = \gamma(1)$ in H. Then identifying tangent vectors in the various tangent spaces to G along γ by means of parallel displacement along γ we have

$$\alpha(h) = \alpha(e) + \int_0^1 (\nabla_{\gamma'(t)}\alpha)(\gamma(t))dt$$

(cf. [Helgason 1978, thm 7.1, page 41).
Now $\gamma'(t) \in T_{\gamma(t)}H$, $\alpha(e) = 0$ and $\nabla_X\alpha = [X,\alpha]$ by (4.26) and $[\alpha, \mathfrak{h}] \subset \mathfrak{h}$ and it follows by the remark made above that $\alpha(h)$ is tangent to H (at $\gamma(1)$), so that $\alpha(h) + \Sigma\, u_i\beta_i(h)$ is in $T_hH \subset T_hG$ for all $h \in H$.

4.29. <u>Another example</u>. Consider the linear Lie group G consisting of all 2 x 2 matrices of the form $\begin{pmatrix} x & z \\ 0 & 1 \end{pmatrix}$, $x,z \in \mathbb{R}$, $x > 0$. The Lie algebra $\mathfrak{g}$ of G consists of all real 2 x 2 matrices of the form $\begin{pmatrix} a & b \\ 0 & 0 \end{pmatrix}$. In the coordinates x,z the invariant vectorfields are linear combinations of

$$x \frac{\partial}{\partial x}, \; x \frac{\partial}{\partial z}$$

and the vectorfields α such that $[\alpha, \mathfrak{g}] \subset \mathfrak{g}$ and $\alpha(e) = 0$ are linear combinations of the three vectorfields

$$x \ln x \frac{\partial}{\partial x}, \; z \frac{\partial}{\partial z}, \; \frac{\partial}{\partial z} - x \frac{\partial}{\partial z}$$

5. FIBRE LINEAR SYSTEMS.

A rather different class of nonlinear systems with enough special structure to make one optimistic is what I like to call fibre linear systems. As an example consider a system whose total state x can be partioned into two parts $x = \begin{pmatrix} x_1 \\ x_2 \end{pmatrix}$ evolving according to

$$(5.1) \qquad \dot{x}_1 = A_1 x_1 + B_1 u_1, \qquad x_1 \in \mathbb{R}^{n_1}$$

$$(5.2) \qquad \dot{x}_2 = A_2(x_1, u_1) + B_2(x_1 u_1) u_2, \qquad x_2 \in \mathbb{R}^{n_2}$$

where A_1 and B_1 are constant matrices and A_2 and B_2 depend only on x_1 and u_1. Thus the total system consists of an ordinary linear system on the base and the state and controls of this influence the systems in the fibre which are also linear given x_1, u_1. One can of course even write down the input-output map of such a system explicitly (more or less).

More generally the first system in the base can itself be nonlinear, perhaps itself a fibre linear system with linear base giving rise so to speak to a three stage tower of linear systems. Generalizations on arbitrary rather than trivial vectorbundles now are easy to define.

5.3. <u>The Heisenberg group again</u>. Consider the Heisenberg group H example of section 4 above again.

Write $x_1 = (x,y)$, $x_2 = z$. Then for all the group linear systems on H, x_1 evolves as a linear system and given x_1 then $z = x_2$ evolves as a slightly generalized linear system

$$\dot{z} = a(x_1,u_1) + A(x_1)z + B(x_1)u_2$$

So that these systems are also fibre linear with linear base. This is a general phenomenon: every group linear system on a unipotent Lie group can be considered as a tower of linear systems in the sense suggested above.

6. REMARKS ON FILTERING FOR GROUP-LINEAR SYSTEMS.

Consider the general nonlinear filtering problem (Ito equations)

$$(6.1) \qquad dx_t = f(x_t)dt + G(x_t)dw_t, \quad dy_t = h(x_t)dt + dv_t,$$

where w_t, v_t are independent Wiener noise processes also independent of the initional random variable x_0. Here h,f,G are vector and matrix valued functions of the appropriate dimensions. Given enough regularity so that the density of the p(x,t) of $\hat{x}_t = E[x| y_s, 0 \leq s \leq t]$, the conditional state at time t given the observations $y^t = \{y_s : 0 \leq s \leq t\}$, exists, a certain unnormalized version $\rho(x,t)$ of p(x,t) satisfies the socalled Duncan-Mortenson-Zakai equation (which is driven by the observations)

$$(6.2) \qquad d\rho = \frac{1}{2} \sum_{i,j} \frac{\partial^2}{\partial x_i \partial x_j}((GG^T)_{ij}\rho) - \sum_i \frac{\partial}{\partial x_i}(f_i \rho) - \\ - \frac{1}{2} \sum_i h_i^2 \rho - \sum_i h_i dy_i$$

(cf. e.g. [Davis-Marcus 1981] for a derivation of this equation). This equation is in Fisk-Stratonovic form. The Lie algebra generated by the differential operator

$$\mathcal{L} = \frac{1}{2} \sum_{i,j} \frac{\partial^2}{\partial x_i \partial x_j}(GG^T)_{ij} - \sum_i \frac{\partial}{\partial x_i} f_i - \frac{1}{2} \sum_i h_i^2$$

(where $(GG^T)_{ij}$ is the (i,j)-th entry of the matrix GG^T, f_i, h_i the i-th component of the vector f,h) and the operators (multiplication with) $h_1, \dots h_p$ is called the estimation algebra. It is likely to be of considerable importance in the analysis of the filtering problem (= building finite dimensional systems driven by the observations which produce $\hat{x}_t$ as outputs), cf. [Brockett 1981], [Hazewinkel-Marcus, 1980] and several more papers in [Hazewinkel-Willems, 1981].

The most general group linear stochastic Ito equation on the Heisenberg group is

$$(6.4)\qquad \begin{pmatrix} dx_1 \\ dx_2 \\ dx_3 \end{pmatrix} = \begin{pmatrix} a_1x_1 + a_4x_2 \\ -a_1x_2 + a_2x_1 + a_6x_2 \\ \tfrac{1}{2}a_2x_1^2 + a_3x_1 + \tfrac{1}{2}a_4x_2^2 + a_5x_2 + a_6x_3 \end{pmatrix} dt + \sum_{i=1}^{m} \begin{pmatrix} b_{1i} \\ b_{2i} \\ x_1b_{2i}+b_{3i} \end{pmatrix} dw_i$$

$a_1,\dots,a_6$; $b_{ji} \in \mathbb{R}$, and the most general observation equations coming from a group homomorphism $H \to \mathbb{R}$ are of the form

$$(6.5)\qquad dy_i = (c_{1i}x_1+c_{2i}x_2)dt + dv_i$$

6.6. <u>Proposition</u>. Consider a system on the Heisenberg group given by a signal equation of type (6.4) with observation equations of type (6.5). Then the observation Lie algebra is always pro-finite dimensional.

A Lie algebra L is pro-finite dimensional if there exists a sequence of ideals $L_1 \supset L_2 \supset \dots$ such that L/L_i is finite dimensional for all i and $\bigcap L_i = 0$. Cf.e.g.[Hazewinkel-Marcus, 1980] for a number of remarks on the relevance of this property for filtering problems.

Indeed writing out the various operators explicitly one observes that they are sums of operators of the type

$$x^\alpha \frac{\partial^\beta}{\partial x^\beta} \frac{\partial^i}{\partial z^i} \; , \; i = 0,1,2\ldots; \; |\alpha|, |\beta| \leq 2,$$

$$x = \binom{x_1}{x_2}, \; z = x_3$$

where α and β are multiindices $|\alpha| = \alpha_1 + \alpha_2$. The operators $x^\alpha \frac{\partial^\beta}{\partial x^\beta}$, $|\alpha|, |\beta| \leq 2$ span a finite dimensional Lie algebra LS_2

(of dimension 15) so that the estimation algebra is a subalgebra of the "current-algebra"

$$LS_2 \otimes \underline{R} \, [\tfrac{\partial}{\partial z}]$$

which is of course profinite dimensional. As a finite dimensional Lie algebra LS_2 can of course be embedded in a Lie algebra of vectorfields on $\underline{R}^N$, some large N) and this then easily gives rise to an inbedding of the current algebra $LS_2 \otimes \underline{R}[\tfrac{\partial}{\partial z}]$. In this case, however, there exists an inbedding of LS_2 modulo its centre in the vectorfields on $\underline{R}^5$ which comes from all Kalman-Bucy filters put together (and is closely related to the Segal-Shale-Weil representation), cf. [Hazewinkel, 1981], which is more likely to be useful.

(A result like proposition 6.6 holds generally also for higher dimensional Heisenberg groups (and hence for all 2-step nilpotent Lie groups) and I would like to pose the question whether it holds for every fibre linear system with linear base (and suitable output maps "linear" in the fibres).

Things change dramatically if instead of using observation like (6.5) one uses an observation equation

(6.7) $dy = x_3 dt + dv$

E.g. the system

(6.8) $dx_1 = dw, \; dx_2 = x_1 dt, \; dx_3 = \frac{1}{2} x_1^2 dt, \; dy = x_3 dt + dv$

has the Weyl algebra $W_1 = \underline{R}\langle x_1, \frac{\partial}{\partial x_1} \rangle$ as a subalgebra. This is perhaps not surprising because the map $(x_1,x_2,x_3) \to x_3$ is not "homogeneous" with respect to H. Indeed there is no action of H on $\mathbb{R}$ which makes this map H-equivariant. There is an action of H on $\mathbb{R}^2$ which makes $(x_1,x_2,x_3) \to (x_2,x_3)$ H-equivariant. This, at first sight, would make an observation equation like

$$(6.9) \qquad \begin{pmatrix} dy_1 \\ dy_2 \end{pmatrix} = \begin{pmatrix} x_2 \\ x_3 \end{pmatrix} dt + \begin{pmatrix} dv_1 \\ dv_2 \end{pmatrix}$$

permissible, and this would also give a subalgebra W_1 in the estimation Lie algebra. However, in (6.9) the noises do not enter in a group-equivariant way. To achieve that one needs observation equations like

$$(6.10) \qquad \begin{pmatrix} dy_1 \\ dy_2 \end{pmatrix} = \begin{pmatrix} x_2 \\ x_3 \end{pmatrix} dt + \begin{pmatrix} dv_1 \\ x_1 dv_1 \end{pmatrix}$$

And this raises the general question of obtaining a D-M-Z type equation for an (unnormalized) conditional density for more general systems

$$(6.11) \qquad dx = f(x)dt + G(x)dw, \quad dy = h(x)dt + J(x)dv$$

With this open question I would like to conclude this paper.

REFERENCES.

1. R.A. Blumenthal 1980, Foliated manifolds with flat basic connection, preprint, St. Louis Univ., St. Louis.
2. R.W. Brockett 1972a, System theory on group manifolds and coset spaces, SIAM J. Control 10, 265-284.
3. R.W. Brockett 1973, Lie theory and control systems defined on spheres, SIAM J. Appl. Math. 25, 213-225.
4. R.W. Brockett 1979, Feedback invariants for nonlinear systems, In Proc. 7th IFAC World congress (Helsinki 1978), Pergamon Press, 1979.

5. R.W. Brockett 1981, Geometrical methods in stochastic control and estimation, In: M. Hazewinkel, J.C. Willems (eds), Stochastic systems: the mathematics of filtering and identification and applications, Reidel Publ. Cy, 1981.

6. M.H.A. Davis, S.I. Marcus 1981, An introduction to nonlinear filtering, In: M. Hazewinkel, J.C. Willems (eds), Stochastic Systems: the mathematics of filtering and identification and applications, Reidel Publ. Cy, 1981.

7. H. Doi 1979, Non-existence of torsion free flat connections on reductive homogeneous spaces, Hiroshima Math. J. 9, 321-322.

8. M. Hazewinkel 1981, The linear systems Lie-algebra, the Segal-Shale-Weil representation and all Kalman-Bucy filters, preprint.

9. M. Hazewinkel, S.I. Marcus 1980, On Lie algebras and finite dimensional filtering, to appear in Stochastics, Report 8019, Econometric Inst., Erasmus Univ. Rotterdam, 1980.

10. M. Hazewinkel, J.C. Willems (eds), 1981, Stochastic systems: the mathematics of filtering and identification and applications, Reidel Publ. Cy, 1981.

11. S. Helgason 1978, Differential geometry, Lie groups and symmetric spaces, Acad. Pr.

12. R. Hermann, A.J. Krener, 1977, Nonlinear controllability and observability, IEEE Trans AC 22, 728-740.

13. R.M. Hirschhorn, 1977, Invertibility of control systems on Lie groups, SIAM J. Control and Opt. 15, 1034-1049.

14. A.J. Krener, 1973, On the equivalence of control systems and the linearization of nonlinear systems, SIAM J. Control 11, 670-676.

15. H. Matsushima, K. Okamoto 1979, Non-existence of torsion free flat connections on a real semisimple Lie group, Hiroshima Math. J. 9, 59-60.

16. R.S. Palais 1957, A global formulation of the Lie theory of transformation groups, Memoirs Amer. Math. Soc. 22.

17. J.A. Wolff 1976, Spaces of constant curvature, 3-rd edition, Publish or Perish.

CONSTRUCTION OF FORMAL AND ANALYTIC REALIZATIONS OF NONLINEAR SYSTEMS

B.Jakubczyk
Institute of Mathematics
Polish Academy of Sciences
00-950 Warszawa, Śniadeckich 8, Poland

1. Introduction. In [7] we gave sufficient and necessary conditions for the existence of realizations of nonlinear input-output maps together with a theoretical construction of a minimal realization. This construction is set theoretical and not suitable for practical computations, however.

In this report we give an explicit construction of a realization of a nonlinear input-output map. This is a power series construction. Namely, we show that the i.o. map of an analytic i.o. system (black box) can be represented by a formal power series of noncommutative variables in the input alphabet. The Taylor expansions of a realization

$$\dot{x} = f(x,u), \quad x(0) = x_0 \qquad (1)$$
$$y = h(x)$$

can be computed from the coefficients of this series. The construction works at formal and analytic levels and should allow for some arithmetization (algebraization) of the realization theory of nonlinear systems.

In our approach we modify a Fliess' idea of using formal power series in noncommuting variables for the input-output description of regular (bilinear) and linear-analytic systems (see [2]-[5]). Our result is also related to an existence theorem in Fliess [5]. The approach presented in the paper (as well as the approach of [5]) is perhaps more in the spirit of the original approach of linear realization theory (cf. Kalman et al.[10]), comparing to the approach of Sussmann [12],[13] and the approach in [6],[7],[8]. We hope that it will produce more computational algorithms of computing nonlinear realizations (or partial realizations).

We give a scetch of proof of our main result (Theorem 1) only. The complete proof together with another method of construction will be given in [9].

2. Formal power series in noncommuting variables.

Let Ω be a nonempty set called an alphabet. The elements of Ω will be treated as noncommuting variables. Denote by Ω^* the free monoid generated by Ω, i.e. the set of words

$$\alpha_1\alpha_2\ldots\alpha_n, \quad n \geq 0, \quad \alpha_i \in \Omega,$$

where the product wv of two words is obtained by writing the word v immediately after the word w. The empty word is the identity e in Ω^*.

Any function $\mathcal{F} : \Omega^* \longrightarrow R$ is called a formal power series in noncommuting variables Ω. The value of $\mathcal{F}$ at $w \in \Omega^*$ is denoted by $\langle\mathcal{F},w\rangle$ and $\mathcal{F}$ is often written in the form of a formal sum

$$\mathcal{F} = \sum_{w\in\Omega^*} \langle\mathcal{F},w\rangle w.$$

If the support of $\mathcal{F}$ is finite, i.e. $\langle\mathcal{F},w\rangle \neq 0$ for finitely many $w \in \Omega^*$, then $\mathcal{F}$ is called polynomial. The degree of a polynomial is the largest lenght of a word w for which $\langle\mathcal{F},w\rangle \neq 0$.

With the product

$$(2) \qquad \mathcal{F}\mathcal{G} = \sum_{w\in\Omega^*} \Big(\sum_{\substack{w',w''\in\Omega^* \\ w'w''=w}} \langle\mathcal{F},w'\rangle\langle\mathcal{G},w''\rangle \Big) w$$

the set of all formal power series in noncommuting variables Ω forms an associative algebra denoted by $R\langle\langle\Omega\rangle\rangle$. Similarly, all polynomials form an algebra $R\langle\Omega\rangle$. The elements of Ω^* are treated as monomials in $R\langle\Omega\rangle$.

3. Input-output map and input-output series.

Assume now that Ω is the input space (input alphabet) which is any set. Let, for simplicity, the output be scalar, i.e. the output space $Y = R$. The input signals are assumed to be piecewise constant functions $u : [0,T] \longrightarrow \Omega$, $u(t) = \alpha_i$ for $t \in [t_1+\ldots+t_{i-1}, t_1+\ldots+t_i)$, $i=1,\ldots,\mu$, where $T = t_1+\ldots+t_\mu$ is not fixed. Such functions will be written in the form

$$(3) \qquad u = (t_\mu\alpha_\mu)\ldots(t_1\alpha_1),$$

where $t_i \in R_+ = [0,\infty)$, $\alpha_i \in \Omega$ and $\mu \geq 0$. The set of all such functions is a semigroup S_Ω with multiplication being the concatenation: $vu = (\tau_\nu\beta_\nu)\ldots(\tau_1\beta_1)(t_\mu\alpha_\mu)\ldots(t_1\alpha_1)$ for $v = (\tau_\nu\beta_\nu)\ldots(\tau_1\beta_1)$ and the identity e being the empty sequence.

The semigroup of inputs S_Ω can be identified with the set of formal sequences (3) with $\mu \geq 0$, $\alpha_i \in \Omega$, where we identify

(4) $\quad (t_1\alpha)(t_2\alpha) = (t_1+t_2)\alpha, \quad (0\alpha_1)\ldots(0\alpha_\mu) = e.$

A subset $U \subset S_\Omega$ is called a neighborhood of identity in S_Ω if for any sequence $\alpha_1,\ldots,\alpha_\mu \in \Omega$, $\mu \geq 1$, the set of parameters $(t_1,\ldots,t_\mu) \in R_+^\mu$ for which $(t_\mu\alpha_\mu)\ldots(t_1\alpha_1)$ is in U is a neighborhood of zero in R_+^μ.

Any map $P : U \longrightarrow R$, where $U \subset S_\Omega$ is a neighborhood of identity in S_Ω, will be called an <u>input-output map</u>. This map will be the input--output representation of our black box system. Practically, this means that we measure the output effect of a control $u \in U$ at the end of the action of u.

The <u>i.o. map</u> P is called <u>analytic</u> if each map

$$(t_1,\ldots,t_\mu) \longrightarrow P((t_\mu\alpha_\mu)\ldots(t_1\alpha_1)), \quad \mu \geq 1, \quad \alpha_1,\ldots,\alpha_\mu \in \Omega,$$

is analytic on its domain of definition in R_+^μ and has an analytic extension to an open subset of R^μ.

Any analytic i.o. map $P : U \longrightarrow R$ defines a noncommutative formal power series $\mathcal{P} \in R\langle\langle\Omega\rangle\rangle$ by the formula

(5) $$\langle\mathcal{P},\alpha_1\ldots\alpha_\mu\rangle = \frac{\partial}{\partial t_1}\ldots\frac{\partial}{\partial t_\mu}P((t_\mu\alpha_\mu)\ldots(t_1\alpha_1))\,|_{t_1=\ldots=t_\mu=0}\,.$$

$\mathcal{P}$ is called the <u>input-output series</u> and is defined uniquely by P

It can be easily seen that the i.o. map can be recovered from the i.o. series by the formula

(6) $$P((t_\mu\alpha_\mu)\ldots(t_1\alpha_1)) = \sum_{i_1,\ldots,i_\mu\geq 0} \langle\mathcal{P},\alpha_1^{i_1}\ldots\alpha_\mu^{i_\mu}\rangle \frac{t_1^{i_1}\ldots t_\mu^{i_\mu}}{i_1!\ldots i_\mu!}\,,$$

where we denote $\alpha^i = \alpha\ldots\alpha$ i-times. In fact, the following can be proved.

<u>Proposition 1</u>. Any analytic i.o. map $P : U \longrightarrow R$ defines an (input--output) series $\mathcal{P}$ in $R\langle\langle\Omega\rangle\rangle$, via (5). Conversely, any series $\mathcal{P} \in R\langle\langle\Omega\rangle\rangle$ for which the series (6) are convergent in a neighborhood of zero in R_+^μ for any $\alpha_1,\ldots,\alpha_\mu \in \Omega$, $\mu \geq 1$, defines via (6) an analytic i.o. map $P : U \longrightarrow R$. This map has an analytic extension to a maximal open neighborhood of identity in S_Ω.

We end this section with a formal definition of a realization. System (1) is called a <u>realization</u> of an i.o. map $P : U \longrightarrow R$ if for any $u \in U$ differential equation (1) has a well defined solution and $P(u) = y(T,u)$, i.e.,

(7) $$P(u) = h \circ \gamma_{t_\mu}^{\alpha_\mu} \circ \dots \circ \gamma_{t_1}^{\alpha_1} (x_o),$$

where $u = (t_\mu \alpha_\mu)\dots(t_1\alpha_1)$ and γ_t^α denotes the local flow of the vector field $f(\cdot,\alpha)$. In general, we assume that the realization is analytic, i.e. x is in a real analytic manifold M and the maps $h : M \longrightarrow R$, $f(\cdot,\alpha) : M \longrightarrow TM$, $\alpha \in \Omega$, are real analytic.

Using the definition of the derivative of a function h along a vector field $f_\alpha(\cdot) = f(\cdot,\alpha)$, $f_\alpha(h)(x) := \frac{d}{dt} h \circ \gamma_t^\alpha(x)\,|_{t=0}$ we can also associate the notion of a realization directly to the i.o. series $\mathcal{P}$.

Proposition 2. If the analytic i.o. map $P : U \longrightarrow R$ has a realization (1), then the i.o. series $\mathcal{P}$ corresponding to P is given by

(8) $$\langle \mathcal{P}, \alpha_1 \dots \alpha_\mu \rangle = (f_{\alpha_1} \dots f_{\alpha_\mu}(h))(x_o).$$

Conversely, if a series $\mathcal{P} \in R\langle\langle\Omega\rangle\rangle$ is defined by (8), where f_α, $\alpha \in \Omega$ and h are given by an analytic system (1), then $\mathcal{P}$ defines via (6) an analytic i.o. map P such that (1) is a realization of P.

Proof. The proof of the first statement follows immediatly from (5) and (7).

The converse follows from the fact that an analytic system (1) defines, via (7) an analytic i.o. map P. This is the desired map which follows from (8).

Proposition 2 suggests a definition of a formal realization of a formal power series $\mathcal{P} \in R\langle\langle\Omega\rangle\rangle$. Let $x_o=0$. A formal power series $h \in R[[t_1,\dots,t_n]]$ of commutative real variables $t_1,\dots,t_n$ and formal vector fields $f_\alpha = \sum_{j=1}^{n} f_{\alpha,j} \frac{\partial}{\partial t_j}$, $f_{\alpha,j} \in R[[t_1,\dots,t_n]]$, $\alpha \in \Omega$ are called a formal realization of $\mathcal{P}$ if (8) holds with $f_\alpha(h) = \sum_j f_{\alpha,j} \frac{\partial}{\partial t_j} h$

4. More about power series.

To construct our realization we introduce certain operations which can be performed on formal power series.

For $\mathcal{P}, \mathcal{P}' \in R\langle\langle\Omega\rangle\rangle$ we define a series $\mathcal{Q} \in R\langle\langle\Omega\rangle\rangle$ by

(9) $$\langle \mathcal{Q}, \alpha_1 \dots \alpha_\mu \rangle = \sum_{\substack{p_i, q_i = 0,1 \\ p_i + q_i = 1}} \langle \mathcal{P}, \alpha_1^{p_1} \dots \alpha_\mu^{p_\mu} \rangle \langle \mathcal{P}', \alpha_1^{q_1} \dots \alpha_\mu^{q_\mu} \rangle,$$

where we denote $\alpha_i^1 = \alpha_i$ and $\alpha_i^o = e$ - empty word. The series $\mathcal{Q}$ is denoted by

$$\mathcal{Q} = \mathcal{P} \sqcup \mathcal{P}'$$

and is called the shuffle product or the Hurwitz product of $\mathcal{S}$ and $\mathcal{S}'$ (equivalent definitions appear in automata and formal languages theories, compare [1], [11]). The role of the shuffle product in our context is explained by the following proposition (cf.[3]).

Proposition 3. If $\mathcal{S}$ and $\mathcal{S}'$ are formal power series which correspond via (8) to functions $h, h' : M \longrightarrow R$ then the shuffle product $\mathcal{S} ш \mathcal{S}'$ corresponds to the product hh'.

Proof. From the Leibnitz rule

$$f_\alpha(hh') = f_\alpha(h)h' + hf_\alpha(h')$$

we get easily by induction with respect to the length of the word $\alpha_1 \dots \alpha_\mu$ that

$$f_{\alpha_1} \dots f_{\alpha_\mu}(hh') = \Sigma f_{\alpha_{i_1}} \dots f_{\alpha_{i_k}}(h) f_{\alpha_{j_1}} \dots f_{\alpha_{j_\ell}}(h')$$

where the sum is taken over all disjoint subsequences $\alpha_{i_1}, \dots, \alpha_{i_k}$ and $\alpha_{j_1} \dots, \alpha_{j_\ell}$ of $\alpha_1, \dots, \alpha_\mu$ which together fill the whole sequence. The above formula together with (8) gives (9).

The set $R\langle\langle\Omega\rangle\rangle$ with the usual addition and the shuffle product forms a commutative algebra with the identity 1 (i.e. the series $\mathcal{S}$ such that $\langle\mathcal{S},e\rangle = 1$, $\langle\mathcal{S},w\rangle = 0$, $w \neq e$). An element $\mathcal{S} = \lambda + \bar{\mathcal{S}} \in R\langle\langle\Omega\rangle\rangle$ is invertible in this algebra (shuffle invertible) if $\lambda = \langle\mathcal{S},e\rangle \neq 0$. The inverse is given by

$$(10) \qquad \mathcal{S}^{ш-1} = \lambda^{-1}(1 - \tilde{\mathcal{S}} + \tilde{\mathcal{S}} ш \tilde{\mathcal{S}} - \tilde{\mathcal{S}} ш \tilde{\mathcal{S}} ш \tilde{\mathcal{S}} + \dots), \quad \text{where } \tilde{\mathcal{S}} = \lambda^{-1}\bar{\mathcal{S}}.$$

We shall also use the following operation. Let $\mathcal{F} = \sum_{w\in\Omega^*} \lambda_w w$ be any polynomial in $R\langle\Omega\rangle$. $\mathcal{F}$ defines a linear operator in $R\langle\langle\Omega\rangle\rangle$ given by the formulas

$$(11) \qquad \mathcal{F}(\mathcal{S}) = \sum_{w\in\Omega^*} \lambda_w w(\mathcal{S}), \quad \text{where}$$

$$w(\mathcal{S}) = \sum_{v\in\Omega^*} \langle\mathcal{S}, vw\rangle v.$$

This operation has the following interpretation. If $w = \alpha_1 \dots \alpha_\mu$ and $\mathcal{S}$ corresponds, via (8), to the function h, then $w(\mathcal{S})$ corresponds to the function $f_{\alpha_1} \dots f_{\alpha_\mu}(h)$.

For $\mathcal{F}, \mathcal{G} \in R\langle\Omega\rangle$ denote

$$[\mathcal{F}, \mathcal{G}] = \mathcal{F}\mathcal{G} - \mathcal{G}\mathcal{F},$$

where $\mathcal{F}\mathcal{G}$ denotes the product (2).
The set of elements in $R\langle\Omega\rangle$ obtained from the generating variables $\alpha \in \Omega$ by taking linear combinations and the products $[\cdot,\cdot]$ forms a Lie algebra $L(\Omega)$ called the free Lie algebra generated by Ω. By the above, each element X of $L(\Omega)$ acts as a linear operator in $R\langle\langle\Omega\rangle\rangle$.

We define also a bigger set of linear operators in $R\langle\langle\Omega\rangle\rangle$. By a shuffle product $\mathcal{Q} \sqcup\!\sqcup X$ of $\mathcal{Q} \in R\langle\langle\Omega\rangle\rangle$ and $X \in L(\Omega)$ we mean the following operator $R\langle\langle\Omega\rangle\rangle \longrightarrow R\langle\langle\Omega\rangle\rangle$

$$(\mathcal{Q} \sqcup\!\sqcup X)(\mathcal{P}) = \mathcal{Q} \sqcup\!\sqcup (X(\mathcal{P})).$$

The linear space spanned by linear combinations of $\mathcal{Q} \sqcup\!\sqcup X$, $\mathcal{Q} \in R\langle\langle\Omega\rangle\rangle$, $X \in L(\Omega)$ is denoted by $L((\Omega))$.

We close this section with a definition of a rank of $\mathcal{P}$ for $\mathcal{P} \in R\langle\langle\Omega\rangle\rangle$. $\mathcal{P}$ defines the following subspace of $R\langle\langle\Omega\rangle\rangle$

$$(12) \qquad \mathcal{V} = \{\mathcal{F}(\mathcal{P}) \mid \mathcal{F} \in R\langle\Omega\rangle\}.$$

We have the following canonical bilinear form on $L(\Omega) \times \mathcal{V}$

$$(X, \mathcal{Q}) \longrightarrow \langle X(\mathcal{Q}), e\rangle ,$$

where $\langle X(\mathcal{Q}), e\rangle$ is the value at identity of the series $X(\mathcal{Q})$. We define a <u>rank of</u> $\mathcal{P}$, called <u>Lie rank</u>, as the rank of this bilinear form. In other words

$$(13) \qquad \operatorname{rank} \mathcal{P} = \sup \operatorname{rank}\{\langle X_i(\mathcal{Q}_j), e\rangle\}_{i,j=1}^{k}$$

where the supremum is taken over all $k \geq 1$ and the sequences $X_1, \ldots, X_k \in L(\Omega)$ and $\mathcal{Q}_1, \ldots, \mathcal{Q}_k \in \mathcal{V}$. For a finite alphabet Ω this definition is equivalent to the definition of the Lie rank given by Fliess [5].

5. <u>Construction of a realization.</u>

Let $\mathcal{P} \in R\langle\langle\Omega\rangle\rangle$ be a given formal power series, for example an input-output series. Assume that $\operatorname{rank} \mathcal{P} = n < \infty$. Then there exist sequences $X_1, \ldots, X_n \in L(\Omega)$ and $\mathcal{Q}_1, \ldots, \mathcal{Q}_n \in \mathcal{V}$ such that $\operatorname{rank} \langle H, e\rangle = n$, where

$$H = \{X_i(\mathcal{Q}_j)\}_{i,j=1}^{n}$$

is a matrix with elements in $R\langle\langle\Omega\rangle\rangle$ and $\langle H, e\rangle$ is the matrix of its first coefficients in R. We claim that there is also a sequence

$\bar{X}_1,\dots,\bar{X}_n \in L((\Omega))$ such that

$$\bar{X}_i(\mathcal{Q}_j) = \delta_{ij} ,$$

where δ_{ij} is the Kronecker symbol: $\delta_{ij} = 1$ if $i = j$ and $\delta_{ij} = 0$ if $i \neq j$.

In fact, by our assumption the matrix H, treated as a matrix over the commutative algebra $R\langle\langle\Omega\rangle\rangle$ with the shuffle product, has an invertible determinant and so it has an inverse which can be computed for example by

$$(14) \qquad H^{-1} = (\det H)^{-1} H^*,$$

where H^* is the adjoint matrix of H (without emphasizing this in notation we use here always the shuffle product). Denote the elements of the matrix H^{-1} by R_{ij} and define

$$(15) \qquad \bar{X}_i = \sum_{j=1}^{n} R_{ij} ш X_j .$$

Then we have

$$\bar{X}_i(\mathcal{Q}_j) = \sum_{k=1}^{n} R_{ik} ш (X_k(\mathcal{Q}_j)) = \delta_{ij} ,$$

i.e., $\bar{X}_1,\dots,\bar{X}_n$ form the desired sequence.

Now we are in a position to state our main theorem.

<u>Theorem 1</u>. Let P be an analytic input-output map. If the formal power series $\mathcal{P}$ of P has a finite Lie rank equal to n then P has an analytic realization (1) and, in certain local coordinates $\tau = (\tau_1,\dots,\tau_n)$ around x_o, the Taylor expansions of $f(\cdot,\alpha) = \sum f_{\alpha,j}\frac{\partial}{\partial\tau_j}$ and h can be constructed as follows. Let $X_1,\dots,X_n \in L((\Omega))$ and $\mathcal{Q}_1,\dots,\mathcal{Q}_n \in \mathcal{U}$ be such that $X_i(\mathcal{Q}_j) = \delta_{ij}$. Then

$$h(\tau) = \sum_i h^{(i)}(0)\frac{\tau^i}{i!} , \qquad f_{\alpha,j}(\tau) = \sum_i f^{(i)}_{\alpha,j}(0)\frac{\tau^i}{i!} ,$$

where $i = (i_1,\dots,i_n)$, $\tau^i = \tau_1^{i_1}\dots\tau_n^{i_n}$, $i! = i_1!\dots i_n!$ and $h^{(i)}(0)$, $f^{(i)}_{\alpha,j}(0)$ are given by

$$(16) \qquad \begin{aligned} h^{(i_1,\dots,i_n)}(0) &= \langle X_n^{i_n}\dots X_1^{i_1}(\mathcal{P}),e\rangle \\ f_{\alpha,j}^{(i_1,\dots,i_n)}(0) &= \langle X_n^{i_n}\dots X_1^{i_1}\alpha(\mathcal{Q}_j),e\rangle . \end{aligned}$$

If we start with any power series $\mathcal{S} \in R\langle\langle\Omega\rangle\rangle$ with rank $\mathcal{S} = n$, then the above formulas give a formal realization of $\mathcal{S}$.

The formulas in Theorem 1 can be also written in the following compact form

$$h(\tau_1,\dots,\tau_n) = \langle e^{\tau_n X_n}\dots e^{\tau_1 X_1}(\mathcal{S}),e\rangle$$

$$f_{\alpha,j}(\tau_1,\dots,\tau_n) = \langle e^{\tau_n X_n}\dots e^{\tau_1 X_1}\alpha(\mathcal{Q}_j),e\rangle ,$$

where we write formally $e^{tX} = \Sigma \frac{t^i}{i!} X^i$.

In view of Theorem 1 the whole procedure of constructing the realization from an i.o. series can be summarized as follows.

Step 1. Compute rank $\mathcal{S}$ and check if rank $\mathcal{S} < \infty$. Find $X_1,\dots,X_n \in L(\Omega)$ and $\mathcal{Q}_1,\dots,\mathcal{Q}_n \in \mathcal{O}$ such that $\mathrm{rank}\{\langle X_i(\mathcal{Q}_j),e\rangle\} = n = \mathrm{rank}\,\mathcal{S}$.

Step 2. Compute H^{-1} over the shuffle algebra (i.e. using the shuffle product as the product) where $H = \{X_i(\mathcal{Q}_j)\}$. If you use formula (14) then compute $(\det H)^{-1}$ by formula (10).

Step 3. Replace $X_1,\dots,X_n$ by $\bar{X}_i = \sum_j R_{ij} \,\text{ш}\, X_j$ where $H^{-1} = \{R_{ij}\}$. The new $X_1,\dots,X_n$ satisfy the formula $X_i(\mathcal{Q}_j) = \delta_{ij}$.

Compute $h^{(i_1,\dots,i_n)}(0)$ and $f_{\alpha,j}^{(i_1,\dots,i_n)}(0)$ using Theorem 1.

Remark 1. If we want to compute the Taylor series of h and $f_{\alpha,j}$ up to a finite order N, then the corresponding computations on formal power series should be done up to some finite order only (this means that the terms of higher order can be neglected). This order depends on the number k which is the highest degree of the polynomials $\mathcal{F}_1,\dots,\mathcal{F}_n$ which define $\mathcal{Q}_1,\dots,\mathcal{Q}_n$, by $\mathcal{Q}_i = \mathcal{F}_i(\mathcal{S})$ (see (12)) and on the number p being the highest degree of $X_1,\dots,X_n$ which appear in Step 1 (treated as polynomials in $R\langle\Omega\rangle$). In particular, if the alphabet Ω is finite then the computation of Taylor expansions of h and $f_{\alpha,j}$ is a finite algorithm in Step 2 and Step 3. However, the computation of rank $\mathcal{S}$ in Step 1 is not a finite algorithm in general. It is an interesting (and open) problem to find finite algorithms which give partial realizations, i.e., realize $\mathcal{S}$ up to some finite order similarly as it was done in the linear case (cf. [10]).

Remark 2. All the results of the paper hold in the multioutput case $Y = R^r$, too. In this case the i.o. map $P : U \longrightarrow R^r$ defines r formal power series $\mathcal{S} = (\mathcal{S}_1,\dots,\mathcal{S}_r)$, $\mathcal{S}_1,\dots,\mathcal{S}_r \in R\langle\langle\Omega\rangle\rangle$ and the definition of the subspace $\mathcal{O} \subset R\langle\langle\Omega\rangle\rangle$ should be changed for

$$\mathcal{O} = \operatorname{span}\{\mathfrak{F}(\mathcal{P}_i) \mid \mathfrak{F} \in R\langle\Omega\rangle,\ i = 1,\ldots,r\}.$$

6. Scetch of the proof.

Below we outline the proof of Theorem 1 in the nonformal case. A detailed proof together with a construction of the realization which omits the computation of the shuffle inverse (14) will be given in [9].

Let P be a given analytic i.o. map. It can be shown that the Lie rank of the corresponding i.o. series $\mathcal{P}$ is equal to the rank of P defined in [6], [7]. Then, analogously as in [7] one can show that P has an analytic realization (1) such that dim M = rank $\mathcal{P}$.

While we know that the realization exists, we can interpret all the operations on the formal power series as operations on functions and vector fields on M, via (8). In particular, the shuffle product corresponds to the product of functions on M (Proposition 3), and the action (11) of the elements in $L(\Omega)$ on the elements of $R\langle\langle\Omega\rangle\rangle$ corresponds to the action of vector fields in the Lie algebra generated by the vector fields f_α, $\alpha \in \Omega$, on functions on M.

Taking this into account it can be seen that the procedure presented in the paper corresponds to the following way of computing the Taylor series of the functions h and $f_{\alpha,j}$, where $f_\alpha = \Sigma f_{\alpha,j} \frac{\partial}{\partial\tau_j}$. Denote the vector fields which correspond to the elements $X_1,\ldots,X_n \in L(\Omega)$ by $g_1,\ldots,g_n$ and the functions corresponding to $\mathcal{Q}_1,\ldots,\mathcal{Q}_n$ by $\tau_1,\ldots,\tau_n$, $\tau_i : M \longrightarrow R$ (by possibly neglecting the constant term we have $\tau_i(x_o) = 0$). The condition $\operatorname{rank}\{g_i(\tau_j)(x_o)\} = \operatorname{rank}\{\langle X_i(\mathcal{Q}_j),e\rangle\} = n$ implies that the gradients of $\tau_1,\ldots,\tau_n$ are linearly independent, so we may choose $\tau_1,\ldots,\tau_n$ as local coordinates in a neighborhood of x_o.

Note now that for a vector field g on M we have $g = \sum_i g(\tau_i)\frac{\partial}{\partial\tau_i}$ ($g(\tau_i)$ is the derivative of τ_i along g). We replace the vector fields $g_1,\ldots,g_n$ by $\bar{g}_1,\ldots,\bar{g}_n$,

$$\bar{g}_i = \sum_{j=1}^{n} h_{ij} g_j ,$$

where $\{h_{ij}\} = \{g_i(\tau_j)\}^{-1}$. In this way we obtain that $\bar{g}_i(\tau_j) = \delta_{ij}$, i.e., $\bar{g}_i = \frac{\partial}{\partial\tau_i}$.

Denote $\bar{g}_i = g_i$. The Taylor series of the functions h and $f_{\alpha,j}$ are now given by

$$h^{(i_1,\dots,i_n)}(0) = g_n^{i_n}\dots g_1^{i_1}(h)(x_o)$$

$$f_{\alpha,j}^{(i_1,\dots,i_n)}(0) = g_n^{i_n}\dots g_1^{i_1}f_\alpha(\tau_j)(x_o),$$

i.e., they correspond to the formulas given by Theorem 1.

References

[1] S.EILENBERG, "Automata, Languages and Machines" Vol.A, Academic Press, New York 1974.

[2] M.FLIESS, Sur la realization des systèmes dynamique bilinéaires, C.R.Acad.Sec.Paris, t.277(1973) ser.A, pp.923-926.

[3] M.FLIESS, Un outil algébrique: les series formelles non commutatives, in "Mathematical Systems Theory, eds. Marchesini, Mitter, Springer 1975, pp.122-148.

[4] M.FLIESS, Realizations of nonlinear systems and abstract transitive Lie algebras, Bull.Am.Math.Soc.Vol.2 No 3(1980),pp.444-446.

[5] M.FLIESS, Fonctionelles causales non linéaires et indéterminées non commutatives, Bull.Soc.Math.Franc.Vol.109(1981),pp.3-40.

[6] B.JAKUBCZYK, Existence and uniqueness of nonlinear realizations, Proc.Conf."Analyse des Systèmes", Bordeaux 1978, Asterisque 75-76(1980), pp.141-147.

[7] B.JAKUBCZYK, Existence and uniqueness of realizations of nonlinear systems, SIAM J.Control and Optimiz. Vol.18(1980), pp.

[8] B.JAKUBCZYK, B.KAŚKOSZ, Realizability of Volterra series with constant kernels, Nonlinear Analysis, Theory, Methods and Appl., Vol.5(1980), pp.167-183.

[9] B.JAKUBCZYK, Realizations of nonlinear systems via formal power series in noncommuting variables, in preparation.

[10] R.KALMAN, P.FALB, M.ARBIB, "Topics in Mathematical System Theory", Mc Graw-Hill, New York 1969.

[11] A.SALOMAA, M.SOITTOLA, "Automata-Theoretic Aspects of Formal Power Series", Springer Verlag, New York 1978.

[12] H.SUSSMANN, Minimal realizations of nonlinear systems, in "Geometric Methods", Proc.Conf.London 1973, ed. Mayne.

[13] H.SUSSMANN, Existence and uniqueness of minimal realizations of nonlinear systems, Math.Syst.Theory, Vol.10(1976/77), pp.263-284.

(Ad f, G) INVARIANT AND CONTROLLABILITY DISTRIBUTIONS

Arthur J. Krener
Department of Mathematics
University of California, Davis
visiting: Department of Electrical Engineering
Imperial College, London

Alberto Isidori
Istituto di Automatica
Università di Roma

1. INTRODUCTION

In the last few years there has been an increasing interest in nonlinear feedback systems, and a systematic work of generalization of Wonham's geometric approach to linear feedback systems is being set up (see [1-6]).

Key tools are those of f invariance and (f,g) invariance for distributions, introduced in [7], [1] and [2]. In this paper, we compare previous definitions of "f invariance" and introduce a new notion, based on Sussmann's results about the integrability of C^∞ distributions [7], which we term (Ad f, G) invariance. Then we also introduce the concept of (Ad f, G) controllability subdistribution (a generalization of the notion of an (A,B) controllability subspace).

2. MATHEMATICAL PRELIMINARIES

Throughout this paper we consider nonlinear systems described by differential equations of the form

$$\text{(2.1 a)} \qquad \dot{x} = f(x,u) = g_0(x) + \sum_{i=1}^{m} g_i(x)u_i$$

$$\text{(2.1 b)} \qquad y = h(x)$$

The state x belongs to an n-dimensional C^∞ manifold M, $u_i \in R$, the vector fields $g_0(x)$, $g_1(x),\dots,g_m(x)$ are complete C^∞ vector fields on M and $h : M \to \mathbb{R}^p$ is a C^∞ function. Occasionally, we shall make an explicit assumption of analyticity.

The following notions are standard. A C^∞ distribution Δ is a mapping assigning to each $x \in M$ a linear subspace $\Delta(x)$ of T_xM, with the property that for all $x \in M$ there exists a neghbourhood U of x and a set of C^∞ vector fields $\{X_i\}_{i\in I}$ defined

---

Research supported in part by the NSF under MCS-8003263 and by a Senior Fellowship from the SRC.

on U such that $\Delta(x)$ is spanned by the set of vectors $\{X_i(x)\}_{i\in I}$. A vector field X belongs to a distributions Δ if $X(x)\in\Delta(x)$ for all $x\in M$. A distribution Δ_1 contains a distribution Δ_2 if $\Delta_1(x) \supseteq \Delta_2(x)$ for all $x \in M$. A distribution Δ is involutive if $X \in \Delta$, $Y \in \Delta$ implies $[X,Y]\in \Delta$. A distribution Δ is nonsingular if the dimension of $\Delta(x)$ is constant over M. An integral submanifold N of Δ is a connected, immersed submanifold $N \subseteq M$ such that, for each $x \in N$, $T_xN = \Delta(x)$. An integral submanifold N of Δ is maximal if every integral submanifold N' of Δ with the property that $N' \supseteq N$ coincides with N. A distribution Δ is integrable if its maximal integral submanifolds define a partition of M.

Let X be a complete vector field on M and let $\Phi^X_t(x)$ denote the corresponding flow, i.e. the C^∞ mapping $\mathbb{R} \times M \to M$ with the property that

$$\frac{d}{dt}\Phi^X_t(x) = X(\Phi^X_t(x))$$

$$\Phi^X_o(x) = x$$

For each t, Φ^X_t defines a diffeomorphism $x \mapsto \Phi^X_t(x)$. Let Y be another vector field on M. For all $t \in \mathbb{R}$, there exists a unique vector field, denoted $Ad^tX(Y)$, which is Φ^X_t-related to Y, i.e. that satisfies the condition

$$(\Phi^X_t)_* Ad^tX(Y) = Y\circ\Phi^X_t(x)$$

for all $x \in M$.

The following two Definitions clarify the concepts of "X invariance" for a distribution Δ.

Definition 2.1. A distribution Δ is Ad X invariant (X-invariant in [7]) if for all $Y \in \Delta$ and for all $t \in \mathbb{R}$

$$Ad^tX(Y) \in \Delta$$

A distribution Δ is ad X invariant (X-invariant in [1]) if for all $Y \in \Delta$

$$[X,Y] \in \Delta$$

Remark 2.1. Clearly, a distribution is Ad X-invariant iff for all $t \in \mathbb{R}$ $(\Phi^X_t)_*$ maps $\Delta(x)$ into $\Delta(\Phi^X_t(x))$, for all $x \in M$ (see [7]).

Remark 2.2. The vector field $Ad^tX(Y)$ can be given a Taylor series expansion via the Campbell-Backer-Hausdorff formula

$$Ad^tX(Y)(x) = \sum_{k=0}^{\infty} \frac{t^k}{k!} ad^kX(Y)(x)$$

where

$$ad^oX(Y) = Y \quad \text{and } ad^kX(Y) = [X,ad^{k-1}X(Y)]$$

Thus, by differentiation, we see that Ad X invariance implies ad X invariance. The converse is clearly true in C^{ω}. In C^{∞}, the two notions are equivalent only under some suitable extra assumption like, e.g., the nonsingularity of Δ.

Remark 2.3. A distribution is involutive iff it is ad X invariant for all $X \in \Delta$.

The basic integrability results are the following

Theorem (Sussmann [7]). A distribution Δ is integrable iff it is Ad X invariant for every $X \in \Delta$.

Corollary (Frobenius) A distribution Δ is integrable only if it is involutive.

Corollary (Frobenius) A nonsingular distribution Δ is integrable iff it is involutive.

Corollary (Hermann-Nagano) A C^{ω} distribution Δ is integrable iff it is involutive.

When referred to a collection of vector fields, like the ones appearing on the right-hand-side of (2.1a), Definition 2.1. is extended as follows (again, see [7] and [1], where the same notions are used, with different notation).

Definition 2.2. A distribution Δ is Ad f invariant (resp. ad f invariant) if for every $u \in \mathbb{R}^m$, Δ is Ad $f(\cdot,u)$ invariant (resp. ad $f(\cdot,u)$ invariant).

If Δ is a given distribution, there is a smallest C^{∞} distribution which contains Δ and is Ad f invariant. This distribution will be denoted with the symbol

$$\langle \mathrm{Ad}\ f | \Delta \rangle$$

It is easy to see [7] that the subspace $\langle \mathrm{Ad}\ f|\Delta \rangle(x)$ of T_xM is the linear hull of all the vectors of the form $(g^{-1})_* X \circ g(x)$, where X is a vector field in Δ and g is a diffeomophism of the form $\Phi_{t_1}^{f(\cdot,u_1)} \circ \Phi^{f(\cdot,u_2)} \circ \dots \circ \Phi_{t_n}^{f(\cdot,u_n)}$, with $n \in \mathbb{Z}$, $t_i \in \mathbb{R}$, $u_i \in \mathbb{R}^m$.

The symbol

$$\langle \mathrm{ad}\ f | \Delta \rangle$$

shall denote the smallest distribution which contains Δ and is ad f invariant. The subspace $\langle \mathrm{ad}\ f|\Delta \rangle(x)$ of T_xM is the linear hull of all vectors of the form $[f(\cdot,u_1), [f(\cdot,u_2),\dots,[f(\cdot,u_n),X]\dots]](x)$, where $X \in \Delta$, $n \in \mathbb{Z}$, $u_i \in \mathbb{R}^m$.

Let $R(F)$ denote the distribution spanned by the set of vector fields $\{g_i\}_{i=0,1,\dots,n}$ and $R(G)$ the distribution spanned by the set of vector fields $\{g_i\}_{i=1,\dots,n}$. The following distributions are of paramount importance in the study of accessibility properties of control systems

(2.2) $\langle \mathrm{Ad}\ f|R(F)\rangle$ (resp. $\langle \mathrm{ad}\ f|R(F)\rangle$)

(2.3) $\langle \mathrm{Ad}\ f|R(G)\rangle$ (resp. $\langle \mathrm{ad}\ f|R(G)\rangle$)

The distribution $\langle Ad\ f|R(F)\rangle$ is integrable (in C^∞, it may properly contain $\langle ad\ f|R(F)\rangle$) and related to the partition of M into equivalence classes with respect to the relation of weak accessibility [7]. In C^ω, the distributions $\langle Ad\ f|R(G)\rangle$ and $\langle ad\ f|R(G)\rangle$ coincide and are related to the partition of M into equivalence classses with respect to the relation of weak accessibility "in zero units of time" [8].

3. (Ad f,G) INVARIANCE

In this and the following section we assume the dynamics (2.1a) be modified by feedback, i.e. that there exists a pair $\alpha(x),\beta(x)$ of m×1 and m×m matrix valued C^∞ functions of x such that

$$u_i = \alpha_i(x) + \sum_{j=1}^{m} \beta_{ij}(x)v_i$$

with $v_i \in \mathbb{R}$. The new dynamics shall be written as

$$\dot{x} = \tilde{f}(x,v) = \tilde{g}_o(x) + \sum_{i=1}^{m} \tilde{g}_i(x)v_i \tag{3.1}$$

For the sake of compactness, we shall introduce the notations

$$G := \mathrm{row}(g_1,\dots,g_m)$$
$$F := \mathrm{row}(g_o,g_1,\dots,g_m)$$
$$\gamma := \begin{pmatrix} 1 & 0 \\ \alpha & \beta \end{pmatrix}$$

and write

$$\tilde{g}_o(x) = g_o(x) + G(x)\alpha(x)$$
$$G(x) = G(x)\beta(x)$$
$$\tilde{F}(x) = F(x)\gamma(x)$$

We say that a distribution Δ separates the controls if there exists a feedback γ with invertible β and a partition of $\beta = (\beta_1\ \beta_2)$ such that

$$\Delta \cap R(G) = R(\tilde{G}_1)$$
$$\Delta \cap R(\tilde{G}_2) = \{0\}$$

where $\tilde{G}_i(x) = G(x)\beta_i(x)$, $i = 1,2$. Such feedback γ is said to be separating.

The following definitions provide nonlinear generalizations of the notion of an (A,B) invariant subspace.

Definition 3.1. A distribution Δ is (Ad f,G) invariant (resp. (ad f,G) invariant) if there exists a feedback γ such that Δ is Ad $\tilde{f}$ invariant (resp. ad $\tilde{f}$ invariant).

Definition 3.2. A distribution Δ is locally (ad f,G) invariant if for every constant $u \in \mathbb{R}^m$

$$X \in \Delta \Rightarrow [f(\cdot,u),X] \in \Delta + R(G)$$

The link between the two Defintions is given by the following Lemma.

Lemma 3.1. If Δ is nonsingular, involutive and separates the controls then the following are equivalent

(a) Δ is locally (ad f,G) invariant,

(b) there exists an open cover $\{U_j\}$ of M and separating feedbacks γ_j defined on U_j such that Δ is (ad $\tilde{f}$,G) invariant on U_j under γ_j.

(c) there exists an open cover $\{U_j\}$ and separating feedbacks γ_j on U_j such that Δ is (Ad $\tilde{f}$,G) invariant on U_j under γ_j.

Proof. The equivalence of (b) and (c) follows from the nonsingularity of Δ. It is trivial to verify that (b) implies (a). In [4] it is shown that (a) implies (b) using the stronger hypothesis that $\Delta \cap R(G)$ and $R(G)$ are nonsingular. But the proof only uses this to show that Δ separates the controls. Moreover the feedback so constructed is easily seen to be separating. Similar results are found in [5]. ◀

4. CONTROLLABILITY DISTRIBUTIONS

In this section we introduce various nonlinear generalizations of the notion of an (A,B) controllability subspace (see also [6]).

Definition 4.1. A distribution Δ is (Ad f,G) controllable (resp. (ad f,G) controllable) if separates the controls and, for some separating feedback γ,

$$\Delta = \langle \mathrm{Ad}\ f | R(\tilde{G}_1) \rangle$$

$$(\text{resp.} \quad \Delta = \langle \mathrm{ad}\ f | R(\tilde{G}_1) \rangle)$$

The local version of this definition is based on a generalization of the controllability subspace algorithm, introduced by Wonham [9].

Controllability subdistribution algorithm. Let Δ be a given distribution. It is possible to prove that the class of all distributions $\hat{\Delta}$ satisfying the condition

$$\hat{\Delta} = \Delta \cap ([f,\hat{\Delta}] + R(G)) \tag{4.1}$$

has a unique minimal element, denoted $\Delta^c(\Delta)$. To this end, define a non-decreasing sequence of distributions Δ_k, by $\Delta_o = \{0\}$ and

$$\Delta_k = \Delta \cap ([f,\Delta_{k-1}] + R(G)) \tag{4.2}$$

Clearly, $\Delta_o \subseteq \Delta_1$ and, by induction, it follows that $\Delta_{k-1} \subseteq \Delta_k$. For, if $\Delta_{k-2} \subseteq \Delta_{k-1}$, then $\Delta_{k-1} = \Delta \cap ([f,\Delta_{k-2}] + R(G)) \subseteq \Delta \cap ([f,\Delta_{k-1}] + R(G)) = \Delta_k$. Let

(4.3) $$\Delta^c(\Delta) := \bigcup_{k\geq 0} \Delta_k$$

Clearly, this distribution satisfies (4.1) for, if on some open subset U of M $\Delta_k = \Delta_{k-1}$, then $\Delta_{k+i} = \Delta_k$ on U for all i>0. On the other hand, if $\hat{\Delta}$ is any distribution satisfying (4.1), then trivially $\hat{\Delta} \supseteq \Delta_o$ and, by induction, it follows that $\Delta \supseteq \Delta_k$ for all k>0. Thus the right-hand-side of (4.3) is the unique minimal element of the class of all distributions $\hat{\Delta}$ satisfying the condition (4.1).

Definition 4.2. A distribution Δ is locally (ad f,G) controllable if Δ is locally (ad f,G) invariant and $\Delta^c(\Delta) = \Delta$.

In order to establish a link between the two defintions, we need the following result , which generalizes a property of (A,B) controllability subspaces.

Lemma 4.1. Suppose Δ is (ad f,G) invariant under invertible feedback γ, then

(4.4) $$\Delta^c(\Delta) = \langle \text{ad } \tilde{f} | \Delta \cap R(G) \rangle$$

Proof. We observe, firstly, that from the equality

$$[\tilde{f}(\cdot,v),X] = [g_o,X] + [G,X](\alpha+\beta v) - G\, X(\alpha+\beta v)$$

we can deduce, because of the nonsingularity of β, that

(4.5) $$[\tilde{f},\Delta] + R(G) = [f,\Delta] + R(G)$$

where Δ is a given distribution.

Now we define a nondecreasing sequence of distributions $\bar{\Delta}_k$, by

$$\bar{\Delta}_1 = \Delta \cap R(G)$$
$$\bar{\Delta}_k = [\tilde{f},\bar{\Delta}_{k-1}] + \Delta_1$$

and we show, by induction, that $\bar{\Delta}_k = \Delta_k$, with Δ_k as defined by (4.2). Clearly, $\bar{\Delta}_1 = \Delta_1$. Assume now that $\Delta_{k-1} = \bar{\Delta}_{k-1}$ and observe that, since Δ is ad $\tilde{f}$ invariant, $[\tilde{f},\Delta_{k-1}] \subseteq \Delta$. By (4.5) we have

$$\Delta_k = \Delta \cap ([f,\Delta_{k-1}] + R(G)) = \Delta \cap ([\tilde{f},\Delta_{k-1}] + R(G)) = [\tilde{f},\Delta_{k-1}] + \Delta \cap R(G) =$$
$$= [\tilde{f},\Delta_{k-1}] + \bar{\Delta}_1 = \bar{\Delta}_k$$

Since

$$\langle \text{ad } f | \Delta \cap R(G) \rangle = \bigcup_{k\geq 1} \bar{\Delta}_k$$

the proof is complete. ◀

At this point it is possible to prove a result analogous to Lemma 3.1.

Lemma 4.2. Suppose Δ is nonsingular, involutive and separates the controls;then the following are equivalent:

(a) Δ is locally (ad f,G) controllable.

(b) there exists an open cover $\{U_j\}$ of M and separating feedbacks γ_j defined on U_j such that Δ is (ad f,G) controllable on U_j under γ_j.

(c) there exists an open cover $\{U_j\}$ of M and separating feedbacks γ_j defined on U_j such that Δ is (Ad f,G) controllable on U_j under γ_j.

Proof. (a) ⇒ (b). If Δ is locally (ad f,G) controllable, then

$$\text{(i)} \qquad [f,\Delta] \subseteq \Delta + R(G)$$

$$\text{(ii)} \qquad \Delta^c(\Delta) = \Delta$$

The first, thanks to Lemma 3.1., implies that there exists an open cover $\{U_j\}$ of M and separating (thus nonsingular) feedbacks γ_j defined on U_j such that Δ is (ad f,G) invariant on U_j under γ_j. Thus, by Lemma 4.1., we have that

$$\Delta^c(\Delta) = \langle \text{ad } \tilde{f} | \Delta \cap R(G) \rangle = \langle \text{ad } \tilde{f} | R(\tilde{G}_1) \rangle$$

on U_j. From this and (ii), the assertion follows.

(b) ⇒ (a). On U_j, under the separting feedback γ_j, we have

$$\Delta = \langle \text{ad } \tilde{f} | R(\tilde{G}_1) \rangle = \langle \text{ad } \tilde{f} | \Delta \cap R(G) \rangle$$

Δ is (ad $\tilde{f}$,G) invariant under invertible feedback and, thus, by Lemma 4.1., $\Delta = \Delta^c(\Delta)$. Moreover, by Lemma 3.1., Δ is locally (ad f,G) invariant. Thus, Δ is locally (ad f,G) controllable.

(b) ⇔ (c). It is a consequence of nonsingularity. ◀

It is easy to show that the family of all locally (ad f,G) controllable distributions is a semilattice with respect to inclusion and distribution addition. Thus the family of all locally (ad f,G) controllable distributions contained in a given distribution Δ has a unique maximal element. Like in the case of linear systems (see [9]), this can be computed via the controllability subdistribution algorithm, applied to the unique maximal locally (ad f,G) invariant distribution contained in Δ.

REFERENCES

[1] A. ISIDORI, A.J. KRENER, C. GORI-GIORGI, S. MONACO - Nonlinear Decoupling via Feedback: a Differential Geometric Approach, IEEE Trans. Aut. Contr., 26(1981), pp. 331-345.

[2] R.M. HIRSCHORN - (A,B)-invariant Distributions and the Disturbance Decoupling of Nonlinear Systems, SIAM J. Contr. Optim., 17 (1981), pp. 1-19.

[3] H. NIJMEIJER, A.J. VAN DER SCHAFT - Controlled Invariance for Nonlinear Control Systems, to appear on IEEE Trans. Aut. Contr.

[4] A. ISIDORI, A.J. KRENER, C. GORI-GIORGI, S. MONACO - Locally (f,g) Invariant Distributions, Systems and Control Letters, 1(1981), pp. 12-15.

[5] H. NIJMEIJER - Controlled Invariance for Affine Control Systems, Int. J. Control, 34(1981), pp. 825-833.

[6] H. NIJMEIJER - Controllability Distributions for Nonlinear Systems, Rep. BW 140/81 (1981), Stichting Math. Centrum (Amsterdam).

[7] H.J. SUSSMANN - Orbits of Families of Vector Fields and Integrability of Distributions, Trans. Am. Math. Soc., 180 (1973), pp. 171-188.

[8] H.J. SUSSMANN, V. JURDIJEVIC - Controllability of Nonlinear Systems, J. Diff. Equations, 12(1972), pp. 95-116.

[9] M. WONHAM - Multivariable Control Systems: a Geometric Approach, Springer Verlag, 1979 (2nd edition).

SCHUR TECHNIQUES FOR RICCATI DIFFERENTIAL EQUATIONS*

Alan J. Laub
Dept. of Elec. Engrg. — Systems
Univ. of Southern California
Los Angeles, CA 90007

1. INTRODUCTION

One of the most intensely studied nonlinear matrix equations arising in mathematics and engineering is the Riccati equation. Despite a voluminous literature, there has been, to date, little numerical analysis and no quality mathematical software for the numerical solution of algebraic or differential Riccati equations.

In this paper we shall outline the numerical solution of certain Riccati differential equations by means of so-called Schur techniques. These methods were first used in [1] to provide a reliable, general-purpose method for solving the algebraic equation for modest-sized problems (say, of order $\lesssim 100$) where there is no usefully exploitable structure in the coefficient matrices. We shall extend those methods here to the solution of Riccati equations of the form

$$\dot{X}(t) = X(t)EX(t) + FX(t) + X(t)G + H \qquad (1)$$

where $E \in \mathbb{R}^{m \times n}$, $F \in \mathbb{R}^{n \times n}$, $G \in \mathbb{R}^{m \times m}$, $H \in \mathbb{R}^{n \times m}$, and $X \in \mathbb{R}^{n \times m}$. An important special case of (1) is the symmetric equation

$$\dot{X}(t) = X(t)EX(t) + FX(t) + X(t)F^T + H \qquad (2)$$

where $m = n$, $E = E^T$, and $H = H^T$.

We shall describe two algorithms for the symmetric problem (2), as that is the one occurring most frequently in control and systems applications. Additional assumptions on E, F, and H are often made to ensure certain properties of the solution X. However, an important feature of Schur techniques is that they apply equally well to the non-symmetric equation which tends to be the more typical situation when solving two-point boundary value problems by invariant imbedding. Similar remarks apply for the discrete-time or difference equation analogues of (1) and (2).

2. SCHUR METHODS FOR ALGEBRAIC RICCATI EQUATIONS

We shall now present a review of the Schur method for solving the algebraic equations

$$XEX + FX + XG + H = 0 \qquad (3)$$

*This research was supported by the U.S. Army Research Office under Contract DAAG29-81-K-0131.

or

$$XEX + FX + XF^T + H = 0. \tag{4}$$

This method, which is described in detail in [1], has been shown to provide a rather efficient and generally reliable method for solving these equations. Before describing the algorithm we shall first review, for completeness, some necessary background material in linear algebra.

Recall that a matrix $A \in \mathbb{R}^{nxn}$ is ORTHOGONAL if $A^T = A^{-1}$ and $A \in \mathbb{C}^{nxn}$ is UNITARY if $A^H = A^{-1}$. Now let $J = \begin{pmatrix} 0 & I \\ -I & 0 \end{pmatrix} \in \mathbb{R}^{2nx2n}$ where I denotes the nth order identity matrix. Then $A \in \mathbb{R}^{2nx2n}$ is HAMILTONIAN if $J^{-1}A^TJ = -A$ and SYMPLECTIC if $J^{-1}A^TJ = A^{-1}$. An easily proved property of a Hamiltonian matrix A which will be useful in the sequel is that if $\lambda \in \Lambda(A)$ then $-\lambda \in \Lambda(A)$ with the same multiplicity ($\Lambda(A)$ denotes the spectrum of A). Similarly, if A is symplectic and $\lambda \in \Lambda(A)$ then $\frac{1}{\lambda} \in \Lambda(A)$ with the same multiplicity. It is also easy to show that if A is either Hamiltonian or symplectic and T is symplectic, then $T^{-1}AT$ is either Hamiltonian or symplectic according as A is. Finally, we need two theorems from classical similarity theory which form the theoretical cornerstone of modern numerical linear algebra.

Theorem 1 (Schur Canonical Form): Let $A \in \mathbb{R}^{nxn}$ have eigenvalues $\lambda_1, \ldots, \lambda_n$. Then there exists a unitary similarity transformation U such that U^HAU is upper triangular with diagonal elements $\lambda_1, \ldots, \lambda_n$ in that order.

Theorem 2 (Real Schur Canonical Form := RSF): Let $A \in \mathbb{R}^{nxn}$. Then there exists an orthogonal similarity transformation U such that U^TAU is quasi-upper-triangular (i.e., upper triangular with the possible exception of 2x2 blocks on the (block) diagonal corresponding to complex conjugate pairs of eigenvalues). Moreover, U can be chosen so that the 2x2 and 1x1 diagonal blocks appear in any desired order.

If in Theorem 2 we partition $S := U^TAU$ into $\begin{pmatrix} S_{11} & S_{12} \\ 0 & S_{22} \end{pmatrix}$ where $S_{11} \in \mathbb{R}^{kxk}$, $0 < k \leq n$, we shall refer to the first k columns of U as the SCHUR VECTORS corresponding to $\Lambda(S_{11}) \subseteq \Lambda(A)$. The Schur vectors corresponding to the eigenvalues of S_{11} span the eigenspace corresponding to those eigenvalues even when some of the eigenvalues are multiple and are more reliably obtainable numerically than the corresponding eigenvectors and principal vectors (see [2]). This property has been exploited in [1] to provide a rather more generally reliable algorithm than eigenvector methods for solving algebraic Riccati equations and we shall now review the relevant results, deferring to [1] for the numerical details such as how to effect the ordering of the diagonal elements of a RSF

Consider first (4) and assume further that $E = E^T \geq 0$, $H = H^T \geq 0$. In the case considered in [1] it is also assumed that (F,B) is a stabilizable pair where B is a full rank factorization (FRF) of E (i.e., $BB^T = E$ and rank (B) = rank (E)) and that (C,F) is a detectable pair where C is a FRF of H (i.e., $C^TC = H$ and rank (C) = rank (H)). Under these assumptions, (4) is known to have a unique nonnegative definite solution. There may be other solutions to (4) and the method described below can be used to find them as well but the nonnegative definite solution is frequently the one of interest in applications.

The Schur method is based on the Hamiltonian matrix

$$Z := \begin{pmatrix} -F^T & -E \\ H & F \end{pmatrix}.$$

Our assumptions guarantee that Z has no pure imaginary eigenvalues. Thus by Theorem 2 we can find an orthogonal transformation $U \in \mathbb{R}^{2n\times 2n}$ which puts Z in RSF:

$$U^TZU = S = \begin{pmatrix} S_{11} & S_{12} \\ 0 & S_{22} \end{pmatrix}$$

with $S_{ij} \in \mathbb{R}^{n\times n}$. It is possible to arrange, moreover, that the real parts of the spectrum of S_{11} are negative while the real parts of the spectrum of S_{22} are positive. The matrix U is conformably partitioned into four nxn blocks $U = \begin{pmatrix} U_{11} & U_{12} \\ U_{21} & U_{22} \end{pmatrix}$. We then have the following theorem.

Theorem 3: U_{11} is nonsingular and $X := U_{21}U_{11}^{-1}$ is the unique symmetric nonnegative definite solution of (4). Moreover, $\Lambda(S_{11}) = \Lambda(F + XE)$.

The solution of (3) is based on the matrix $Z = \begin{pmatrix} -G & -E \\ H & F \end{pmatrix}$ where, again, certain assumptions are needed to provide a criterion for which Schur vectors of Z are to be used to construct an appropriate Riccati solution.

One further refinement may also be employed for the solution of (4). There the symmetric matrix E was simply assumed to be nonnegative definite but in many applications (such as optimal control or filtering) E is of the form $E_1E_2^{-1}E_1^T$ where $E_2 = E_2^T > 0$. In [3] it is shown that one may consider the matrix pencil

$$\lambda \begin{pmatrix} I & 0 & 0 \\ 0 & I & 0 \\ 0 & 0 & 0 \end{pmatrix} - \begin{pmatrix} F^T & 0 & E_1 \\ -H & -F & 0 \\ 0 & E_1^T & -E_2 \end{pmatrix} \tag{5}$$

which involves no inverses. An equivalent 2n x 2n generalized eigenvalue problem can then be derived from (5) and the Riccati solution is determined from a basis for the stable eigenspace of the reduced problem as suggested in [4].

3. SCHUR METHODS FOR RICCATI DIFFERENTIAL EQUATIONS

We shall now outline two Schur methods for the numerical solution of the symmetric problem (2). Remarks similar to those in the previous Section provide the appropriate generalization necessary for the nonsymmetric problem (1).

3.1 First Method

The first algorithm to be described is essentially a Schur vector formulation, with a few other numerical improvements, of an algorithm of Vaughn which was applied in [5] to discrete-time problems and in [6] to (2).

The Riccati initial value problem to be discussed in detail is

$$\dot{X}(t) = X(t)EX(t) + FX(t) + X(t)F^T + H; \quad X(0) = A \tag{6}$$

where $E = E^T \geq 0$, $H = H^T \geq 0$, and $A = A^T \geq 0$. The Schur algorithm is based on the well-known relationship between (6) and the following $2n^{th}$ order linear initial value problem:

$$\frac{d}{dt}\begin{pmatrix} P(t) \\ Q(t) \end{pmatrix} = \begin{pmatrix} -F^T & -E \\ H & F \end{pmatrix}\begin{pmatrix} P(t) \\ Q(t) \end{pmatrix}; \quad \begin{pmatrix} P(0) \\ Q(0) \end{pmatrix} = \begin{pmatrix} I \\ A \end{pmatrix}. \tag{7}$$

Specifically, we have $X(t) = Q(t)P^{-1}(t)$.

Under suitable hypotheses, the Hamiltonian matrix $Z = \begin{pmatrix} -F^T & -E \\ H & F \end{pmatrix}$ may be reduced by an orthogonal similarity to the form

$$U^TZU = \begin{pmatrix} S_{11} & S_{12} \\ 0 & S_{22} \end{pmatrix}$$

with $\Lambda(S_{11}) \subseteq RHP$, $\Lambda(S_{22}) \subseteq LHP$. This dichotomy of the eigenvalues is precisely what precludes a straightforward direct integration of (7) since the solution would involve terms with both $e^{\lambda_i t}$, $\lambda_i \in \Lambda(S_{11})$ and $e^{\lambda_j t}$, $\lambda_j \in \Lambda(S_{22})$. These terms may rapidly diverge in magnitude causing numerical difficulties. The problem is alleviated as follows. Perform a change of variables:

$$\begin{pmatrix} P(t) \\ Q(t) \end{pmatrix} = U\begin{pmatrix} \hat{P}(t) \\ \hat{Q}(t) \end{pmatrix}. \tag{8}$$

Then

$$\frac{d}{dt}\begin{pmatrix}\hat{P}(t)\\ \hat{Q}(t)\end{pmatrix} = \begin{pmatrix}S_{11} & S_{12}\\ 0 & S_{22}\end{pmatrix}\begin{pmatrix}\hat{P}(t)\\ \hat{Q}(t)\end{pmatrix} \tag{9}$$

so that

$$\begin{aligned} X(t) &= Q(t)P^{-1}(t) \\ &= (U_{21}\hat{P}(t) + U_{22}\hat{Q}(t))(U_{11}\hat{P}(t) + U_{12}\hat{Q}(t))^{-1} \\ &= (U_{21} + U_{22}\hat{Q}(t)\hat{P}^{-1}(t))(U_{11} + U_{12}\hat{Q}(t)\hat{P}^{-1}(t))^{-1}. \end{aligned}$$

We now derive a convenient expression for

$$\hat{R}(t) := \hat{Q}(t)\hat{P}^{-1}(t). \tag{10}$$

First note that from (9),

$$\hat{Q}(t) = e^{tS_{22}}\hat{Q}(0). \tag{11}$$

Substituting (11) in the equation for $\hat{P}$ we find

$$\frac{d}{dt}\hat{P}(t) = S_{11}\hat{P}(t) + S_{12}e^{tS_{22}}\hat{Q}(0) \tag{12}$$

which, when integrated, gives

$$\hat{P}(t) = e^{tS_{11}}\hat{P}(0) + \int_0^t e^{(t-\tau)S_{11}}S_{12}e^{\tau S_{22}}\hat{Q}(0)d\tau. \tag{13}$$

Now (13) can be solved for $\hat{P}(0)$ to yield

$$\begin{pmatrix}\hat{P}(0)\\ \hat{Q}(t)\end{pmatrix} = \begin{pmatrix}e^{-tS_{11}} & -\int_0^t e^{-\tau S_{11}}S_{12}e^{\tau S_{22}}d\tau\\ 0 & e^{tS_{22}}\end{pmatrix}\begin{pmatrix}\hat{P}(t)\\ \hat{Q}(0)\end{pmatrix}. \tag{14}$$

Now notice that

$$I = P(0) = U_{11}\hat{P}(0) + U_{12}\hat{Q}(0)$$

$$A = Q(0) = U_{21}\hat{P}(0) + U_{22}\hat{Q}(0)$$

whence

$$\hat{Q}(0) = R\hat{P}(0)$$

where

$$R := (AU_{12} - U_{22})^{-1}(U_{21} - AU_{11}). \tag{15}$$

Then we have

$$\hat{Q}(t) = e^{tS_{22}}\hat{Q}(0)$$

$$= e^{tS_{22}}R\hat{P}(0)$$

$$= e^{tS_{22}}R\left(e^{-tS_{11}}\hat{P}(t) - \int_0^t e^{-\tau S_{11}}S_{12}e^{\tau S_{22}}d\tau\hat{Q}(0)\right). \tag{16}$$

By uniqueness, we must have

$$\hat{Q}(0) = Re^{-tS_{11}}\hat{P}(t) - R\int_0^t e^{-\tau S_{11}}S_{12}e^{\tau S_{22}}d\tau\hat{Q}(0)$$

so that

$$\hat{Q}(0) = \left(I + R\int_0^t e^{-\tau S_{11}}S_{12}e^{\tau S_{22}}d\tau\right)^{-1}Re^{-tS_{11}}\hat{P}(t).$$

Substituting in (11) we have

$$\hat{Q}(t) = \hat{R}(t)\hat{P}(t)$$

where

$$\hat{R}(t) := e^{tS_{22}}\left(I + R\int_0^t e^{-\tau S_{11}}S_{12}e^{\tau S_{22}}d\tau\right)^{-1}Re^{-tS_{11}} \tag{17}$$

To summarize, we have

$$X(t) = (U_{21} + U_{22}\hat{R}(t))(U_{11} + U_{12}\hat{R}(t))^{-1}$$

where $R(t)$ is given by (17) and R is given by (15).

The computation of $\hat{R}(t)$ involves exponentials of the matrices S_{22} and $-S_{11}$ both of which are quasi-upper-triangular and have spectra in the LHP which alleviates the previous difficulty. The calculation of these exponentials is greatly expedited by exploiting their upper triangular structure and efficient algorithms can be constructed from either Padé approximation (see, for example [7]) or the recursion formulas of [8]. The integral term involving matrix exponentials in (17) can be computed efficiently using the formulas developed in [9].

3.2 Second Method

The second algorithm to be described is a Schur vector formulation of a standard solution technique in which the solution of (6) is determined as a sum of the solution of the steady-state equation (4) and a term involving the solution of a Lyapunov equation, some (stable) matrix exponentials, and certain matrix factorizations.

Let us consider then the solution of (6) with the stabilizability and detectability assumptions of Section 2 for E, F, and H. Let $\overline{X}$ be the unique, nonnegative definite solution of the steady-state Riccati equation (4), i.e.,

$$\overline{X}E\overline{X} + F\overline{X} + \overline{X}F^T + H = 0. \tag{18}$$

We now need the following:

Lemma 1: Assume $(A-\overline{X})^{-1}$ exists $(A = X(0))$ and define $Y(t) := (X(t)-\overline{X})^{-1}$. Then $Y(t)$ exists for all $t \geq 0$ and satisfies

$$-\frac{d}{dt}Y(t) = \overline{F}^T Y(t) + Y(t)\overline{F} + E \;; \quad Y(0) = (A-\overline{X})^{-1} \tag{19}$$

where $\overline{F} := F + \overline{X}E$.

Proof: First note that by subtracting (18) from the differential equation (6) we have

$$\begin{aligned}\frac{d}{dt}Y^{-1}(t) &= \frac{d}{dt}(X(t)-\overline{X}) \\ &= XEX - \overline{X}E\overline{X} + F(X-\overline{X}) + (X-\overline{X})F^T \\ &= (X-\overline{X})E(X-\overline{X}) + \overline{F}(X-\overline{X}) + (X-\overline{X})\overline{F}^T \\ &= Y^{-1}EY^{-1} + \overline{F}Y^{-1} + Y^{-1}\overline{F}^T.\end{aligned} \tag{20}$$

Now from the relation

$$-\frac{d}{dt}Y(t) = Y(t)\left(\frac{d}{dt}Y^{-1}(t)\right)Y(t)$$

we find, upon substitution of (20), that

$$-\frac{d}{dt}Y(t) = \overline{F}^T Y + Y\overline{F} + E \tag{21}$$

which is (19). Note that since $Y(0) = (A-\overline{X})^{-1}$ exists, $Y(t)$ exists for all $t \geq 0$ as a solution of (21).

By Theorem 3, $\overline{F}$ defined in Lemma 1 is stable. Thus there exists a unique, positive definite solution $\overline{Y}$ to the Lyapunov equation

$$\bar{F}^T\bar{Y} + \bar{Y}\bar{F} + E = 0. \tag{22}$$

Then subtracting (21) from (22) we get the Lyapunov differential equation

$$\frac{d}{dt}(Y(t) - \bar{Y}) = -\bar{F}^T(Y(t) - \bar{Y}) - (Y(t) - \bar{Y})\bar{F} \tag{23}$$

with initial condition $Y(0) - \bar{Y} = (A - \bar{X})^{-1} - \bar{Y}$. Now the solution of (23) can be written as

$$Y(t) - \bar{Y} = e^{-t\bar{F}^T}[(A - \bar{X})^{-1} - \bar{Y}]e^{-t\bar{F}}$$

from which $X(t)$ can then be recovered as follows:

$$Y(t) = (X(t) - \bar{X})^{-1} = \bar{Y} + e^{-t\bar{F}^T}[(A - \bar{X})^{-1} - \bar{Y}]e^{-t\bar{F}}$$

whence

$$X(t) = \bar{X} + [\bar{Y} + e^{-t\bar{F}^T}[(A - \bar{X})^{-1} - \bar{Y}]e^{-t\bar{F}}]^{-1}$$

$$= \bar{X} + e^{t\bar{F}}[(A - \bar{X})^{-1} - \bar{Y} + e^{t\bar{F}^T}\bar{Y}e^{t\bar{F}}]^{-1}e^{t\bar{F}^T} . \tag{24}$$

Notice that only stable matrix exponentials appear in (24).

The symmetry in the term $e^{t\bar{F}}[\ldots]^{-1}e^{t\bar{F}^T}$ of (24) can be further exploited as follows. Since $(A - \bar{X})$ is symmetric (but possibly indefinite) it has a diagonal pivoting factorization

$$A - \bar{X} = C_1 D_1 C_1^T \tag{25}$$

where C_1 is a product of elementary unit upper triangular and permutation matrices and D_1 is a symmetric block diagonal matrix with blocks of order 1 or 2; see [10], [11], [12] for numerical details. This factorization is much more cheaply computed than the spectral factorization and can be implemented in a numerically stable way.

Now do a diagonal pivoting factorization of $D_1^{-1} + C_1(e^{t\bar{F}^T}\bar{Y}e^{t\bar{F}} - \bar{Y})C_1^T$ (which we assume is nonsingular):

$$D_1^{-1} + C_1(e^{t\bar{F}^T}\bar{Y}e^{t\bar{F}} - \bar{Y})C_1^T = C_2 D_2 C_2^T. \tag{26}$$

Recall that we assumed $(A - \bar{X})$ was nonsingular so that C_1 and D_1 are nonsingular. Similarly, we have also assumed above that C_2 and D_2 are nonsingular.

Finally, solve the linear system

$$C_2\Phi = C_1 e^{t\bar{F}^T} \tag{27}$$

for Φ. Then it may be checked that

$$e^{t\bar{F}}[(A-\bar{X})^{-1} - \bar{Y} + e^{t\bar{F}^T}\bar{Y}e^{t\bar{F}}]^{-1}e^{t\bar{F}^T} = \Phi^T D_2^{-1}\Phi \ . \tag{28}$$

Let us now summarize the algorithm:

1. Solve the algebraic Riccati equation (18) for $\bar{X}$ by the Schur method described in Section 2.
2. Perform a diagonal pivoting factorization of $(A-\bar{X})$; see (25).
3. Solve the Lyapunov equation (22) by the RSF-based Bartels-Stewart algorithm [13] for $\bar{Y}$.
4. With $\bar{F} = F + \bar{X}E$ compute $e^{t\bar{F}}$ by, say Padé approximation; see [7]. Note that $\bar{F}$ is stable.
5. Perform the diagonal pivoting factorization (26).
6. Solve the linear system (27) for Φ.
7. Then $X(t) = \bar{X} + \Phi^T D_2^{-1}\Phi$.

One further modification is possible in Step 5 above. There the quantity

$$W(t) := e^{t\bar{F}^T}\bar{Y}e^{t\bar{F}} - \bar{Y}$$

had to be computed. This could either be done directly in the obvious way or it can be noticed that $W(t)$ solves the Lyapunov differential equation

$$\frac{d}{dt} W(t) = \bar{F}^T W(t) + W(t)\bar{F} - E \ ; \quad W(0) = 0. \tag{29}$$

Moreover, if (29) is solved then solution of the algebraic Lyapunov equation (22)(Step 3) is no longer necessary. However, many methods for solving (29) involve a solution of the algebraic equation

$$\bar{F}^T W + W\bar{F} = E \tag{30}$$

but there do also exist methods such as [14] which avoid (30) through the use of other efficient time-stepping formulas.

4. REFERENCES

[1] Laub, A.J., A Schur Method for Solving Algebraic Riccati Equations, *IEEE Trans. Aut. Contr.*, AC-24(1979), 913-921.

[2] Golub, G.H., and J.H. Wilkinson, Ill-Conditioned Eigensystems and the Computation of the Jordan Canonical Form, *SIAM Rev.*, 18(1976), 578-619.

[3] Van Dooren, P., A Generalized Eigenvalue Approach for Solving Riccati Equations, *SIAM J. Sci. Stat. Comp.*, 2(1981), 121-135.

[4] Pappas, T., A.J. Laub, and N.R. Sandell, On the Numerical Solution of the Discrete Time Algebraic Riccati Equation, *IEEE Trans. Aut. Contr.*, AC-25(1980), 631-641.

[5] Vaughn, D., A Nonrecursive Algebraic Solution for the Discrete Riccati Equation, *IEEE Trans. Aut. Contr.*, AC-15(1970), 597-599.

[6] Vaughn, D., A Negative-Exponential Solution to the Matrix Riccati Equation, *IEEE Trans. Aut. Contr.*, AC-14(1969), 72-75.

[7] Moler, C.B., and C. Van Loan, Nineteen Dubious Ways to Compute the Exponential of a Matrix, *SIAM Rev.*, 20(1978), 801-836.

[8] Parlett, B., A Recurrence Among the Elements of Functions of Triangular Matrices, *Lin. Alg. & Applics.*, 14(1976), 117-121.

[9] Van Loan, C., Computing Integrals Involving the Matrix Exponential, *IEEE Trans. Aut. Contr.*, AC-23(1978), 395-404.

[10] Bunch, J.R., Analysis of the Diagonal Pivoting Method, *SIAM J. Numer. Anal.*, 8(1971), 656-680.

[11] Bunch, J.R., and B.N. Parlett, Direct Methods for Solving Symmetric Indefinite Systems of Linear Equations, *SIAM J. Numer. Anal.*, 8(1971), 639-655.

[12] Bunch, J.R., and L. Kaufman, Some Stable Methods for Calculating Inertia and Solving Symmetric Linear Systems, *Math. Comp.*, 31(1977), 163-179.

[13] Bartels, R.H., and G.W. Stewart, Solution of the Matrix Equation $AX + XB = C$, *Comm. ACM*, 15(1972), 820-826.

[14] Serbin, S.M., and C.A. Serbin, A Time-Stepping Procedure for $\dot{X} = A_1X + XA_2 + D$, $X(0) = C$, Dept. of Mathematics, Univ. of Tennessee, Dec. 1979.

TOWARD A THEORY OF NONLINEAR STOCHASTIC REALIZATION

Anders Lindquist*, Sanjoy Mitter†, and Giorgio Picci§

1. INTRODUCTION

The following is a central problem in stochastic systems theory: Given a stationary stochastic process $\{y(t);t\in\mathbb{R}\}$, find a (possibly infinite-dimensional) vector Markov process $\{x(t);t\in\mathbb{R}\}$, called the *state process*, and a function f so that $y(t) = f(x(t))$ for all $t\in\mathbb{R}$. Moreover, find a stochastic differential equation driven by a Wiener process and having the state process x as its unique solution. The problem of characterizing the family of all such representations is known as the *stochastic realization problem.*

There is by now a rather comprehensive theory of stochastic realization for the case that $\{y(t);t\in\mathbb{R}\}$ is *Gaussian* [1-3], in which case the representations can be taken to be *linear*, i.e. both f and the stochastic differential equation are linear. This linear theory can be applied to *non-Gaussian* processes also, but then we need to give up the requirement that x is Markov and that it is generated by a Wiener process, replacing these concepts by "wide sense Markov" [4] and "orthogonal increment process" respectively. If we are not willing to do so, a *nonlinear* stochastic realization theory is needed. That is the topic of this paper.

In this paper we shall apply Wiener's theory of homogeneous chaos [5,6] to the nonlinear stochastic realization problem. For simplicity and ease of notation we shall assume that the process y is scalar, although the machinery which we develop is sufficient to accommodate also the vector case. Other assumptions, such as y admitting an innovation representation, are however crucial to our approach. (In this respect, it might be more appropriate to consider a process y with stationary increments, and indeed with minor modifications we could have done so.) In the extension of this work we see the possibility of making contact with *nonlinear filtering* [7,8] and that is partially a motivation for this work.

2. PROBLEM FORMULATION

Let $\{y(t);t\in\mathbb{R}\}$ be a non-Gaussian stationary stochastic process which is mean-square continuous, purely nondeterministic, and centered, and let $\mathcal{Y}$ be the sigma-field

---

* Department of Mathematics, University of Kentucky, Lexington, Kentucky 40506. This research was supported partially by the National Science Foundation under grant ECS-7903731 and partially by the Air Force Office of Scientific Research under grant AFOSR 78-3519.

† Laboratory for Information and Decision Systems, Massachusetts Institute of Technology, Cambridge, Massachusetts 02139. This research was supported by the Air Force Office of Scientific Research under grant AFOSR 77-3281D.

§ LADSEB-CNR, Corso Stati Uniti 4, 35100 Padova, Italy.

generated by y. Then define H to be the Hilbert space of all centered $\mathcal{Y}$-measurable random variables, having inner product $\langle\xi,\eta\rangle = E\{\xi\eta\}$. Since y is stationary there is a strongly continuous group of unitary operators $\{U_t; t\in\mathbf{R}\}$ on H, called the *shift*, such that $y(t+s) = U_t y(s)$ for all t and s [9]. Let $\mathcal{Y}_t^-$ and $\mathcal{Y}_t^+$ be the sigma-fields generated by $\{y(s); s\le t\}$ and $\{y(s); s\ge t\}$ respectively.

Next assume that y has an innovation process $\{\nu(t); t\in\mathbf{R}\}$, by which we shall here mean a Wiener process such that $\sigma\{\nu(\tau)-\nu(\sigma); \tau,\sigma\le t\} = \mathcal{Y}_t^-$. (Here $\sigma\{\cdot\}$ denotes the sigma-field generated by the random variables inside the curly brackets.) Then, by symmetry, it also has a backward innovation process $\{\bar\nu(t); t\in\mathbf{R}\}$, i.e. another Wiener process such that $\sigma\{\bar\nu(\tau)-\bar\nu(\sigma); \tau,\sigma\ge t\} = \mathcal{Y}_t^+$. Now, since $\mathcal{Y} = \sigma\{\nu(t); t\in\mathbf{R}\} = \sigma\{\bar\nu(t); t\in\mathbf{R}\}$ we can apply Wiener's homogeneous chaos theory [5,6]. Let H_1 denote the Gaussian space [5] generated by $\{\nu(t); t\in\mathbf{R}\}$ or, which is equivalent, by $\{\bar\nu(t); t\in\mathbf{R}\}$. Since y is mean-square continuous, H_1 is a separable space, and therefore H_1 has a countable orthonormal basis $\{\xi_i\}_{i=0}^{\infty}$. Now, let P_n be the (closed) linear subspace of all polynomials in $\{\xi_i\}_{i=0}^{\infty}$ of degree not exceeding n. Next define $H_n = P_n \ominus P_{n-1}$, i.e. the orthogonal complement of P_{n-1} in P_n. Then it can be shown [5,6] that

$$H = H_1 \oplus H_2 \oplus H_3 \oplus \dots \tag{1}$$

where $\oplus$ denotes orthogonal direct sum. The space H_n is called the nth *homogeneous chaos* of H. Since $y(0)\in H$, there is an orthogonal decomposition

$$y(0) = y_1(0) + y_2(0) + y_3(0) + \dots \tag{2}$$

where $y_n(0)\in H_n$. It is easy to see that each chaos H_n is invariant under the shift U_t, and consequently, for any $t\in\mathbf{R}$, we have a decomposition such as (2) for y(t) in terms of $y_n(t) := U_t y_n(0)$, n=1,2,3...

In order to obtain a state space description we introduce a *past space* H^- and a *future space* H^+ as follows. Let $H^-(H^+)$ be the subspace of all centered $\mathcal{Y}^-$-measurable ($\mathcal{Y}^+$-measurable) random variables. Then, defining H_1^- and H_1^+ to be the Gaussian spaces generated by $\{\nu(\tau)-\nu(\sigma); \tau,\sigma\le 0\}$ and $\{\bar\nu(\tau)-\bar\nu(\sigma); \tau,\sigma\ge 0\}$ respectively, we obtain the chaos expansions

$$\begin{cases} H^- = H_1^- \oplus H_2^- \oplus H_3^- \oplus \dots & \text{(3a)} \\ H^+ = H_1^+ \oplus H_2^+ \oplus H_3^+ \oplus \dots & \text{(3b)} \end{cases}$$

where clearly $H_n^-\subset H_n$ and $H_n^+\subset H_n$ for all n. Note that $H_1^-\cap H_1^+ \ne \emptyset$ and $H_1^-\vee H_1^+ = H_1$, but $H^-\vee H^+ \ne H$.

Now, if y were a Gaussian process, y would have a component only in the first chaos, i.e. $y = y_1$, and consequently state spaces for y could be constructed along the lines of [1,2] by finding the minimal Markovian (H_1^-,H_1^+)-splitting subspaces in H_1 [We recall that, for two subspaces A and B, X is an (A,B)-*splitting subspace* if $\langle E^X\alpha, E^X\beta\rangle = \langle\alpha,\beta\rangle$ for all $\alpha\in A$ and $\beta\in B$, where E^X denotes orthogonal projection on the subspace X.] However, for a non-Gaussian process y, there will be some nontrivial component y_n, $n>1$, and consequently the state space construction will have to involve at least those higher chaoses in which y has a component. To this end define the index set

$N := \{n \mid y_n(0) \neq 0\} \cup \{1\}$. For reasons which will soon be evident, we shall have to always include the first chaos in our analysis. (In particular, see Section 7.)

Hence we call X a *state space for* y if

$$X = \bigoplus_{n\in N} X_n , \tag{4}$$

where $X_n \subset H_n$ is an (H_n^-, H_n^+)-splitting subspace, and X is Markovian in the sense that, if $\mathcal{X} := \sigma\{X\}$, $\mathcal{X}^- := \sigma\{\bigvee_{t\leq 0} U_t X\}$ and $\mathcal{X}^+ := \sigma\{\bigvee_{t\geq 0} U_t X\}$, $\mathcal{X}^-$ and $\mathcal{X}^+$ are conditionally independent given $\mathcal{X}$; we shall write this $\mathcal{X}^- \perp\!\!\!\perp \mathcal{X}^+ \mid \mathcal{X}$. We say that X is *minimal* if there is no other state space X' for which $\mathcal{X}' := \sigma\{X'\}$ is properly contained in $\mathcal{X}$.

The problem at hand is now to construct all minimal state spaces for y and to obtain a dynamical representation (realization) for each of them.

3. THE STRUCTURE OF H

According to Itô's Theorem [10]

$$H_n = \{I_n(f;\nu) \mid f \in \hat{L}_2(\mathbb{R}^n)\} \tag{5}$$

where I_n is the multiple Wiener integral

$$I_n(f;\nu) = \int_{-\infty}^{\infty}\int_{-\infty}^{t_1}\cdots\int_{-\infty}^{t_{n-1}} f(t_1,t_2,\ldots,t_n)\,d\nu(t_1)\ldots d\nu(t_n) \tag{6}$$

and $\hat{L}_2(\mathbb{R}^n)$ are the symmetric functions in $L_2(\mathbb{R}^n)$. Although the region of integration is such that (5) remains the same if $\hat{L}_2(\mathbb{R}^n)$ is exchanged for $L_2(\mathbb{R}^n)$, we prefer the former since we have a one-one correspondence between elements in H_n and $\hat{L}_2(\mathbb{R}^n)$. In fact, we can establish an isometric isomorphism between these spaces [5,6,10]. Now, for i=1,2,...,n, let $\eta_i \in H_1$ be arbitrary. Then there exist unique functions $f_i \in L_2(\mathbb{R})$ such that $\eta_i = \int_{-\infty}^{\infty} f_i(t)\,d\nu(t)$. Next define

$$\eta_1 * \eta_2 * \ldots * \eta_n = n!\; I_n(f;\nu) \tag{7}$$

where

$$f(t_1,t_2,\ldots,t_n) = \frac{1}{n!}\sum_{\pi\in G} f_{\pi_1}(t_1) f_{\pi_2}(t_2)\ldots f_{\pi_n}(t_n) , \tag{8}$$

G being the symmetric group of permutations of n letters. Since finite linear combinations of functions of type (8) are dense in $\hat{L}_2(\mathbb{R}^n)$, Itô's Theorem implies that

$$H_n = \overline{\text{sp}}\{\eta_1 * \eta_2 * \ldots * \eta_n \mid \eta_1,\eta_2,\ldots,\eta_n \in H_1\} \tag{9a}$$

where sp denotes closed linear hull. We shall write this as

$$H_n = H_1 * H_1 * \ldots * H_1 . \tag{9b}$$

By Itô's formula [11; p.38]

$$\eta_1 * \eta_2 * \ldots * \eta_n = (\eta_2 * \eta_3 * \ldots * \eta_n)\cdot\eta_1 - \sum_k (\eta_2 * \ldots * \eta_{k-1} * \eta_{k+1} * \ldots * \eta_n)\cdot\langle\eta_1,\eta_k\rangle \tag{10}$$

which can be solved recursively. For example,

$$\eta_1 * \eta_2 = \eta_1\eta_2 - \langle\eta_1,\eta_2\rangle$$

$$\eta_1 * \eta_2 * \eta_3 = \eta_1\eta_2\eta_3 - \eta_1 \cdot \langle\eta_2,\eta_3\rangle - \eta_2 \cdot \langle\eta_1,\eta_3\rangle - \eta_3 \cdot \langle\eta_1,\eta_2\rangle \ .$$

The $*$-operation is obviously commutative. In particular,

$$\underbrace{\eta * \eta * \dots * \eta}_{n \text{ times}} = h_n(\eta, \langle\eta,\eta\rangle^{\frac{1}{2}}) \tag{11}$$

(in the sequel we shall write this η^{n*}) where

$$h_n(x,\sigma) = \frac{(-\sigma)^n}{n!} \exp\left(\frac{x^2}{2\sigma}\right) \frac{\partial^n}{\partial x^n} \exp\left(-\frac{x^2}{2\sigma}\right) , \tag{12}$$

n=0,1,2,..., are the *Hermite polynomials* (cf [11; p.37]). Analogously to (9) we have

$$\begin{cases} H_n^- = H_1^- * H_1^- * \dots * H_1^- & \text{(13a)} \\ H_n^+ = H_1^+ * H_1^+ * \dots * H_1^+ & \text{(13b)} \end{cases}$$

Let $H_1^{n\circledast} = H_1 \circledast H_1 \circledast \dots \circledast H_1$ denote the symmetric tensor-product Hilbert space of H_1 by itself taken n times. Then for arbitrary $\xi_i,\eta_i \in H_1$, i=1,2,...,n, with $\langle\cdot,\cdot\rangle_{n\circledast}$ the inner product in $H_1^{n\circledast}$, we have

$$\langle\xi_1 \circledast \xi_2 \circledast \dots \circledast \xi_n, \eta_1 \circledast \eta_2 \circledast \dots \circledast \eta_n\rangle_{n\circledast} = \frac{1}{n!} \sum_{\pi\in G} \langle\xi_{\pi_1},\eta_1\rangle\langle\xi_{\pi_2},\eta_2\rangle \dots \langle\xi_{\pi_n},\eta_n\rangle \tag{14}$$

where $\xi_1 \circledast \xi_2 \circledast \dots \circledast \xi_n$ is the symmetric tensor product [5,6]. Since finite linear combinations of such tensor products are dense in $H_1^{n\circledast}$, it is now easy to see that $H_1^{n\circledast}$ is isometrically isomorphic to $\hat{L}_2(\mathbb{R}^n)$ and hence to H_1^{n*}. For n=2 we can illustrate this by factoring the symmetric bilinear map $(\eta_1,\eta_2) \to \eta_1 * \eta_2$ as follows

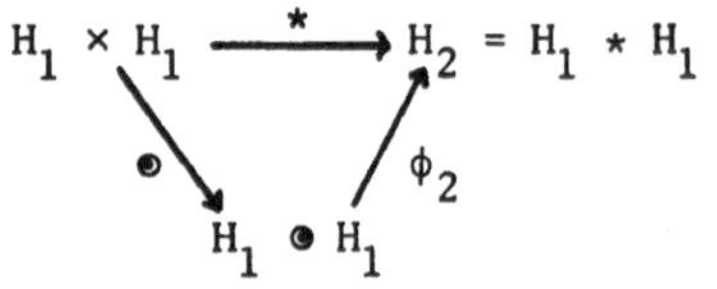

where ϕ_2 is the unique linear map which makes the diagram commute; ϕ_2 is unitary. Similar unitary maps ϕ_n are defined for n=3,4,...

If $A_1, A_2, \dots,$ are linear operators in H_1, we define $A_1 * A_2 * \dots * A_n : H_n \to H_n$ via

$$(A_1 * A_2 * \dots * A_n)(\eta_1 * \eta_2 * \dots * \eta_n) = (A_1\eta_1) * (A_2\eta_2) * \dots * (A_n\eta_n)$$

on a dense set in H_n and then extend it continuously to all of H_n. We define $A_1 \circledast A_2 \circledast \dots \circledast A_n : H_1^{n\circledast} \to H_1^{n\circledast}$ analogously. For n=2 we have then the following picture:

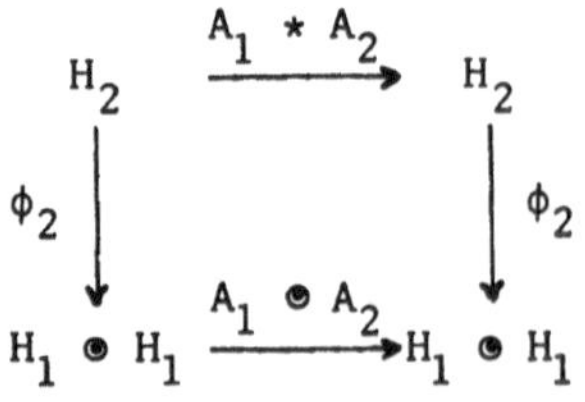

and analogously for n > 2.

4. STATE SPACE CONSTRUCTION

THEOREM 1. *The subspace* $X \subset H$ *is a minimal state space for* y *if and only if*

$$X = \bigoplus_{n \in N} X_n \tag{15a}$$

where X_1 *is a minimal Markovian* (H_1^-, H_1^+) *-splitting subspace and*

$$X_n = X_1^{n*} := X_1 * X_1 * \dots * X_1 \tag{15b}$$

(n times). Then each X_n *is a minimal* (H_n^-, H_n^+) *-splitting subspace.*

The proof of this theorem is based on the following lemmas.

LEMMA 1. *Let* $\eta = \eta_1 \circledast \eta_2 \circledast \dots \circledast \eta_n$ *where* $\eta_i \in H_1$ *for* $i=1,2,\dots,n$. *Let* X *be a subspace of* H_1. *Then*

$$E^{X \circledast X \circledast \dots \circledast X}\eta = (E^X \eta_1) \circledast (E^X \eta_2) \circledast \dots \circledast (E^X \eta_2) .$$

PROOF. Let $\hat{\eta}_i := E^X \eta_i$, and let $\xi = \xi_1 \circledast \xi_2 \circledast \dots \circledast \xi_n$ where $\xi_1, \xi_2, \dots, \xi_n$ are arbitrary elements in X. Then, by (14),

$$\langle \eta_1 \circledast \eta_2 \circledast \dots \circledast \eta_n - \hat{\eta}_1 \circledast \hat{\eta}_2 \circledast \dots \circledast \hat{\eta}_n \rangle_{n\circledast} =$$

$$\langle (\eta_1 - \hat{\eta}_1) \circledast \eta_2 \circledast \dots \circledast \eta_n, \xi \rangle_{n\circledast} + \langle \hat{\eta}_1 \circledast (\eta_2 - \hat{\eta}_2) \circledast \dots \circledast \eta_n, \xi \rangle_{n\circledast} + \dots + \langle \hat{\eta}_1 \circledast \hat{\eta}_2 \circledast \dots \circledast (\eta_n - \hat{\eta}_n), \xi \rangle_{n\circledast} =$$

$$\frac{1}{n!} \sum_{\pi \in G} \{ \langle \eta_1 - \hat{\eta}_1, \xi_{\pi_1} \rangle \langle \eta_2, \xi_{\pi_2} \rangle \dots \langle \eta_n \xi_{\pi_n} \rangle + \langle \eta_1, \xi_{\pi_1} \rangle \langle \eta_2 - \hat{\eta}_2, \xi_{\pi_2} \rangle \dots \langle \eta_n, \xi_{\pi_n} \rangle +$$

$$\dots + \langle \eta_1, \xi_{\pi_1} \rangle \langle \eta_2, \xi_{\pi_n} \rangle \dots \langle \eta_n - \hat{\eta}_n, \xi_{\pi_n} \rangle \} ,$$

which equals zero since $\eta_i - \hat{\eta}_i \perp X$. □

LEMMA 2. *Let* X_1 *be a subspace of* H_1, *and let* X_n *be defined by* (15b). *Then* X_1 *is an* (H_1^-, H_1^+)*-splitting subspace if and only if* X_n *is an* (H_n^-, H_n^+)*-splitting subspace.*

PROOF. Due to isomorphism, we can identify H_n^-, H_n^+ and X_n with $(H_1^-)^{n\circledast}$, $(H_1^+)^{n\circledast}$ and $X_1^{n\circledast}$ respectively. But Lemma 1, (14), and the definition of splitting subspace imply that $H_1^- \perp H_1^+ \mid X$ if and only if $(H_1^-)^{n\circledast} \perp (H_1^+)^{n\circledast} \mid X_1^{n\circledast}$. Hence the lemma follows. □

LEMMA 3. *Let* X_1 *and* X_n *be the splitting subspaces of Lemma 2. Then* X_n *is minimal if and only if* X_1 *is minimal.*

PROOF. By Proposition 1 in [12] it suffices to show that the condition $\bar{E}^{X_1}H_1^- = X_1$ is equivalent to $\bar{E}^{X_n}H_n^- = X_n$ (bar over E stands for closure) and that $\bar{E}^{X_1}H_1^+ = X_1$ is equivalent to $\bar{E}^{X_n}H_n^+ = X_n$. By isomorphism $\bar{E}^{X_n}H_n^- = X_n$ can be identified with $\bar{E}^{X_1^{n\circledast}}(H_1^-)^{n\circledast} = X_1^{n\circledast}$, which, by Lemma 1, holds if and only if $\bar{E}^{X_1}H_1^- = X$. A similar argument establishes the other equivalence. □

PROOF OF THEOREM 1. Let the X described in the theorem be denoted $\hat{X}$, and set $\hat{\mathcal{X}} := \sigma(\hat{X}_1)$. Then $\hat{X}$ is the space of all centered $\hat{\mathcal{X}}$-measurable random variables in H [5,6], and

$\sigma(\hat{X}) = \hat{\mathcal{X}}$. Moreover, if $X^- := \bigvee_{t\le 0} U_t X$, it is not hard to see that $\hat{\mathcal{X}}^- := \sigma(\hat{X}^-) = \sigma(\hat{X}_1^-)$, since $U_t \hat{X}_n = (U_t X_1)^{n*}$. An analogous relation holds for summation over the future.

(if): Show that $\hat{X}$ is a minimal state space. Since $\hat{X}_1$ is Markovian, so is $\hat{X}$. Hence, in view of Lemma 2, $\hat{X}$ is a state space for y. Now assume that $\hat{X}$ is not minimal. Then there is a state space X such that $\mathcal{X} := \sigma(X)$ is properly contained in $\hat{\mathcal{X}}$. Then, since all $\xi \in X$ are $\hat{\mathcal{X}}$-measurable and $\hat{X}$ is the space of all centered $\hat{\mathcal{X}}$-measurable random variables, $X \subset \hat{X}$. Therefore X_1 must be a proper subset of $\hat{X}_1$, or else $\mathcal{X} \supset \sigma(X_1) = \sigma(\hat{X}_1) = \hat{\mathcal{X}}$. This contradicts the minimality of $\hat{X}_1$.

(only if): Let X be a minimal state space for y. First let us assume that X_1 is not minimal. Then there is another (H_1^-, H_1^+)-splitting subspace $\hat{X}_1$ which is a proper subspace of X_1. Let $\hat{\mathcal{X}} = \sigma(\hat{X}_1)$, and let $\hat{X}$ be the space of all $\hat{\mathcal{X}}$-measurable elements in H. Clearly $\hat{\mathcal{X}} \subset \mathcal{X} := \sigma(X)$. We want to show that this inclusion is proper, contradicting minimality of X. But this is the case, for there is a $\xi \in X_1$ such that $\xi \perp \hat{X}_1$. Consequently, by the Gaussian property, $\sigma\{\xi\}$ and $\hat{\mathcal{X}}$ are independent, while both are subfields of $\mathcal{X}$. Hence X_1 must be minimal, and $X_1 = \hat{X}_1$. Next assume that X_n is not of the form (15b), i.e. $X_n \ne \hat{X}_n$. Then since $\hat{X}_n$ is minimal (Lemma 3), $X_n \not\subset \hat{X}_n$, i.e. there is a $\xi \in X$ which does not belong to $\hat{X}$ and consequently is not $\hat{\mathcal{X}}$-measurable. Hence $\hat{\mathcal{X}}$ is a proper subfield of $\mathcal{X}$ contradicting minimality of X. Therefore $X = \hat{X}$. Finally $\hat{X}$ is Markovian only if $\hat{X}_1$ is Markovian. The last statement of the theorem follows from Lemma 3. □

5. THE STATE SPACE COMPONENT OF THE FIRST CHAOS

Thus it remains to determine the minimal Markovian (H_1^-, H_1^+)-splitting subspaces X_1. This is *almost* the problem solved in [1-3]. To explain how it differs, let $\zeta \in H_1^- \cap H_1^+$ be defined in the following manner. If $y_1(0) \ne 0$, set $\zeta := y_1(0)$, otherwise let it be arbitrary. (Remember that $H_1^- \cap H_1^+ \ne \emptyset$.) Next define the process $z(t) := U_t \zeta$. Then $z(t) \in H_1$ for all t. Moreover,

$$\begin{cases} H_1^-(z) \subset H_1^- & (16a) \\ H_1^+(z) \subset H_1^+ & (16b) \end{cases}$$

where $H_1^-(z)$ and $H_1^+(z)$ are the closed linear hulls of the random variables $\{z(t); t \le 0\}$ and $\{z(t); t \ge 0\}$ respectively. Since y is purely nondeterministic and mean-square continuous, so is z. Therefore z has a spectral density $\Phi(i\omega)$. A scalar solution W of the equation

$$W(s)W(-s) = \Phi(s) \tag{17}$$

will be called a (full-rank) *spectral factor* of z. Now, if y is Gaussian as assumed in [1,2], $z = y$ and we have equality in each of relations (16). Then there is a procedure in [1,2] to determine X_1 from a certain pair $(W, \bar{W})$ of spectral factors. However, in the non-Gaussian case, $z \ne y$, and we cannot assume that relations (16) hold with equality, not even when $z = y_1$. Hence there is a "mismatch" between the process z and the geometry in H_1, and consequently the procedure of [1,2] will have to be modified.

Fortunately the basic results of [1,2] depend in no crucial way on the spectral factor construction. The following result found in [1,2] is a consequence of the geometry in H_1 only. The theorem requires some new notation: For any Wiener process $\{u(t);t\in\mathbb{R}\}\subset H_1$, let $H_1^-(du)$ and $H_1^+(du)$ be the Gaussian spaces generated by the increments $\{u(\tau)-u(\sigma);\tau,\sigma\le 0\}$ and $\{u(\tau)-u(\sigma);\tau,\sigma\ge 0\}$ respectively. In particular, we have $H_1^-(d\nu)=H_1^-$ and $H_1^+(d\bar{\nu})=H_1^+$. Here and in the sequel, when we talk of a "Wiener process," we shall always mean a centered Gaussian process *defined on the whole real line* by a spectral representation

$$u(t) = \int_{-\infty}^{\infty} \frac{e^{i\omega t}-1}{i\omega}\, d\hat{u}(i\omega) \;, \tag{18}$$

where $d\hat{u}$ is a Gaussian orthogonal stochastic measure such that $E|d\hat{u}|^2 = \frac{1}{2\pi}\, d\omega$.

THEOREM 2. *A subspace* $X_1\subset H_1$ *is a minimal Markovian* (H_1^-,H_1^+)*-splitting subspace if and only if*

$$X = H_1^-(du) \ominus H_1^-(d\bar{u}) \tag{19}$$

for some pair $(u,\bar{u})$ *of Wiener processes in* H_1 *such that*

$$H_1^-(d\bar{u}) \subset H_1^-(du) \tag{20a}$$

$$H_1^- \subset H_1^-(du) \tag{20b}$$

$$H_1^+ \subset H_1^+(d\bar{u}) \tag{20c}$$

$$H_1^+(d\bar{u}) = H_1^+ \vee H_1^+(du) \tag{20d}$$

$$H_1^-(du) = H_1^- \vee H_1^-(d\bar{u}) \;. \tag{20e}$$

The processes u and $\bar{u}$ (which are essentially unique) are called respectively the *forward* and the *backward generating processes* of X. (Condition (20a) is equivalent to $H_1^-(du)$ and $H_1^+(d\bar{u})$ intersecting perpendicularly. Moreover, (20d) is an observability and (20e) a constructibility condition [1,2].)

The Gaussian space of any Wiener process u in H_1 coincides with H_1 [9], and consequently any $\eta\in H_1$ can be written

$$\eta = \int_{-\infty}^{\infty} f(-t)\,du(t) \tag{21a}$$

where $f\in L_2(\mathbb{R})$, or equivalently,

$$\eta = \int_{-\infty}^{\infty} \hat{f}(i\omega)\,d\hat{u}(i\omega) \tag{21b}$$

where $\omega\to\hat{f}(i\omega)$ is the Fourier transform [9]. [We shall refer also to the function $\hat{f}$ as the Fourier-transform, although it properly should be called the (double-sided) Laplace-transform.] Relations (22) establishes an isometric isomorphism between H_1 and $L_2(\mathbb{I})$, where $\mathbb{I}$ is the imaginary axis. Let $T_u : H_1\to L_2(\mathbb{I})$ be the map $T_u\eta=\hat{f}$. Then it can be seen that T_u is unitary. Let T_u^* denote the adjoint, i.e. $\eta = T_u^*\hat{f}$, which is relation (21b). The shift U_t corresponds to $e^{i\omega t}$ under the isomorphism T_u.

LEMMA 4. *There is a one-one correspondence between Wiener processes* u *in* H_1 *and spectral factors* W *of* z *described by the following rule. For each* u, $W:=T_u\zeta$ *is a spectral factor. For each spectral factor* W, u *defined by* (18) *and* $d\hat{u}=W^{-1}Gd\hat{v}$, *where* $G:=T_v\zeta$, *is a Wiener process.*

PROOF. Let $W:=T_u\zeta$. Then

$$z(t) = \int_{-\infty}^{\infty} e^{i\omega t} W d\hat{u} , \tag{22}$$

from which it is easy to see that the inverse Fourier transform of $E\{z(t)z(0)'\}$ is $W(i\omega)W(-i\omega)$, establishing W as a spectral factor. In particular G is a spectral factor. Therefore, for any spectral factor W, $d\hat{u}:=W^{-1}Gd\hat{v}$ is a Gaussian orthogonal stochastic measure such that $E|d\hat{u}|^2=\frac{1}{2\pi}d\omega$, for $d\hat{v}$ is. Hence (18) is a Wiener process. □

Next we introduce the *Hardy spaces* H_2^+ and H_2^-: Let $H_2^+(H_2^-)$ be the subspace of $L_2(\Pi)$ of functions whose inverse Fourier-transforms vanish on the negative (positive) real line. From (21) it follows that $T_u H^-(du) = H_2^+$ and $T_u H^+(du) = H_2^-$. A function K which is bounded and analytic in the open left half-plane and has modulus one on the imaginary axis is called *inner*. Define $K^*(i\omega):=K(-i\omega)$; K^* is the inverse of K. If $f\in H_2^+$ and K is inner, $fK\in H_2^+$, and H_2^+K is a subspace of H_2^+. Let $H(K)$ denote the orthogonal complement of H_2^+K in H_2^+.

THEOREM 3. *Let* $\zeta\in H_1^-\cap H_1^+$ *be arbitrary, and set* $G:=T_v\zeta$ *and* $\bar{G}:=T_{\bar{v}}\zeta$. *Let* $\Gamma:=G/\bar{G}$. *Then* X_1 *is a minimal Markovian* (H_1^-,H_1^+)*-splitting subspace if and only if there is a pair of inner functions* $(Q,\bar{Q}^*)$ *such that* $K:=\Gamma Q\bar{Q}^*$ *is also inner,* K *and* Q *are coprime,* K *and* $\bar{Q}^*$ *are coprime, and*

$$X_1 = T_u^* H(K) , \tag{23}$$

where u *is the Wiener process* (18) *with* $d\hat{u}=Q^*d\hat{v}$.

PROOF. We present an appropriately modified version of the proof in [1,2]. The idea is to translate conditions (20) to the Hardy space setting and apply Beurling's Theorem [13]. To this end, first note that if u_1 and u_2 are two Wiener processes in H_1, and W_1 and W_2 are their corresponding spectral factors,

$$T_{u_2}\eta = (T_{u_1}\eta)(W_2/W_1) \tag{24}$$

for any $\eta\in H_1$, as is easily seen from Lemma 4. Then, if W and $\bar{W}$ are the spectral factors corresponding to u and $\bar{u}$ respectively, applying the map T_u to (20a), (20b) and (20e) and $T_{\bar{u}}$ to (20c) and (20d) yields

$$H_2^+K \subset H_2^+ \quad \text{where} \quad K := W/\bar{W} \tag{25a}$$

$$H_2^+Q \subset H_2^+ \quad \text{where} \quad Q := W/G \tag{25b}$$

$$H_2^-\bar{Q} \subset H_2^- \quad \text{where} \quad \bar{Q} := \bar{W}/\bar{G} \tag{25c}$$

$$H_2^- = (H_2^-\bar{Q}) \vee (H_2^-K^*) \tag{25d}$$

$$H_2^+ = (H_2^+Q) \vee (H_2^+K) . \tag{25e}$$

Now, since $H_1^-(d\bar{u})$ is invariant under the shift $\{U_t; t \le 0\}$, H_2^+K is invariant under $\{e^{i\omega t}; t \le 0\}$. Therefore, by Beurling's Theorem [13], (25a) holds if and only if K is inner. In the same way we see that (25b) is equivalent to Q being inner and (25c) to $\bar{Q}$ being inner with respect to H_2^-, i.e. $\bar{Q}^*$ inner. Moreover, (25d) and (25e) are valid if and only if the stated coprimeness conditions hold [13], and (19) is equivalent to $T_u X = H_2^+ \ominus (H_2^+K) =: H(K)$, i.e. (23). The statement about u follows from Lemma 4. □

REMARK. Let us pinpoint in what way this theorem differs from the corresponding result in [1,2]. In the case studied in [1,2], the pairs $(W,\bar{W})$ which generate splitting subspaces are precisely those for which $W \in H_2^+$, $\bar{W} \in H_2^-$, and K is inner. In the present setting these three conditions must also hold, but in addition we must have $W = GQ$ and $\bar{W} = \bar{G}\bar{Q}$. These factorizations correspond to the inner-outer factorizations of [1,2], but the difference is that now G and $\bar{G}$ are not outer. Consequently, some of the pairs $(W,\bar{W})$ mentioned above will be excluded. Note that the innovation process does not correspond to an outer spectral factor of z, since $\bar{E}^{H^-}H^+$ is not the predictor space of z. □

6. THE STATE PROCESS

We recall from Section 2 that

$$y(t) = \sum_{n \in N} y_n(t) \tag{26a}$$

where

$$y_n(t) = \int_{-\infty}^{t}\int_{-\infty}^{t_1}\cdots\int_{-\infty}^{t_{n-1}} g_n(t-t_1,t-t_2,\ldots,t-t_n)\,d\nu(t_1)\,d\nu(t_2)\ldots d\nu(t_n) \tag{26b}$$

for some $g_n \in L_2(\mathbb{R}^n)$. Let us assume that this innovation representation is given, i.e. that the functions $\{g_n; n \in N\}$ are known.

Let us now consider a minimal state space X with forward generating process u. Then, since $H_1^- \subset H_1^-(du)$,

$$y_n(0) \in H_n^-(du) := H_1^-(du) * H_1^-(du) * \ldots * H_1^-(du) \tag{27}$$

(n times) and consequently there is a representation

$$y_n(0) = \int_{-\infty}^{0}\int_{-\infty}^{t_1}\cdots\int_{-\infty}^{t_{n-1}} w_n(-t_1,-t_2,\ldots,-t_n)\,du(t_1)\,du(t_2)\ldots du(t_n) \tag{28}$$

for some $L_2(\mathbb{R}^n)$. Defining w_n to be zero whenever some argument is zero, we may write this $y_n(0) = I_n(w_n;u)$. By the same recipe we write $y_n(0) = I_n(g_n;\nu)$. We need to determine w_n from g_n. To this end, let $\hat{f} \in L_2(\mathbb{II}^n)$ be the n-fold Fourier-transform

$$\hat{f}(i\omega_1,\ldots,i\omega_n) = \int_{-\infty}^{\infty}\cdots\int_{-\infty}^{\infty} e^{-i\omega_1 t_1 - \ldots - i\omega_n t_n} f(t_1,\ldots,t_n)\,dt_1\ldots dt_n \,.$$

Let $F_n : L_2(\mathbb{R}^n) \to L_2(\mathbb{II}^n)$ be the operator defined by $\hat{f} = F_n f$. The following is a multidimensional version of (21).

LEMMA 5. *Let* $f \in L_2(\mathbb{R}^n)$ *and set* $\hat{f} := F_n f$. *Let* u *be a Wiener process* (18). *Then*

$$I_n(f;u) = I_n(\hat{f};\hat{u}) \,. \tag{29}$$

PROOF. First let f be of the form (8). Then $\hat{f}$ has this form too, and $\hat{f}_i = F_1 f_i$. From (21) we have

$$\eta_i = \int_{-\infty}^{\infty} f_i(-t)\,du(t) = \int_{-\infty}^{\infty} \hat{f}_i(i\omega)\,d\hat{u}(i\omega) .$$

Then, by Itô's formula, i.e. (11) with ν exchanged for u or $\hat{u}$, each member of (29) can be reduced to the same expression in $\eta_1,\eta_2,\dots,\eta_n$. Hence (29) holds for functions of type (8). Then, since finite linear combinations of functions of type (8) are dense in $L_2(\mathbb{R}^n)$ or $L_2(\mathbb{I}^n)$, (20) holds in general. □

Consequently, defining $W_n := F_n w_n$ and $G_n := F_n g_n$, w_n can be determined from g_n via the relation

$$W_n(i\omega_1,\dots,i\omega_n) = G_n(i\omega_1,\dots,i\omega_n)Q(i\omega_1)\dots Q(i\omega_n) , \tag{30}$$

for $d\hat{\nu} = Qd\hat{u}$ (Theorem 3).

It is well-known [6], and we have already used this fact in Section 2, that $\eta = I_n(\hat{f};\hat{u})$ defines an isomorphism between H_n and $L_2(\mathbb{I}^n)$. More precisely $T_u^{(n)}:H_n \to L_2(\mathbb{I}^n)$, defined by $T_u^{(n)}\eta = \hat{f}$, is a map with the property that $(n!)^{\frac{1}{2}}T_u^{(n)}$ is unitary. The space of Fourier-transforms of functions (such as w_n) in $L_2(\mathbb{R}^n)$ which vanish whenever an argument is negative, can be identified with $(H_2^+)^{n\circledast}$ so that $T_u^{(n)}H_n^-(du) = (H_2^+)^{n\circledast}$. In the sequel we shall use precisely this realization of the tensor-product Hilbert space $(H_2^+)^{n\circledast}$. Then the tensor product $f_1 \circledast f_2 \circledast \dots \circledast f_n$ is given by (8). Also, for subspaces $A_1,A_2,\dots,A_n$ in H_n,

$$T_n^{(n)}\{A_1 * A_2 * \dots * A_n\} = (T_uA_1) \circledast (T_uA_2) \circledast \dots \circledast (T_uA_n) \tag{31}$$

so that in particular

$$T_u^{(n)}X_n = H(K) \circledast H(K) \circledast \dots \circledast H(K) \tag{32}$$

(n times). Then, since $y_n(0) \in X_n$, $W_n \in H(K)^{n\circledast}$.

Following [1,2] we say that X is *regular* if $H(K)$ contains only Fourier-transforms of continuous functions. All X with $\dim X_1 < \infty$ are clearly regular. It can be shown [1,2] that if X is regular the functional

$$V\hat{f} = \frac{1}{(2\pi)^n}\int_{\mathbb{R}^n} \hat{f}(i\omega_1,\dots,i\omega_n)\,d\omega_1\dots d\omega_n \tag{33}$$

is bounded on $H(K)^{n\circledast}$. Hence, since $V\hat{f} = f(0)$ where $f = F_n^{-1}\hat{f}$, there is a $B_n \in H(K)^{n\circledast}$ such that

$$f(0) = \langle \hat{f},B_n\rangle_{H(K)^{n\circledast}} \tag{34}$$

(Riesz Theorem). Next, as in [1,2], we define a strongly continuous semigroup $\{e^{At};t \geq 0\}$ on $H(K)$ by

$$e^{At}\hat{f} = P^{H(K)}e^{-i\omega t}\hat{f} \tag{35}$$

where $P^{H(K)}$ denotes the orthogonal projection on the subspace $H(K)$. Moreover define the linear bounded operator $C_n : H(K)^{n\circledast} \to \mathbb{R}$ given by

$$C_n\hat{f} = \langle W_n,\hat{f}\rangle_{H(K)^{n\circledast}} \ . \tag{36}$$

Then the following lemma is a multilinear version of the construction in [14,15] which is being used in [1,2].

LEMMA 6. *The integrand in* (28) *admits the factorization*

$$w_n(t_1,t_2,\dots,t_n) = C_n\left(e^{At_1}\circledast e^{At_2}\circledast\dots\circledast e^{At_n}\right)B_n \tag{37}$$

for $t_1,t_2,\dots,t_n \geq 0$.

PROOF. In view of (34)

$$w_n(t_1,t_2,\dots,t_n) = \left\langle e^{i\omega_1 t_1+i\omega_2 t_2+\dots+i\omega_n t_n}W_n,B_n\right\rangle \ .$$

Since $W_n \in H(K)^{n\circledast}$, we have

$$w_n(t_1,t_2,\dots,t_n) = \left\langle W_n, P^{H(K)^{n\circledast}} e^{-i\omega_1 t_1-i\omega_2 t_2-\dots-i\omega_n t_n}B_n\right\rangle$$

which is the required result. □

Consequently

$$y_n(t) = C_n x_n(t) \tag{38}$$

where $x_n(t)$ is the $H(K)^{n\circledast}$-valued process

$$x_n(t) = \int_{-\infty}^{t}\int_{-\infty}^{t_1}\dots\int_{-\infty}^{t_{n-1}}\left[e^{A(t-t_1)}\circledast\dots\circledast e^{A(t-t_n)}\right]B_n du(t_1)\dots du(t_n) \ . \tag{39}$$

If $H(K)$ is infinite dimensional, $\{x_n(t);t\in\mathbb{R}\}$ is not an ordinary stochastic process but must be defined in a weak sense [16]. Then the *state process* $\{x(t);t\in\mathbb{R}\}$ is defined as the (possibly weakly defined) $\oplus_{n\in N} H(K)^{n\circledast}$-valued process with components x_n; $n\in N$. This terminology is motivated by the following result developed along the lines in [1,2].

PROPOSITION 1. *Let* X *be a regular state space and let* x_n *be given by* (39). *Then*

$$\{\langle\hat{f},x_n(0)\rangle_{H(K)^{n\circledast}} \mid \hat{f}\in H(K)^{n\circledast}\} = X_n \ . \tag{40}$$

Moreover, for each $n\in N$,

$$X_n \perp H_n^+(du) \ . \tag{41}$$

PROOF. Let $\xi\in X_n$ be arbitrary, and let $\hat{f} := T_u^{(n)}\xi$ and $f := F_n^{-1}\hat{f}$. Then, by Lemma 5,

$$\xi = \int_{-\infty}^{0}\int_{-\infty}^{t_1}\dots\int_{-\infty}^{t_{n-1}} f(t_1,\dots,t_n)du(t_1)\dots du(t_n) \ . \tag{42}$$

By exchanging w_n for f in the proof of Lemma 6, we obtain

$$f(t_1,\dots,t_n) = \left\langle\hat{f},\left(e^{-At_1}\circledast\dots\circledast e^{-At_n}\right)B_n\right\rangle_{H(K)^{n\circledast}} \ . \tag{43}$$

Then (42) and (43) together yield (40). In view of (19), $X_1 \subset H_1^-(du) \perp H_1^+(du)$, from which (41) follows. □

In partular, for n=1 we can write (38) and (39) in the following suggestive form

$$\begin{cases} dx_1 = Ax_1 dt + B_1 du \\ y_1 = C_1 x_1 \end{cases} \tag{44}$$

The higher-chaos subsystems are nonlinear. In the next section we shall illustrate this with an example.

Note that a backward realization for X generated by $\bar{u}$ is obtained by developing the above analysis in $(H_2^-)^{n\circledast}$ rather than in $(H_2^+)^{n\circledast}$. Whereas the forward property is characterized by (41), the backward one is determined by $X_n \perp H_n^-(d\bar{u})$ for each $n \in N$.

Finally, in the case that X is not regular, other constructions involving rigged Hilbert spaces are possible [19].

7. THE FINITE-DIMENSIONAL BILINEAR CASE

To illustrate our point let us consider the simplest possible nonlinear problem. Let the process y have the innovation representation

$$y(t) = \int_{-\infty}^{t} g_1(t-\sigma)d\nu(\sigma) + \int_{-\infty}^{t}\int_{-\infty}^{\tau} g_2(t-\tau,t-\sigma)d\nu(\sigma)d\nu(\tau) \tag{45a}$$

and the backward innovation representation

$$y(t) = \int_{t}^{\infty} \bar{g}_1(t-\sigma)d\bar{\nu}(\sigma) + \int_{t}^{\infty}\int_{\tau}^{\infty} \bar{g}_2(t-\tau,t-\sigma)d\bar{\nu}(\sigma)d\bar{\nu}(\tau) . \tag{45b}$$

Assume that $G_1 := F_1 g_1$ is a rational function which is not identically zero. Then $\bar{G}_1 := F_1\bar{g}_1$ has the same properties, and $y_1 \neq 0$. Moreover y_1 has a rational spectral density, namely $\Phi(s) := G_1(s)G_1(-s)$.

Now, setting $\Gamma := G_1/\bar{G}_1$, find all pairs $(Q,\bar{Q}^*)$ of inner functions such that $K := \Gamma Q\bar{Q}^*$ is inner and coprime with Q and $\bar{Q}^*$. For each such solution form

$$X_1 = \int_{-\infty}^{\infty} H(K)Q^* d\nu . \tag{46}$$

Theorem 3 states that the X_1-spaces obtained in this way are precisely the minimal Markovian (H_1^-,H_1^+)-splitting subspaces. In particular, $Q_1 = 1$ yields $X_1 = E^{H_1^-}H_1^+$, and $\bar{Q}_1 = 1$ yields $X_1 = E^{H_1^+}H_1^-$. Since Γ is rational, it can be shown that K must be rational, and consequently X_1 is finite-dimensional [17]. In fact, all X_1 have the same dimension n [1,2]. By using the procedure described in Section 7 of [18] we can determine an nxn-matrix A_1 and an nx1-matrix B_1 from K and a 1xn-matrix C_1 from $W := GQ$ so that

$$\begin{cases} dx_1 = A_1 x_1 dt + B_1 du \\ y_1 = C_1 x_1 , \end{cases} \tag{47}$$

where $sp\{x_1(t),\dots,x_n(t)\} = U_t X_1$, $H_1^+(du) \perp X_1$ and

$$u(t) = \int_{-\infty}^{\infty} \frac{e^{i\omega}-1}{i\omega} Q^*(i\omega)d\hat{\nu} \ . \tag{48}$$

To each X_1 there corresponds a minimal state space, namely

$$X = X_1 \oplus X_2 \ , \tag{49a}$$

where

$$X_2 = X_1 * X_1 \ . \tag{49b}$$

Hence, for each t, the $\frac{1}{2}n(n+1)$ random variables $\{x_1^{ij} := x_1^i(t) * x_1^j(t); j \le i\}$ span $U_t X_2$. (Remember that $x_1^{ij} = x_1^{ji}$.) Let $\{x_2(t); t \in \mathbb{R}\}$ be the $\frac{1}{2}n(n+1)$-dimensional stationary vector process with components $x^{ij}(t)$. Applying Itô's differentiation rule [6] to

$$x_1^{ij}(t) = x_1^i(t)x_1^j(t) - E\{x_1^i(t)x_1^j(t)\}$$

we obtain

$$dx_1^{ij}(t) = \sum_{k=1}^{n} [a_{ik}x_1^{kj}(t) + a_{jk}x_1^{ik}(t)]dt + (b_i x_1^j(t) + b_j x_1^i(t))du$$

where a_{ik} and b_i are the components of A_1 and B_1 respectively. Defining the $\frac{1}{2}n(n+1)\times\frac{1}{2}n(n+1)$-matrix A_2 and the $\frac{1}{2}n(n+1)\times n$-matrix B_2 appropriately, this can be written

$$dx_2 = A_2x_2dt + B_2x_1du \ . \tag{50}$$

Integrating this bilinear equation we get an expression of type

$$x_2(t) = \int_{-\infty}^{t}\int_{-\infty}^{\tau} f(t-\tau,t-\sigma)du(\sigma)du(\tau) \ ,$$

where f is a vector-valued function. Moreover

$$y_2(t) = \int_{-\infty}^{t}\int_{-\infty}^{\tau} w_2(t-\tau,t-\sigma)du(\sigma)du(\tau) \ ,$$

where w_2 is obtained from g_2 via formula (30). Now, since $y_2(0) \in X_2$, there are real numbers $\{c_k; k=1,2,\dots,\frac{1}{2}n(n+1)\}$ such that

$$w_2(\tau,\sigma) = \sum_k c_k f_k(\tau,\sigma)$$

and these numbers can be determined by known methods. Let C_2 be the $\frac{1}{2}n(n+1)$-dimensional row vector with components c_k. Then

$$y_2(t) = C_2x_2(t) \ . \tag{51}$$

Since $y = y_1 + y_2$,

$$\begin{cases} dx_1 = A_1x_1dt + B_1du \\ dx_2 = A_2x_2dt + B_2x_1du \\ y = C_1x_1 + C_2x_2 \end{cases} \tag{52}$$

is a realization of y, for $x = \begin{bmatrix} x_1 \\ x_2 \end{bmatrix}$ is a Markov process. Note that even if y_1 were zero we would need to include x_1 is the state process x, for x_2 by itself is not Markov.

Let $\hat{x}$ be the state process corresponding to $X_1 = E^{H_1^-}H_1^+$ (in the coordinate-system of (52)). It is shown in [1-3] that, for any X_1, $E^{H_1^-}x_1(0) = \hat{x}_1(0)$. Therefore, in view of the definition of x_2 and the fact that $E^{H_2^-} = E^{H_1^-} * E^{H_1^-}$, $E^{H_2^-}x_2(0) = \hat{x}_2(0)$. ($E^{H_1^-}$ and $E^{H_2^-}$ applied to a vector means that the projection is performed componentwise.) Consequently the conditional expectation of $x(t)$ given Y_t^- is

$$E^{Y_t^-}x(t) = \hat{x}(t) \tag{53}$$

for any realization (52). For this reason, remembering that the forward generating process of $E^{H_1^-}H_1^+$ is the innovation ν, we may call the system

$$\begin{cases} d\hat{x}_1 = A_1\hat{x}_1dt + B_1d\nu \\ d\hat{x}_2 = A_2\hat{x}_2dt + B_2\hat{x}_1d\nu \\ \quad y = C_1\hat{x}_1 + C_2\hat{x}_2 \end{cases} \tag{54}$$

the *steady state non-linear filter* of (52), and we have shown that this filter is invariant over the class (52) of minimal realizations. A similar result can be obtained for backward realizations in terms of $\bar{\nu}$.

8. CONCLUDING REMARKS

The purpose of this paper is to investigate the structural aspects of the nonlinear stochastic realization problem and to clarify basis concepts. This is a first step toward a nonlinear realization theory. Hence we have not concerned ourselves with algorithmic aspects of the problem, and our analysis is based on the availability of an innovation representation, the actual determination of which is a nontrivial problem in itself (see [20]).

The question of state space construction needs to be further studied. It could be argued that condition (4) is too restrictive since there could well be $(\oplus_{n\in N}H_n^-, \oplus_{n\in N}H_n^+)$-splitting subspaces which are not of the form (4), having a nonzero angle with some (or even all) H_n. Hence, if we can do without realizations of individual y_n but only need their sum y, it is possible that we are missing state spaces of smaller size.

Our interest in the nonlinear realization problem emanates from its potential value as a conceptual framework for certain classes of nonlinear filtering problems. This will be the topic of a future study.

REFERENCES

1. A. Lindquist and G. Picci, State space models for Gaussian stochastic processes, *Stochastic Systems: The Mathematics of Filtering and Identification and Applications*, M. Hazewinkel and J.C. Willems, Eds., Reidel Publ. Co., 1981.
2. A. Lindquist and G. Picci, Realization theory for multivariate Gaussian processes (to appear).
3. G. Ruckebusch, Theorie géometrique de la représentation markovienne, Thèse de doctorat d'état, Univ. Paris VI, 1980.
4. J.L. Doob, *Stochastic Processes*, John Wiley, New York, 1953.

5. J. Neveu, *Processus Aléatoire Gaussiens*, Presses de L'Université de Montréal, 1968.
6. G. Kallianpur, *Stochastic Filtering Theory*, Springer-Verlag, 1980.
7. D. Ocone, *Topics in Nonlinear Filtering Theory*, Ph.D. thesis, M.I.T., December 1980.
8. S.K. Mitter and D. Ocone, Multiple integral expansions for nonlinear filtering, *Proc. 18th Conf. Decision and Control*, Ft. Lauderdale, Florida, December 1979.
9. Yu. A. Rozanov, *Stationary Random Processes*, Holden-Day, 1967.
10. K. Ito, Multiple Wiener integrals, *J. Math. Soc. Japan* 3, 1951, 157-169.
11. H.P. McKean, *Stochastic Integrals*, Academic Press, 1969.
12. A. Lindquist, G. Picci and G. Ruckebusch, On minimal splitting subspaces and Markovian representations, *Math. Systems Theory* 12, 271-279 (1979).
13. H. Helson, *Lecture Notes on Invariant Subspaces*, Academic Press, 1964.
14. J.S. Baras and R.W. Brockett, H^2 functions and infinite dimensional realization theory, *SIAM J. Control* 13(1975), 221-241.
15. J.S. Baras and P. Dewilde, Invariant subspace methods in linear multivariable-distributed systems and lumped-distributed network synthesis, *Proc. IEEE* 64(1976), 160-178.
16. A.V. Balakrishnan, *Applied Functional Analysis*, Springer-Verlag, 1976.
17. P.A. Fuhrmann, *Linear Systems and Operators in Hilbert Space*, McGraw-Hill, 1981.
18. A. Lindquist and G. Picci, Realization theory for multivariate Gaussian processes II: State space theory revisited and dynamical representations of finite-dimensional state spaces, *Proc. 2nd Intern. Conf. on Information Sciences and Systems*, Patras, Greece, July 1979, Reidel Publ. Co., 108-129.
19. J.W. Helton, Systems with infinite dimensional state space: The Hilbert space approach, *Proc. IEEE* 64 (January 1976), 145-160.
20. H.P. McKean, Wiener's theory of nonlinear noise, *Stochastic Differential Equations*, SIAM-AMS Proceedings, American Math. Society, 1973, 191-209.

BURNING GRASS
and
FLOATING CORKS

C. Lobry

INTRODUCTION :

This is a tentative paper. My objective is to show how "Non linear Feedback Techniques" and "Non Standard Analysis" can fit together and provide an alternative to Reaction Diffusion Equations in the modelling of excitable media. I show by an example : "The propagation of fire in the grass" , that the kind of discrete models I propose seems more suitable for analysis than classical Hodgkin-Huxley and FitzHug Nagumo type partial differential equations.

But I emphasize here that the proposed model is not based on any realistic law. It is purely speculative ! I want to focus on the method rather than the pertinence of the model compared to the real object. This is the reason for which I chose to present the Fire problem in place of some serious scientific question like propagation of excitation in the cardiac muscle or pattern formation in the Belousov reaction.

In the paper everything related to differential equations is trivial for people with little training in the area. Everything is also trivial for those with little training in Non Standard Analysis. But, about partial differential equations, nothing is easy, even for specialists.

1- THE PROBLEM.

Consider an atoll covered by grass. See picture below.

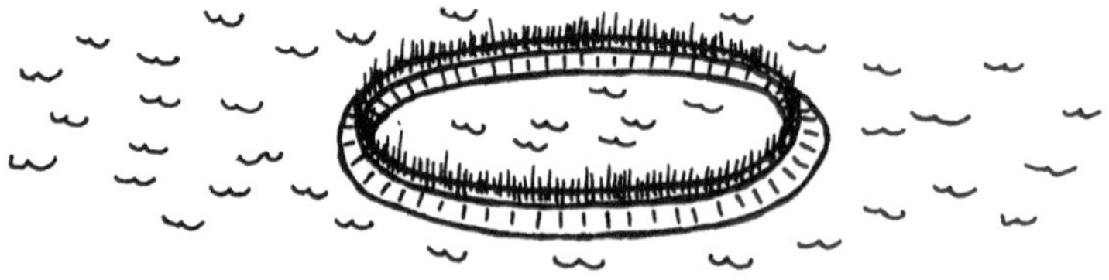

An atoll is supposed to be one dimensional and thus is modelled by the mathematical object S^1, the circle. Suppose that we cut the grass on some length and we put the fire as it is indicated on the picture next page.

The fire front propagates along the atoll and if the time it takes to make one turn is larger than the time it takes for the grass to grow up we have a fire turning indefinitely around the atoll.

Comment 1 : Nobody requires a mathematical model to understand what happens. Nevertheless we hope that if we are able to handle this simple situation with mathematics, then they may be useful in more complex ones.

Comment 2 : Grass is an example of an excitable medium. Membrane of a non myelinated axone is another one. In the final discussion we shall see that these two examples bear some analogy.

2 - P.D.E. MODELS .

An excitable medium has the following property. Consider a piece of the medium, small enough to be assimilated to a homogeneous one, and imagine some stimulus. There is a threshold: Below the threshold value nothing is obser--ved, the state comes back smoothly to the rest value; above there is a large excursion, large amplification before the state comes back to the rest value. The differential equation whose phase plane is pictured below has this property.

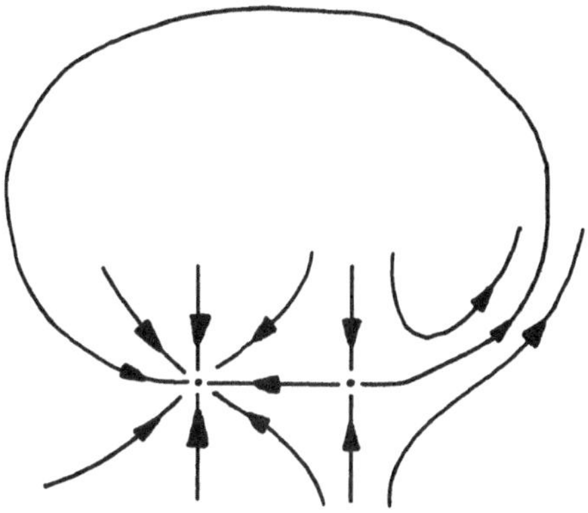

Such an equation may be purely phenomenological or based on physical laws. We do not discuss this point here.

Hypothesis: We assume that a "homogeneous" set of grass has its temperature and height governed by the system of two differential equations:

$$\begin{cases} \dfrac{dH}{dt} = F(H,T) \\ \dfrac{dT}{dt} = G(H,T) \end{cases}$$

where H stands for Height, T for Temperature. The phase portrait is given below.

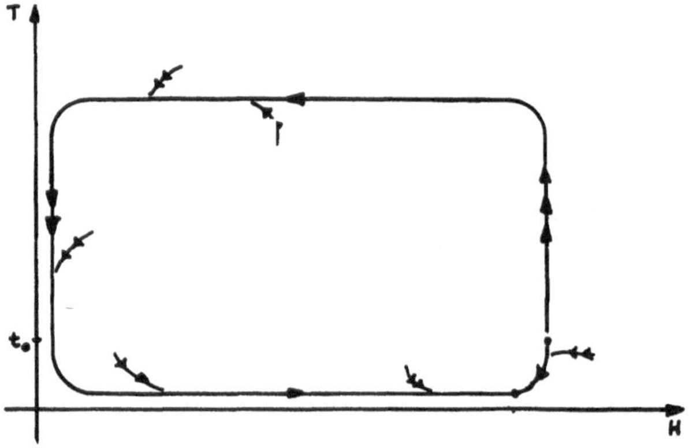

One sees that there are two time scales: Fast elevation of temperature when all the grass is firing, fast decay when the grass is burnt, comparatively slow dynamics for the burning and the growing phase. The picture below shows how this fits with a "cubic shaped fast-slow differential equation".

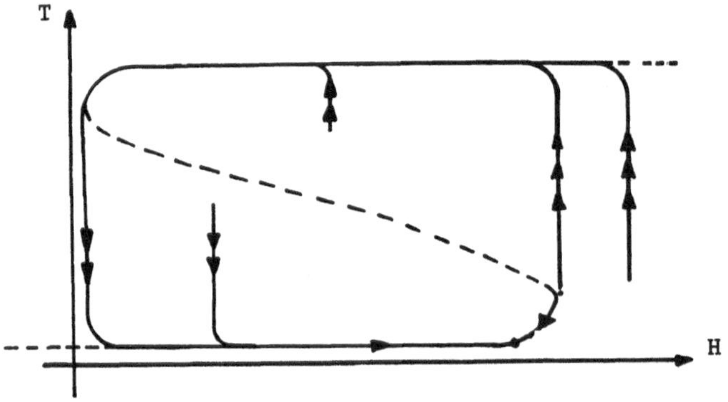

Thus after a change of variables:

$$U = U(H,T)$$
$$V = V(H,T)$$

we can replace this cubic by the standard cubic of $\mathbb{R}^2$:

$$V = \frac{U^3}{3} - U$$

and our system of differential equations becomes the system C.C.S., which means Canonical Cubic System, and is defined on the next page.

$$\text{(C.C.S.)}\quad \begin{cases} \dfrac{dU}{dt} = \dfrac{1}{\varepsilon}\left(V - \left(\dfrac{U^3}{3} - U \right)\right) = f(U,V) \\ \dfrac{dV}{dt} = a - U \qquad\qquad = g(U,V) \quad a > 1 \end{cases}$$

We show below the phase portrait of this system when ε is a small positive real.

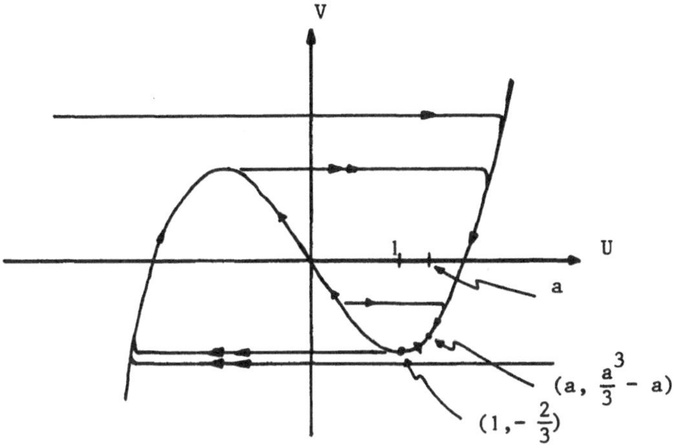

When the medium is distributed in space, U and V depend on a space variable x and the "temperature is diffused". In the new variables U and V the temperature is of the form - U + a constant, and thus we introduce a Laplacian in the first equation.

$$\text{(R.D.E.)}\quad \begin{cases} \dfrac{\partial U}{\partial t}(x,t) = f(U(x,t),V(x,t)) + k^2 \dfrac{\partial^2 U}{\partial x^2}(x,t) \\ \dfrac{\partial V}{\partial t}(x,t) = g(U(x,t),V(x,t)) \end{cases} \quad x \in S^1$$

The space variable is in S^1 and thus we have no boundary conditions to specify. An initial condition with index 0 with respect to the origin corresponds to a "burning initial condition". A periodic firing around the atoll corresponds to a stable solution which is of index one for every t. To my knowledge nobody has proved the existence of such "index preserving solutions" for this kind of partial differential equations. Most of the qualitative results are concerned with the invariance of convex domains, which is not relevant for this purpose. What we need is the invariance of a corona which seems to be a difficult question. See [2] [8].

3 - DISCRETE MODELS

The idea to use discrete models is certainly not new! The discre--tized version of R.D.E. is D.E. (D.E. = Discrete Equation).

(D.E.) $$\begin{cases} \dfrac{dU_i}{dt} = f(U_i,V_i) + \dfrac{k^2}{h^2}(U_{i+1} - 2U_i + U_{i-1}) \\ \dfrac{dV_i}{dt} = g(U_i,V_i) \\ i \in \mathbb{N} \pmod N \qquad N \times h = L \end{cases}$$

where: L is the length of the atoll

h is the size of the mesh

U_i stands for $U(hi,t)$

V_i stands for $V(hi,t)$

First we give a new interpretation of this system of differential equations.

Let X denote a vector field on $\mathbb{R}^n$, here n = 2. Consider N copies of $\mathbb{R}^n$ and denote by C_i the i-th coordinate vector of a point C in $(\mathbb{R}^n)^N$. Consider:

$$\frac{dC_i}{dt} = X(C_i) \qquad i \in \mathbb{N} \pmod N.$$

This system of N identic decoupled differential equations can be interpreted as the movement of N corks, $(C_1,C_2,\ldots\ldots C_N)$ on a flow whose velocity is given by X. Introduce now some feedback law for each cork, based on the observation of the two neighbouring corks (connection with a rubber is a possibility).
Then we get the following system:

(C.C.) $$\frac{dC_i}{dt} = X(C_i) + \Phi(C_{i+1},C_i,C_{i-1}) \qquad i \in \mathbb{N} \pmod N .$$

The symbol C.C. stands here for "Coupled Corks" on a flow.
Remark : Notice that if we also discretize the time then the system (C.C.) bears strong analogies with cellular automata. It was noticed that one way to undersand E.D.R. is to study their cellular automata analogues (see [1] [3] [5]).
From the point of view of feedback systems the equation (C.C.) is a perfect object to look at. If the feedback is a linear one:

(L.F.) $$\Phi(C_{i-1},C_i,C_{i+1}) = \begin{cases} \dfrac{k^2}{h^2}(U_{i-1}-2U_i+U_{i+1}) \\ 0 \end{cases} \quad ; \quad C_i = (U_i,V_i)$$

then the system (C.C.) is equivalent to the system (D.E.), but, and this is essential, from a system theoretic point of view there is no "a priori" reason to choose a linear feedback.

Let us come back to the question of fire propagation. Imagine that our phenomenological system of equations (C.C.S.) is a good model for the burning of the grass in the open air. It is reasonable to postulate (I do not say it is true !) that the elevation of temperature of a non burning blade of grass is caused by the brightness of the neighbouring burning blades. In which case we can assume a phenomenological law of the type below:

(N.L.F.) $$\Phi(C_{i-1}, C_i, C_{i+1}) = \begin{cases} K(U_{i-1}) + K(U_{i+1}) \\ 0 \end{cases}$$

where the function K if defined by the graph below:

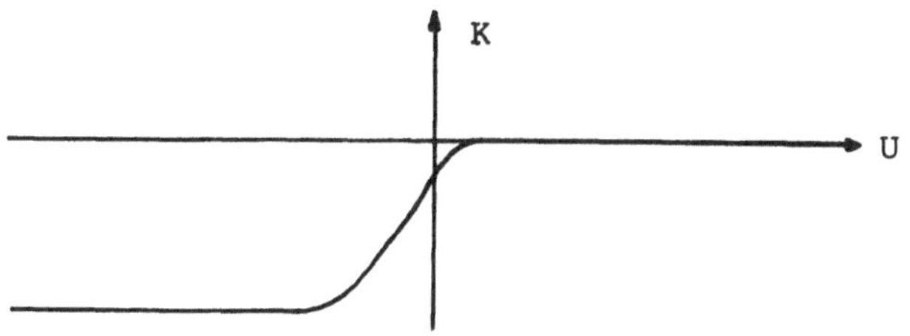

To understand the meaning of the shape of this graph recall that, up to a constant, the temperature is $-U$.

The important point in the forthcoming discussion is that below a certain temperature the grass is not burning an hence has no brightness, which explains the dissymetry in the graph of K. This dissymetry, which at least is as much realistic as linear feedback, is much more easier to manipulate in the mathematical developments. Thus if we have no serious physical reason to choose a linear feedback one should prefer "a priori" the class of nonlinear ones.

4- COUPLED CORKS ON TWO DIMENSIONAL FLOWS .

Let us denote by X the vector field $\begin{Bmatrix} f \\ g \end{Bmatrix}$ of the Canonical Cubic System. Recall that :

$$f(U,V) = \frac{1}{\varepsilon}\left(V - \left(\frac{U^3}{3} - U\right)\right)$$

and now choose ε infinitesimal but strictly positive.

For the first time we use a term from Non Standard Analysis. I renounce of any attempt to explain here what Non Standard Analysis is. The reader who is interested in foundations must read Robinson [9] or the paper of Nelson [7] from which we take our notations and results. The reader who is interested in Non-Standard Analysis treatment of singularly perturbed van der Pol equation (which is equivalent to our Canonical Cubic System) is refered to the paper by Benoit E., F. and M. Diener, J.L. Callot [4]. There is also some philosophy on the subject in [6]. For the reader who does not want to look at these references the best to do is to take the word infinitesimal in the sense it is used by scientists like physicists, chemists, biologists..... It works perfectly and it turns out that everything can be formalised in a mathematical theory.

Let h be an infinitesimal and N be an integer such that $Nh = L$. Thus N must be infinitely large. (We say unlimited).
Consider:

$$(C.C.) \quad \begin{cases} \frac{dC_i}{dt} = X(C_i) + K_\lambda(C_{i-1}) + K_\lambda(C_{i+1}) \\ i \in \mathbb{N} \quad (\text{mod } N) \text{ ; } h \text{ infinitesimal, } h > 0 \text{ ; } Nh = L . \end{cases}$$

and to be more specific the feedback K_λ is defined by the graph below:

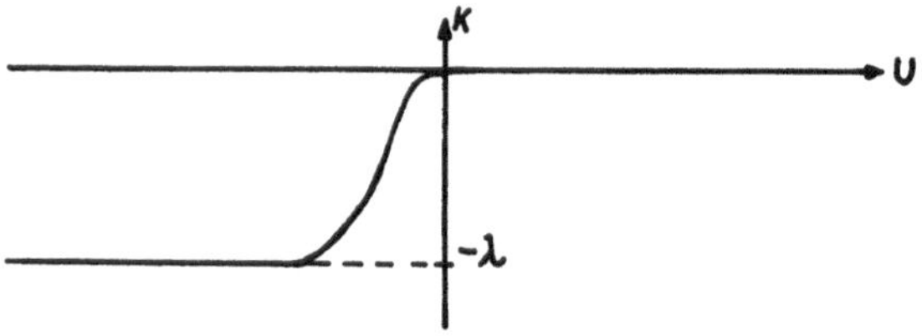

Definition: A rotating solution of (C.C.) is a solution :

$$t \longrightarrow C_i(t) \qquad i \in \mathbb{N} \ (\text{mod } N)$$

with the following properties:

i) For every pair i , i+1 the line joining C_i to C_{i+1} in the plane does not contain (0,0)

ii) The piecewise linear curve defined in the plane by joining C_i to C_{i+1} has index one with respect to the origin.

iii) Every $C_i(t)$ "turns around the origin" in a sense which is evident to formalise.

Theorem: For every pair of strictly positive infinitesimals ε and h there exist real numbers λ_o and L_o such that for $\lambda > \lambda_o$ and $L > L_o$ the differential system (C.C.) has a stable rotating solution.

Demonstration: A rigorous proof of this result needs some notations and few technicalities which are not difficult but are too long to be exposed here. Nevertheless it is easy to understand how it works.

i) Description of the dynamics of an isolated cork.

The equation describing the dynamics of an isolated cork is:

$$\frac{dC_i}{dt} = X(C_i)$$

According to [4] the phase portrait is given by the picture below. Notice that the point (a , $\frac{a^3}{3}$ - a) is the unique stable rest point.

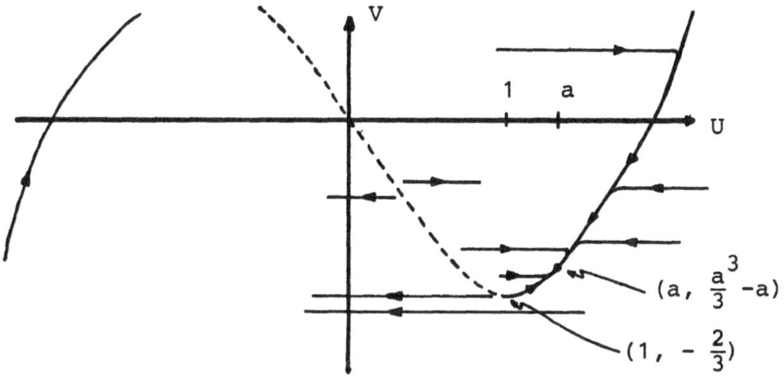

ii) Description of the dynamics of a cork which has a neighbouring cork on the left of the V axis.

Assume that C_{i-1} is fixed on the left of the V axis and C_{i+1} is fixed at the rest point. In this case one has $K(C_{i-1}) = \lambda$ and thus the equations of C_i are :

$$\begin{cases} \dfrac{dU_i}{dt} = \dfrac{1}{\varepsilon} \left(V_i - \left(\dfrac{U_i^3}{3} - U_i + \varepsilon\lambda \right)\right) \\[2ex] \dfrac{dV_i}{dt} = a - U_i \end{cases}$$

It turns out that the slow manifold is the canonical cubic lifted by $\varepsilon\lambda$. We choose $\varepsilon\lambda_o = \frac{a^3}{3} - a + \frac{2}{3} + \rho$, $\rho > 0$ in order to have the minimum of the cubic higher than the rest point of the Canonical Cubic System.

iii) Description of the trajectory of C_i when $C_{i-1}(0) = (0, \frac{a^3}{3} - a)$; $C_i(0)$ is close to the rest point of the Canonical Cubic System, C_{i+1} is fixed at the rest point.

It takes an infinitesimal time α for the cork C_{i-1} to jump to the left branch of the cubic. At time 0 the dynamics for C_i are given by the Canonical Cubic System, at time greater than α by the equations of ii) above. Because $\frac{dV_i}{dt}$ is always finite (by a trivial adaptation of [8] , for instance, one sees that any convex region large enough is invariant, and thus $U_i(t)$ is always finite) from classical majorization we deduce that $V_i(\alpha)-V_i(0)$ is an infinitesimal. After time α , the cork C_{i-1} stays for a finite time on the left branch of the canonical cubic or the lifted one, it depends on C_i which is moving, and thus we are sure that C_{i-1} is on the left of the V axis. This is enough for C_i to jump to the left branch of the cubic.

iiii) Assume now that we have a chain of corks starting at C_0 and ending at C_N , which means that C_0 and C_N have no action on each other, assume that C_0 is on the left of V axis, the other ones being near the rest point of the Canonical Cubic System. It is clear that all the successive corks will jump to the left.

iiiii)Rotating solutions.Suppose it takes a finite time T for a cork C_i which is on the left branch of the cubic to come back near the rest point, namely to be such that $V_i \leq (\frac{a^3}{3} - a) + \rho$ and it takes an infinitesimal τ to jump to the left in the procedure described at iiii). Then during time T, $\omega = \frac{T}{\tau}$ corks will jump from right to left. So if N is smaller than ω there will be no longer corks in position to jump before the first cork comes back near the rest point. All the corks will be on the right branch of the cubic after some time and the process will stop. But if N is greater than ω one sees that the process will not take an end and thus we have a rotating solution. Because N depends on L we conclude that we have a rotating solution if L is large enough.

Formal redaction of point i) to iiiii) proves the theorem.

5 - DISCUSSION

First notice that a weaker version of the previous theorem is:

Theorem bis: There exist ε_0 and h_0 such that for every $0 < \varepsilon < \varepsilon_0$ and every $0 < h < h_0$ there exist λ_0 and L_0 such that :

$$\left.\begin{array}{l}\lambda > \lambda_0\\ L > L_0\end{array}\right\} \Rightarrow \text{(C.C.) has a stable rotating solution.}$$

Proof: Choose for ε_0 and h_0 some infinitesimals. If ε and h are smaller they are also infinitesimals and then the previous theorem applies.

Notice that the Theorem bis in its statement makes no reference to non standard words . One says that this statement is "internal" and it turns out that it is also true if read in terms of Standard Mathematics. See Nelson [7] . No doubt that this theorem has a standard proof, but it will be technical and not particularly illuminating. One sees that Non Standard Language is one way to prove some standard results. This is well known for long time now.

But I think that the theorem of paragraph 4, read in Non Standard Mathematics, contains much more informations. Let us come back to point iiiii). The condition on L is :

$$L > \omega h$$

No mention is made of the fact that ωh may be limited or not. In Standard Mathe--matics this has no meaning, but physically it has. Non Standard Language is able to express it. The condition ωh unlimited says that the number of blades of grass which are set on fire during the limited time T multiplied by the dis--tance between two blades is unlimited, which means that the propagation of the flame has an unlimited, (infinite) velocity. This happens if ω or h are too large. This is coherent with our assumption that fire is communicated instantane--ously from one blade to another provided the temperature is large enough and, as soon as the fire is set, it takes some time to reach this critical value. One may argue that the number λ must be a decreasing function of h because the aptitude of the brightness to set fire decreases with the distance. This is quite true and if we apply this remark to our model it turns out that the velocity of the flame front increases with h up to a certain limit where it stops to propa--gate (when λ is too small). There is a little paradox here; it comes from the fact that we supposed that the brightness has influence just on the two neighbou--ring blades, which is at least discutable....

Just one word about a possibility which has not been exploited in this paper. The solution of our coupled corks system is by essence discrete with respect to the space variable. It is possible to come back to a continuous repre--sentation, not as in Standard Mathematics by a limit procedure ($h \longrightarrow 0$), but by the consideration of the mean of the disctrete solution on infinitely small neighbourhoods of standard points. This point of view will be developed else--where.

Let us stop here these considerations to point out what is, in my opinion, essential.

1) It is quite possible to make reasonable models of propagation of fire in the grass on the basis of "Coupled Corks", the coupling of neighbou--ring corks being essentially non linear, as opposed to the linear interaction when one considers discretization of a diffusion process like a Reaction Diffu-

-sion equation.

2) The use of Non Standard Analysis is very much adapted to describe macroscopic effects (here the flame front) of microscopic causes (here the law of propagation of fire from one blade to another).

3) Even if they are trivial the above considerations on relevance of the model with respect to the real object have a great merit: They are possible ! Try to have a similar discussion with a Reaction Diffusion Equation .

There is one point which remains unclear : Even if "Coupled Corks" models are justified in the case of the fire propagation in the grass this problem seems to have a very special structure . It seems to be discrete by essence. Let me say few words about a classical problem: The propagation of electrical pulses along an axone. The model of Hodgkin-Huxley is a system of three differential equations coupled with a P.D.E. It is a description of a distributed electro - chemical system which explains to some extent how electrical waves propagate in a neurone. By simulation the model shows reasonable accordance with observed beha- -viours. But it is rather difficult to understand it at a mathematical level and even its simplification, the FitzHug Nagumo equation is still a challenge for Mathematicians. Recent experiments show that the membrane of an axone looks like a wall with many small specialised holes which are able to open or to close depending on the difference of the potential through the membrane, and conversely the difference of potential depends of the migration of ions, and thus from the number of holes which are open or closed. See [10] for instance . It is perfecly reasonable to try to modelise this question with the same tools we used to mode- -lise flame propagation.

Acknoledgments:

To G. REEB who convinced me that Non Standard Analysis is a very good tool for mathematical modelling, J.M. LASRY who introduced me to the fire propagation problem (and other related topics), C. REDER and F. MAZAT who helped me to improve successive versions of Cork Floating models.

REFERENCES :

1 J.P. ALLOUCHE, C. REDER : *"Oscillations d'un automate cellulaire en milieu excitable"* Colloque sur les rithmes en biologie, chimie, physique, et autres champs d'applications. Marseille 14-18 septembre 1981. COSNARD, DEMONGEOT, LEBRETON organisateurs, à paraitre dans Lecture notes in biomathematics.

2 K.N. CHUEH, C.C. CONLEY, J.A. SMOLLER : *"Positively invariant regions for sys- -tems of non linear diffusion equations"* Indiana Univ. Math. J. Vol 26,n°2 (1977)

3 P. CIPIERE, C. LOBRY, C. REDER : *"A propos de reaction chimiques oscillantes"* Publications A.A.I. Université de Bordeaux , Octobre 1979.

[4] E. BENOIT, JL. CALLOT, F. et M. DIENER : *"Chasse au Canard"* Publications IRMA Université de STRASBOURG (rue René Descartes 67000 Strasbourg) (1980) 98 p 53 .

[5] J.M. GREENBERG, S.P. HASTINGS : *"Spatial patterns for discrete models of diffusion in excitable media"* SIAM Journal Appl. Math. 34 (1978)

[6] C. LOBRY : *"Mathématiques non classiques, Mathématiques aplliquées ?"* Publications A.A.I. Université de Bordeaux, n° 81 - 06 (Mai 1981).

[7] E. NELSON : *"Internal Set Theory"* B.A.M.S. 83 (1977) pp 1165-1198.

[8] C. REDER : *"Familles de convexes invariantes et équations de diffusion réaction"* A paraitre aux annales de l'Institut Fourier, Publications A.A.I. Université de Bordeaux, n° 80 - 07 (Avril 1980) .

[9] A. ROBINSON :*"Non Standard Analysis"* American Elsevier, N.Y. 1974 .

[10] C. STEVENS : *"Le Neurone"* Pour la Science n° 25, Nov. 1979.

U.E.R. MATHEMATIQUES ET INFORMATIQUE

Université de BORDEAUX

351 Cours de la Libération

33405 TALENCE (France)

SUPERVISORY CONTROL OF DISCRETE EVENT PROCESSES

P.J. Ramadge and W.M. Wonham

Systems Control Group
Dept. of Electrical Engineering
University of Toronto
Toronto, Ontario
CANADA M5S 1A4

ABSTRACT

A discrete event process is defined in algebraic terms and its behavior is given by an appropriate formal language.

For a set of asynchronous processes we examine the problem of synthesizing a centralized supervisor to ensure a desired collective behavior. Our main result is that every supervisor which solves the centralized supervisor problem is the projection of a grammar for the coordinated behavior of the given processes.

1.0 INTRODUCTION

A complex system may consist of many interacting components which operate concurrently. A typical high level control problem for such systems is the supervision of the various components in order to ensure their harmonious interaction and a resultant orderly flow of events.

Simple examples of supervisory control are provided by the start-up and shut-down procedures of industrial processes and the coordination of activities in automated production lines.

In this paper we model sequential discrete event processes using finite graphs and regular languages. Our main interest is in a set of such processes each of which operates asynchronously. Each process communicates with a central supervisor which can be considered as a reference station from which the interaction of all the processes may be observed. This observation takes the form of a shuffling of the incoming communication sequences.

The central supervisor problem is to synthesize the dynamics of the supervisor, as well as its responses to the incoming communications, so as to achieve a desired coordinated behavior. Our main result for this problem is that every supervisor which solves the centralized supervisor problem is a quotient of a grammar for the resultant coordinated behavior. If we think of projections as providing "coarse models" then our result states that every successful supervisor contains a model of the coordinated behavior.

2.0 PRELIMINARIES

We let $\underline{n}^+$ denote the subset of the natural numbers $\{1,2,\ldots,n\}$. If Σ is a set then we denote the free monoid over Σ by Σ^*. Elements of Σ^* are called <u>strings</u> or <u>words</u> and elements of Σ are called <u>symbols</u>. The function $\ell:\Sigma^* \to \Sigma^*$ is defined by $\ell(\varepsilon) := \varepsilon$ and for $w \varepsilon \Sigma^*$ and $\sigma \varepsilon \Sigma$, $\ell(w\sigma) := \sigma$. Let $L \subset \Sigma^*$. We say a string $u \varepsilon \Sigma^*$ is a <u>pre-fix</u> in L if there exists a word $w \varepsilon L$ with $w = uv$ for some string $v \varepsilon \Sigma^*$. Let Pre(L) denote the set of prefixes in L. A string $u \varepsilon$ Pre(L) is a <u>proper prefix</u> in L if $u \notin L$. Let $\Omega \subset \Sigma$, and let ε denote the empty string. The ε-<u>projection</u> of Σ^* onto Ω^* is the monoid homomorphism $h:\Sigma^* \to \Omega^*$ with $h(\omega) = \omega$ if $\omega \varepsilon \Omega$ and $h(\sigma) = \varepsilon$ otherwise.

A <u>directed graph</u> (or simply <u>graph</u>) G is a two-sorted algebra $(N;E; \delta_0,\delta_1)$ with N the node set, E the edge set and the functions $\delta_0:E \to N$, $\delta_1:E \to N$ giving the initial and final nodes respectively of each edge.

A <u>graph morphism</u> from $G = (N,E; \delta_0,\delta_1)$ to $H = (M,D; \delta_0,\delta_1)$ is a pair of functions $f = (f_n,f_e)$ such that the following diagram commutes.

$$
\begin{array}{ccccc}
N & \xleftarrow{\ \delta_0\ } & E & \xrightarrow{\ \delta_1\ } & N \\
\downarrow{f_n} & & \downarrow{f_e} & & \downarrow{f_n} \\
M & \xleftarrow{\ \delta_0\ } & D & \xrightarrow{\ \delta_1\ } & M
\end{array}
$$

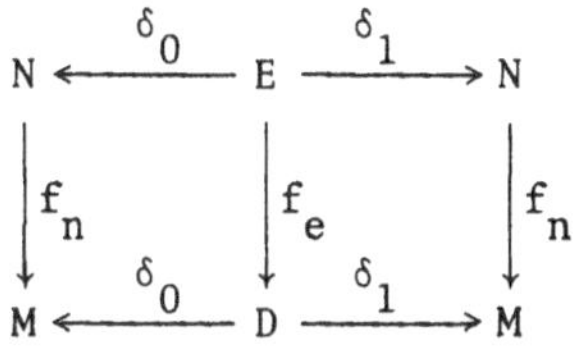

Graphs together with their morphisms form a category which we call <u>Gph</u>.

Let 1 denote the one element set $\{1\}$. Given any set Σ we can construct the special graph $\widetilde{\Sigma} = (1,\Sigma; \delta_0,\delta_1)$. This has edge set Σ and for each $\sigma \varepsilon \Sigma$, $\delta_0(\sigma) = \delta_1(\sigma) = 1$.

A <u>labelled graph</u> G_Σ is a triple (G,g,Σ) with G a graph, Σ a set and g a graph morphism from G to $\widetilde{\Sigma}$. The morphism g labels the edges of G by the elements of the set Σ.

A <u>labelled graph morphism</u> from $G_\Sigma = (G,g,\Sigma)$ to $H_\Omega = (H,h,\Omega)$ is a pair $f = (f_1,f_2)$ of graph morphisms such that the following diagram commutes in <u>Gph</u>.

$$
\begin{array}{ccc}
G & \xrightarrow{\ g\ } & \widetilde{\Sigma} \\
\downarrow{f_1} & & \downarrow{f_2} \\
H & \xrightarrow{\ h\ } & \widetilde{\Omega}
\end{array}
\qquad
\begin{array}{c}
G_\Sigma \\
\downarrow{f} \\
H_\Omega
\end{array}
$$

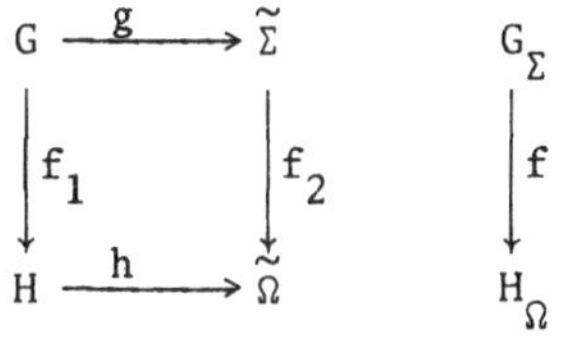

Labelled graphs and their morphisms form a category.

Let $G_\Sigma = (G,g,\Sigma)$, $G = (N,E; \delta_0,\delta_1)$ and $g = (g_n,g_e)$. Consider the following commutative diagram.

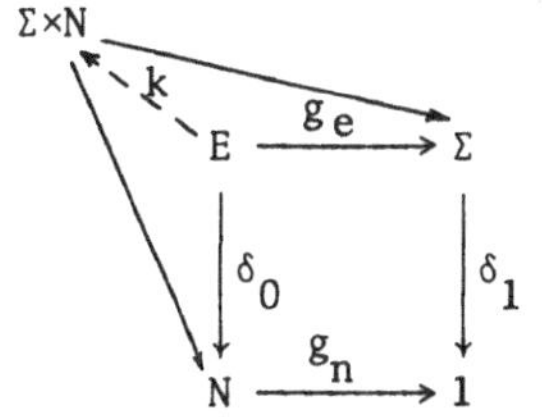

Here unlabelled arrows are natural projections and k is the unique map for which the diagram commutes. We say G_Σ is *deterministically labelled* if k is injective.

3.0 SEQUENTIAL PROCESSES

A *sequential process* (SP) P is represented by a labelled graph G_Σ together with two nonempty subsets S and T of the nodes of G. Elements of the set S are called *initial states* and elements of T are called *final* or *halting states*. P is a *finite state sequential process* iff the graph G is finite.

A *successful path* in G is a path from S into T. The *label* of a path p in G is defined as the concatenation of the labels of the constituent edges of p. The *behavior* of P is then the set $|P| \subset \Sigma^*$ of all labels of successful paths in G.

P is said to be *accessible* if for every state x of P there is a path in G from some $s \in S$ to x and *coaccessible* if from every state x of P there is a path from x into T. A SP which is both accessible and coaccessible is said to be *trim*.

The behavior of a finite state SP is a regular language and every regular language is the behavior of a deterministically labelled SP with only one initial node.

Let $S = (G_\Sigma, S, T)$ and $P = (H_\Omega, U, V)$ be sequential processes. A *morphism* $F: S \to P$ *of sequential processes* consists of:

1. A labelled graph morphism $f = (f_1, f_2)$ from G_Σ to H_Ω.
2. A pair of maps $a: S \to U$ and $b: T \to V$ such that, with f_n the node map of f_1, N the node set of G and M the node set of H, the following diagram commutes.

$$\begin{array}{ccccc} S & \longrightarrow & N & \longleftarrow & T \\ \downarrow{\scriptstyle a} & & \downarrow{\scriptstyle f_n} & & \downarrow{\scriptstyle b} \\ U & \longrightarrow & M & \longleftarrow & V \end{array}$$

Sequential processes and their morphisms form a category which we call *Seq*.

A *deterministic sequential machine* (SM) without outputs is a two-sorted partial algebra $A = (\Sigma, Q; q_0, T, \delta)$. Here q_0 is the initial state of A, T is the set of terminal states of A and $\delta: \Sigma \times Q \to Q$ is a partial function called the state transition map of A. There are numerous possible definitions for the morphisms of SMs. Here we define a morphism $h: A \to B$ with $B = (\Omega, X; x_0, U, \beta)$, as a pair of maps $h_1: \Sigma \to \Omega$ and $h_2: Q \to X$ such that $h_2(q_0) = x_0$, $h_2(T) \subset U$ and $h_2\delta \subset \beta(h_1 \times h_2)$. Trim finite state SMs and their morphisms form a category which we call *Aut^t*.

We extend $\delta: \Sigma \times Q \to Q$ to a partial function $\delta^*: \Sigma^* \times Q \to Q$ by recursion:

1. $\delta^*(\varepsilon, q) := q$ for each $q \in Q$
2. For $u \in \Sigma^*$ and $\sigma \in \Sigma$, $\delta^*(u\sigma, q) := \delta(\sigma, \delta^*(u, q))$ whenever $\delta^*(u,q)$ and $\delta(\sigma, \delta^*(u,q))$ are defined.

We shall abbreviate $\delta^*(w,q)$ to $(q)w$.

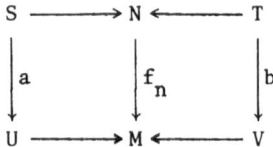

The behavior of a SM A is the set $|A|$ of all finite strings $w \in \Sigma^*$ such that $(q_0)w \in T$. Thus objects of *Aut^t* are *acceptors* of regular languages. Indeed it is easily shown that there is a functor $F:\underline{Aut}^t \to \underline{Seq}$ which maps each SM $\mathcal{A}$ into its "state graph" and $|\mathcal{A}| = |F(\mathcal{A})|$.

A *generalized sequential machine* (GSM) $\mathcal{A}$ is a SM $(\Sigma,Q;\ q_0,T,\delta)$ together with an output map $\tau:\Sigma\times Q \to \Gamma^*$, with the domain of definition of τ equal to the domain of definition of δ. Let $\mathcal{A} = (\Sigma,Q,\Gamma;\ q_0,T,\delta,\tau)$ and $\mathcal{B} = (\Omega,X,\Phi;\ x_0,U,\beta,\nu)$ be GSMs. A *morphism* $h:\mathcal{A} \to \mathcal{B}$ is a triple of maps $h_1:\Sigma \to \Omega$, $h_2:Q \to X$ and $h_3:\Gamma \to \Phi$ such that $h_2(q_0) = x_0$, $h_2(T) \subset U$, $h_2\delta \subset \beta(h_1\times h_2)$ and $h_3\tau \subset \nu(h_1\times h_2)$. GSMs and their morphisms form a category which we call $\underline{Gsm}$.

We extend $\tau:\Sigma\times Q \to \Gamma^*$ to a partial function $\tau^*:\Sigma^*\times Q \to \Gamma^*$ by recursion:

1. $\tau^*(\varepsilon,q) := \varepsilon$
2. For $u\in\Sigma^*$ and $\sigma\in\Sigma$, $\tau^*(u\sigma,q) := \tau^*(u,q)\ \tau(\sigma,(q)w)$ whenever $\tau^*(u,q)$ and $\tau(\sigma,(q)w)$ are defined.

We shall abbreviate $\tau^*(w,q_0)$ to $\tau(w)$.

For each GSM $\mathcal{A}$ there is an underlying SM which is obtained from $\mathcal{A}$ by "forgetting" its output map. In this paper we define the behavior $|\mathcal{A}|$ of a GSM as the behavior of its underlying SM (this is nonstandard).

Isomorphic to the category $\underline{Aut}^t$ is the category $\underline{Grm}$ of deterministic regular grammars. A *deterministic regular grammar* G is a trim finite state SP with only one initial node and with a deterministically labelled graph. The nodes of G are usually called nonterminal symbols, the edges of G productions and the labels of G terminal symbols. The category $\underline{Grm}$ is the resultant full subcategory of $\underline{Seq}$.

Since the categories $\underline{Grm}$ and $\underline{Aut}^t$ are isomorphic, we can regard a grammar as an acceptor or an acceptor as a grammar, whichever happens to be more convenient.

3.1 Operations on Regular Languages

Let $G = (G_\Sigma,n_0,T_1)$ and $H = (H_\Sigma,m_0,T_2)$ be two regular grammars over the alphabet Σ. The *intersection* $G\cap H$ of G and H is defined using the following pullback construction in $\underline{Gph}$.

$$\begin{array}{ccc} K & \xrightarrow{\ b\ } & H \\ {\scriptstyle a}\downarrow & & \downarrow{\scriptstyle h} \\ G & \xrightarrow{\ g\ } & \tilde{\Sigma} \end{array}$$

Here the morphisms a and b and the graph K are the pullback [MacLane, p. 71] in $\underline{Gph}$ of the pair of morphisms h and g. $G\cap H$ is defined to be the trim part of the SP $(K_\Sigma,(n_0,m_0),T_1\times T_2)$. The behavior of $G\cap H$ is of course $|G|\cap|H|$. Thus there exists a finite procedure to construct a grammar for $|G| \cap |H|$, given grammars G and H.

Let Σ and Ω be two disjoint alphabets. We define the *shuffle projection* $p:(\Sigma\cup\Omega)^* \to \Sigma^*\times\Omega^*$ as the unique monoid homomorphism generated by:

1. $p(\sigma) = (\sigma,\varepsilon)$ if $\sigma\in\Sigma$
2. $p(\omega) = (\varepsilon,\omega)$ if $\omega\in\Omega$.

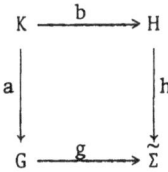

Thus p maps a string s over $\Sigma \cup \Omega$ onto its pair of component strings (x,y) with $x \in \Sigma^*$ and $y \in \Omega^*$. We say that s is a shuffling of x and y.

The shuffle product [Eilenberg, p. 20] of regular languages $L_1 \subset \Sigma^*$ and $L_2 \subset \Omega^*$ is defined by: $L_1 \otimes L_2 := p^{-1}(L_1 \times L_2)$.

If G is a grammar for L_1 and H is a grammar for L_2, then the shuffle product grammar $G \otimes H$ is defined as follows.

Let Q(X) be the node set, $q_0(x_0)$ be the start node and $T_1(T_2)$ be the terminal nodes for $G(H)$. Let $G_{\Sigma \cup \Omega}$ be the labelled graph with node set $Q \times X$ and with an edge $(q,x) \xrightarrow{\sigma} (q',x)$ for each edge $q \xrightarrow{\sigma} q'$ in G, and an edge $(q,x) \xrightarrow{\omega} (q,x')$ for each edge $x \xrightarrow{\omega} x'$ in H. Then $G \otimes H$ is the trim part of the SP $(G_{\Sigma \cup \Omega}, (q_0,m_0), T_1 \times T_2)$. The behavior of $G \otimes H$ is of course $L_1 \otimes L_2$.

4.0 CONTROLLED DISCRETE EVENT PROCESSES

A controlled discrete event process (CDEP) P consists of:

1. A trim SP $P = (G_{\Sigma \cup \varepsilon}, S, T)$
2. A set of input events $\Gamma = \Gamma_e \cup \Gamma_d$ with $\Gamma_e \cap \Gamma_d = \emptyset$ and $\Gamma \cap \Sigma = \emptyset$.
3. A pair of partial functions $f_e : E \to \Gamma_e$ and $f_d : E \to \Gamma_d$ with the domain of definition of f_e equal to the domain of definition of f_d. Here E is the edge set of P. f_e specifies the enabling event and f_d the disabling event for each controlled edge.

A CDEP P is given the following interpretation. The nodes of G represent the states of P and edges in G (called events) represent allowed state transitions. The states in S are allowed initial states and states in T are allowed halting states. Each state transition of P is either controlled or uncontrolled. If a state transition is controlled then it is said to have a status, which can take the value enabled or disabled; otherwise the state transition is always enabled.

The "occurrence" of an enabling event for a controlled transition enables the transition while the "occurrence" of its disabling event disables the transition.

At any time in which P is in state q, P may decide to execute any state transition from q to some other state q' of P. If the state transition is enabled then P executes the transition instantaneously. If the transition is disabled then P waits in state q until the chosen transition becomes enabled; P then executes the chosen transition.

Σ is a set of output names. Each state transition of P carries an external output. If $\sigma \in \Sigma$ is the label of an event e of P, then when e is executed the output σ occurs simultaneously. The behavior $|P|$ of P is thus the set of all possible output strings which may be generated during the operation of the process. We restrict our attention to the case when P is a finite state SP. Then the behavior of P is a regular language.

Let $g : E \to \Sigma \cup \varepsilon$ be the label map of P, let $i : \Sigma \to \Sigma \cup \varepsilon$ be the inclusion map and let $g' : E \to \Sigma$ be the unique maximal partial function with $ig' \subset g$. P is output controlled if there exists a pair of injective partial functions $h_e : \Sigma \to \Gamma_e$ and $h_d : \Sigma \to \Gamma$ such that

$f_e = h_e g'$ and $f_d = h_d g'$. In this paper we assume all CDEPs are output controlled. Thus to specify an initial condition for P we must give the initial state of P and the initial status of each controlled output of P.

5.0 SUPERVISION OF DISCRETE EVENT SYSTEMS

A discrete event system (DES) is a finite set Ⓗ $= \{P_i,\ i\in \underline{n}^+\}$ of CDEPs with disjoint input alphabets and disjoint output alphabets. We define the behavior of Ⓗ as the shuffle product $L_{Ⓗ} := \bigotimes_{i\in\underline{n}^+} |P_i|$ of the behaviors of the constituent processes.

We interpret Ⓗ as a set of independent asynchronous processes. We assume the processes have interacting effects on their shared environment and that this interaction gives rise to the need for supervision of their collective behavior. The objective of supervision is to ensure that the processes interact to achieve harmonious coexistence or to carry out some collective task.

5.1 The Sequential Supervisor

In the remainder of the paper we let Ⓗ $= \{P_i \mid i\in\underline{n}^+\}$ be a DES with Σ_i the output alphabet, $\Gamma_i = \Gamma_{ei} \cup \Gamma_{di}$ the input alphabet and $\Lambda_i \subset \Sigma_i$ the set of controlled outputs of P_i, $i\in\underline{n}^+$. We let $\Sigma = \bigcup_{i\in\underline{n}^+} \Sigma_i$, $\Gamma = \bigcup_{i\in\underline{n}^+} \Gamma_i$ and $\Lambda = \bigcup_{i\in\underline{n}^+} \Lambda_i$.

A sequential supervisor S for Ⓗ is a trim, finite state, deterministic GSM with input set Σ and output set Γ^*. We shall assume that all states of S are allowed halting states. Then $\mathrm{Pre}(|S|) = |S|$.

Figure 5.1 depicts the central supervision of the DES Ⓗ by a sequential supervisor.

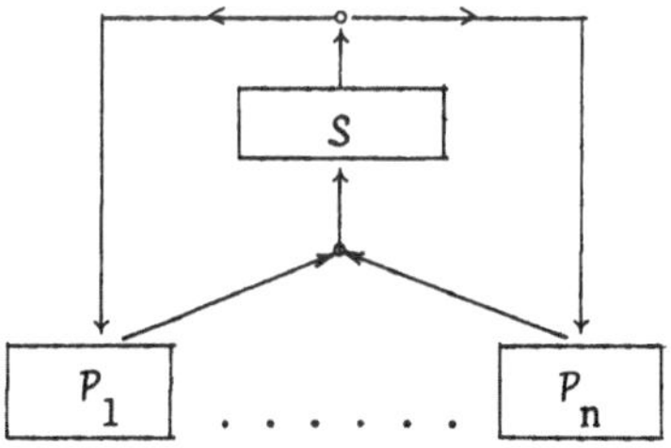

Fig. 5.1

Since S is sequential, its "observation" of the activity of the processes is modelled as a shuffling of the incoming communications into a single sequence. The output sequence of S is a shuffling of the command strings for the processes of Ⓗ. The controls are separated by a shuffle projection and transmitted to the corresponding processes. In this paper we assume that the delay in communication between Ⓗ and S is negligible. This assumption is not essential; however, space limitations preclude a treatment of the more general situation here. Many of the algebraic results which we present are also applicable to the situation when there is a delay in the communication between Ⓗ and S.

Let S be a supervisor for the DES Ⓗ. For each $\sigma\in\Lambda$ let $\langle\sigma\rangle$ denote the enabling event and $\langle\bar{\sigma}\rangle$ the disabling event for σ. Let $\Gamma_\sigma = \{\langle\sigma\rangle, \langle\bar{\sigma}\rangle\}$ and $p_\sigma:\Gamma^* \to \Gamma_\sigma^*$ be the ε-projection of Γ^* onto Γ_σ^*. Let $\alpha:\Lambda \to \{0,1\}$ be the map which specifies the initial status of each controlled output. We extend α to a map $\phi:|S|\times\Lambda \to \{0,1\}$ as follows: 1. $\phi(\varepsilon,\sigma) := \alpha(\sigma)$, 2. for $u\in|S|$, $\phi(u,\sigma) = 1$ if $\alpha(\sigma) = 1$ and $p_\sigma\cdot\tau(u) = \varepsilon$ or if $\ell\cdot p_\sigma\cdot\tau(u) = \langle\sigma\rangle$, otherwise $\phi(u,\sigma) = 0$. If $\phi(u,\sigma) = 1$ (0), then we say σ is <u>enabled (disabled) after</u> u.

The set $L_p \subset \Sigma^*$ of <u>controlled output strings</u> of $(\textcircled{H},S)$ is defined by: $w\in L_p$ iff $w\in|S| \cap \mathrm{Pre}(L_{\textcircled{H}})$ and for each factorization $u\sigma v$ of w, with $\sigma\in\Lambda$, $\phi(u,\sigma) = 1$. We define the <u>controlled behavior</u> of $(\textcircled{H},S)$ as the set $L_c := L_p \cap L_{\textcircled{H}}$. Clearly $\mathrm{Pre}(L_p) = L_p$ and $\mathrm{Pre}(L_c) \subseteq L_p$. If $w\in L_p - \mathrm{Pre}(L_c)$ then w is a possible output string of $(\textcircled{H},S)$ which cannot legally be completed to form a word in $L_{\textcircled{H}}$.

<u>Proposition 5.1</u>

The controlled behavior of $(\textcircled{H},S)$ is a regular language.

□

A supervisor S is <u>functionally complete</u> if for each $u\in L_p$, if $u\sigma \in \mathrm{Pre}(L_{\textcircled{H}})$ with $\sigma\in\Sigma$, and σ is enabled after u, then $u\sigma\in|S|$. We abbreviate functionally complete to "f-complete". A supervisor S is <u>functionally trim</u> if for each state $x\in X$ of S and each $\sigma\in\Sigma$ for which $(x)\sigma$ is defined, there exists a word $u\sigma v \in L_c$ such that $(x_0)u = x$. We abbreviate functionally trim to "f-trim". If S is f-trim, then S has no redundant states or state transitions which play no role in the supervision of $\textcircled{H}$. We assume henceforth that all supervisors for $\textcircled{H}$ are f-complete and f-trim.

Let X be the state set of S. For each $\sigma\in\Lambda$ we say a state $x\in X$ is σ-<u>consistent</u> if for each pair $u,v \in |S|$ with $x = (x_0)u = (x_0)v$, $\phi(u,\sigma) = \phi(v,\sigma)$. We say S is <u>control-consistent</u> if every state of S is σ-consistent for each $\sigma\in\Lambda$. The state of a control-consistent supervisor uniquely determines the status of each controlled output.

Let S be a control-consistent supervisor. The status of σ at x is given by the surjection $\pi_\sigma:X \to \{0,1\}$ with $\pi_\sigma(x) = 1$ iff $\phi(u,\sigma) = 1$ for some (and therefore every) $u\in|S|$ with $x = (x_0)u$. We also let π_σ denote the equivalence kernel of this surjection and define the <u>control partition</u> of X by $\pi = \wedge\{\pi_\sigma | \sigma\in\Lambda\}$.

Two supervisors S and R for $\textcircled{H}$ are said to be <u>control equivalent</u> if they both result in the same set of controlled output strings for $\textcircled{H}$.

<u>Lemma 5.1</u>

For each supervisor S of $\textcircled{H}$ there exists a supervisor S^* and an epimorphism $f:S^* \to S$ in <u>Gsm</u> with S^* a control-consistent supervisor for $\textcircled{H}$ which is control equivalent to S.

Proof. Let $S = (\Sigma,X,\Gamma;\ x_0,T,\delta,\tau)$. If S is control-consistent we may take $S^* = S$. Otherwise we introduce a dynamic extension of S, as follows. Let $\Lambda' \subset \Lambda$ be the set of controlled outputs for which there exist states of S which are not consistent. Let $\sigma\in\Lambda'$ and $X_\sigma = \{0,1\}$. Define a SM $X_\sigma = (\Gamma,X_\sigma;\ x_0,\delta_\sigma)$ by: $x_0 = \phi(\sigma)$ and $\delta_\sigma(\gamma,x) = 1(0)$ if $\gamma = \langle\sigma\rangle(\langle\bar{\sigma}\rangle)$, otherwise $\delta_\sigma(\gamma,x) = x$. Thus X_σ is a SM which records the status of σ. Let m be the cardinality of Λ' and let $Y = \{0,1\}^m$. Define the SM $Y = (\Gamma,Y;\ y_0,\alpha)$ by $y_0 = (x_0^1,\ldots,x_0^m)$ and $\alpha(\gamma,(x^1,\ldots,x^m)) = ((x^1)\gamma,\ldots,(x^m)\gamma)$. Then the dynamic extension of S is the GSM $S_e = (\Sigma,X\times Y,\Gamma;\ (x_0,y_0),\delta_e,\tau_e)$ with $\delta_e(\sigma,x,y) := (\delta(\sigma,x),\ \alpha^*(\tau(\sigma,x),y))$ if $\delta(\sigma,x)$ is defined and σ is enabled in the state y of Y; otherwise $\delta_e(\sigma,x,y)$ is undefined. $\tau_e(\sigma,x,y) := \tau(\sigma,x)$. Clearly S_e is control equivalent to S, every state of S_e is σ-consistent for each $\sigma\in\Lambda$, and S is a quotient of S_e in Gsm. Let $S^* = S_e$.

□

Lemma 5.2

If S is a control-consistent supervisor for Ⓗ, then $L_p = |S| \cap \mathrm{Pre}(L_Ⓗ)$ and $L_c = |S| \cap L_Ⓗ$.

Proof. Let $w\in|S|\cap L_Ⓗ$ and $u\sigma v$ be a factorization of w with $\sigma\in\Lambda$. Since $w\in|S|$, then σ is an allowed input in S at $(x_0)u$, and since S is f-trim, σ must be enabled at $(x_0)u$. Thus $w\in L_c$. The result for L_p follows similarly.

□

5.2 Event Disabling and Deadlock

We impose the restriction that if $\sigma\in\Lambda_i$ is an enabled output of the process $P_i\in$ Ⓗ, then a supervisor cannot disable σ when P_i is already in a state from which σ can occur. Formally we say that a supervisor satisfies the disabling restriction if for each $\sigma\in\Sigma$ and each prefix $u\omega\sigma$ in $L_Ⓗ$ with $\omega\in\Sigma$, if $u\omega$ is a prefix in L_c and σ is disabled after $u\omega$ then σ is disabled after u. If a supervisor S satisfies the disabling restriction then we say that S is a restricted supervisor.

In the case when the communication channels of Fig. 5.1 have finite delay, the disabling restriction is a natural consequence of the delay in communication between S and Ⓗ. In the limiting case when this delay is zero, the restriction can be viewed as a "well-posedness" condition on the disabling of events.

Let S be a control-consistent supervisor for Ⓗ. For each $\sigma\in\Sigma$, define subsets D_σ, E_σ, $\bar{E}_\sigma$ and $\bar{D}_\sigma$ of the state set X of S as follows:

1. $x\in D_\sigma$ iff $(x)\sigma$ is defined.
2. $x\in E_\sigma(\bar{E}_\sigma)$ iff σ is enabled (disabled) at x.
3. $x\in\bar{D}_\sigma$ iff σ is disabled at x and for some prefix u in L_c with $x = (x_0)u$, $u\sigma$ is a prefix in $L_Ⓗ$.

$\bar{D}_\sigma$ is the subset of states of S where σ is disabled and Ⓗ may be in a state from which σ, if enabled, could occur.

Let $S\subset X$. The set $(S)\Sigma^{-1}\subset X$ is defined as follows: $x\in(S)\Sigma^{-1}$ iff there exists

$\sigma\epsilon\Sigma$ such that $(x)\sigma$ is defined and $(x)\sigma\epsilon S$.

Proposition 5.2

For a control-consistent supervisor S the following conditions are equivalent:

1. S is a restricted supervisor.
2. $(\bar{D}_\sigma)\Sigma^{-1} \subset \bar{E}_\sigma$ for all $\sigma\epsilon\Sigma$.
3. $(E_\sigma)\Sigma \cap \bar{D}_\sigma = \emptyset$ for all $\sigma\epsilon\Sigma$.

Proof. (1 → 2) Suppose $\bar{D}_\sigma = \emptyset$. Then $(\bar{D}_\sigma)\Sigma^{-1} = \emptyset \subset \bar{E}_\sigma$. Suppose $\bar{D}_\sigma \neq \emptyset$ and $(\bar{D}_\sigma)\Sigma^{-1} \not\subset \bar{E}_\sigma$. Then there exist $x\epsilon\bar{D}_\sigma$, $y\epsilon(\bar{D}_\sigma)\Sigma^{-1}$, $\omega\epsilon\Sigma$ and a prefix u in L_c such that $(y)\omega = x$, $y = (x_0)u$ and $y\epsilon E_\sigma$. So $u\omega\sigma$ is a prefix in $L_{Ⓗ}$, $u\omega$ is a prefix in L_c, σ is disabled after $u\omega$ and σ is enabled after u. This contradicts the assumption that S is a restricted supervisor.

(2 → 3) $(\bar{D}_\sigma)\Sigma^{-1} \subset \bar{E}_\sigma$ so $(\bar{D}_\sigma)\Sigma^{-1} \cap E_\sigma = \emptyset$. If $x\epsilon(E_\sigma)\Sigma \cap \bar{D}_\sigma$ then there exist $y\epsilon E_\sigma$ and $\omega\epsilon\Sigma$ such that $y\epsilon E_\sigma$ and $(y)\omega\epsilon\bar{D}_\sigma$. Then $y\epsilon E_\sigma \cap (\bar{D}_\sigma)\Sigma^{-1}$. This is a contradiction. Hence $(E_\sigma)\Sigma \cap \bar{D}_\sigma = \emptyset$ for each $\sigma\epsilon\Sigma$.

(3 → 1) Suppose $(E_\sigma)\Sigma \cap \bar{D}_\sigma = \emptyset$ for each $\sigma\epsilon\Sigma$. Let $u\omega\sigma$ be a prefix in $L_{Ⓗ}$, $u\omega$ be a prefix in L_c and let σ be disabled after $u\omega$. Let $x = (x_0)u\omega$ and $y = (x_0)u$. Then $x\epsilon\bar{D}_\sigma$ and hence $y\not\epsilon E_\sigma$. Thus σ is disabled after u and S is a restricted supervisor.

□

A regular language $L \subset L_{Ⓗ}$ is said to be partially invariant if for each $\sigma,\omega\epsilon\Sigma$, if $u\sigma$ and $u\omega$ are prefixes in L and $u\omega\sigma$ is a prefix in $L_{Ⓗ}$ then $u\omega\sigma$ is a prefix in L.

Let A be a deterministic acceptor for L and let Q be the state set of A. For each $\sigma\epsilon\Sigma$, define subsets A_σ, $\bar{\bar{A}}_\sigma$ and $\bar{A}_\sigma$ of Q as follows:

1. $q\epsilon A_\sigma$ iff $(q)\sigma$ is defined.
2. $q\epsilon\bar{\bar{A}}_\sigma$ iff $(q)\sigma$ is undefined and for each prefix u in L with $q = (q_0)u$, $u\sigma$ is not a prefix in $L_{Ⓗ}$.
3. $q\epsilon\bar{A}_\sigma$ iff $(q)\sigma$ is undefined and for some prefix u in L with $q = (q_0)u$, $u\sigma$ is a prefix in $L_{Ⓗ}$.

Clearly A_σ, $\bar{\bar{A}}_\sigma$ and $\bar{A}_\sigma$ are pairwise disjoint and $A_\sigma \cup \bar{\bar{A}}_\sigma \cup \bar{A}_\sigma = Q$.

Proposition 5.3

$L \subset L_{Ⓗ}$ is partially invariant iff for each $\sigma\epsilon\Sigma$, $(\bar{A}_\sigma)\Sigma^{-1} \subseteq \bar{A}_\sigma \cup \bar{\bar{A}}_\sigma$.

□

Proposition 5.4

$L \subset L_{Ⓗ}$ is partially invariant iff for each $\Sigma_i, i\epsilon\underline{n}^+$, and for each $\sigma\epsilon\Sigma_i$:

1. $(A_\sigma)(\Sigma-\Sigma_i) \subset A_\sigma$ i.e. if $q\epsilon A_\sigma$ then for each $\omega\epsilon\Sigma-\Sigma_i$, $(q)\omega\epsilon A_\sigma$ whenever $(q)\omega$ is defined; and
2. $(A_\sigma)\Sigma_i \subset A_\sigma \cup \bar{\bar{A}}_\sigma$ i.e. if $q\epsilon A_\sigma$ then for each $\sigma'\epsilon\Sigma_i$, $(q)\sigma'\epsilon A_\sigma \cup \bar{\bar{A}}_\sigma$ whenever $(q)\sigma'$ is defined.

Proof. (If) Let $\sigma\in\Sigma_i$ and $\omega\in\Sigma$. Suppose $u\sigma$ and $u\omega$ are prefixes in L and $u\omega\sigma$ is a prefix in $L_{\textcircled{H}}$. If $\omega\notin\Sigma_i$ then $q = (q_0)u\omega \in A_\sigma$. Hence $u\omega\sigma$ is a prefix in L. If $\omega\in\Sigma_i$ then $q\in A_\sigma \cup \bar{\bar{A}}_\sigma$. But since $u\omega\sigma$ is a prefix in $L_{\textcircled{H}}$, $q\notin\bar{\bar{A}}_\sigma$. Hence $q\in A_\sigma$ and $u\omega\sigma$ is a prefix in L. Thus L is partially invariant.
(Only if) Suppose L is partially invariant. Let $\sigma\in\Sigma_i$ and $q = (q_0)u\in A_\sigma$ for some prefix u in L. Suppose $(q)\omega$ is defined with $\omega\notin\Sigma_i$. Then $u\sigma$ and $u\omega$ are prefixes in L and by definition of $L_{\textcircled{H}}$, $u\omega\sigma$ is a prefix in $L_{\textcircled{H}}$. Hence $u\omega\sigma$ is a prefix in L. Thus $(q)\omega\in A_\sigma$ as required. Suppose $(q)\omega$ is defined with $\omega\in\Sigma_i$. Then $u\sigma$ and $u\omega$ are prefixes in L. If $u\omega\sigma$ is a prefix in $L_{\textcircled{H}}$ then $(q)\omega\in A_\sigma$, otherwise $(q)\omega\in\bar{\bar{A}}_\sigma$. Hence $(q)\omega\in A_\sigma \cup \bar{\bar{A}}_\sigma$ as required.

□

We have the following restriction on the controlled behavior of $\textcircled{H}$.

Proposition 5.5

If S is an f-complete restricted supervisor for $\textcircled{H}$ then the behavior of $(\textcircled{H},S)$ is partially invariant.

Proof. Assume S is control consistent. Then by Lemma 5.2 $L_c = L_{\textcircled{H}} \cap |S|$. Let $u\sigma$ and $u\omega$ be prefixes in L_c with $\sigma\in\Sigma$ and $\omega\notin\Sigma_i$. Let $u_k := \rho_k(u)$ be the th component of u. Then $u_i\sigma$ is a prefix in $|P_i|$ and if $\omega\in\Sigma_j$ then $u_j\omega$ is a prefix in $|P_j|$. Thus, by the definition of $L_{\textcircled{H}}$, $u\omega\sigma$ is a prefix in $L_{\textcircled{H}}$. Let $x = (x_0)u$ be the state of S after u. Since $u\sigma$ and $u\omega$ are prefixes in L_c then $(x)\sigma$ and $(x)\omega$ are both defined and $x\in E_\sigma$. Since S is a restricted supervisor $(x)\omega\notin\bar{D}_\sigma$; hence $(x)\omega\in E_\sigma$. Further, since S is f-complete we must have $(x)\omega\in D_\sigma$. Thus $u\omega\sigma$ is a prefix in $|S| \cap L_{\textcircled{H}} = L_c$.

Let $\sigma,\sigma'\in\Sigma_i$. Suppose $u\sigma$ and $u\sigma'$ are prefixes in L_c and $u\sigma'\sigma$ is a prefix in $L_{\textcircled{H}}$. Again let $x = (x_0)u$. Then by the same argument as above we conclude $(x)\sigma'\in D_\sigma$ and hence $u\sigma'\sigma$ is a prefix in L_c. The result now follows by Lemma 5.1.

□

Let S be a supervisor for $\textcircled{H}$ and let the controlled behavior of $(\textcircled{H},S)$ be L_c. Heuristically we say that S "permits deadlock" of $\textcircled{H}$ if there exists a string u in L_p after which all processes which have not halted are blocked and cannot proceed.

Let $\Sigma_i^u \subseteq \Sigma_i$ be the subset of outputs of process P_i defined by : $\sigma\in\Sigma_i^u$ iff $u\sigma$ is a prefix in $L_{\textcircled{H}}$. A language $L \subset L_{\textcircled{H}}$ is <u>partially blocking</u> if for each $u\in\mathrm{Pre}(L)$ with $\Sigma_j^u \neq \emptyset$ for some $j\in\underline{n}^+$, there exists $i\in\underline{n}^+$ with $\Sigma_i^u \neq \emptyset$ such that:

1. $u\sigma \in \mathrm{Pre}(L)$ for each $\sigma\in\Sigma_i^u$

and 2. if u is a proper prefix in L then $u_i = p_i(u)$ is a proper prefix in $|P_i|$.

Formally we say that $(\textcircled{H},S)$ is <u>deadlock free</u> if $\mathrm{Pre}(L_c) = L_p$ and L_c is partially blocking.

Proposition 5.6

If $L \subset L_{\textcircled{H}}$ is partially blocking then $\mathrm{Pre}(L) \cap L_{\textcircled{H}} = L$.

□

5.3 Coordination

Let $\textcircled{H}$ be a DES with behavior $L_{\textcircled{H}} \subset \Sigma^*$. A coordination task for $\textcircled{H}$ is specified as an admissible output behavior $L_a \subset L_{\textcircled{H}}$. We shall not discuss how such an admissible behavior is determined, this being a separate problem which will depend on the specific form of the coordination task.

Let G_i be a grammar for $|P_i|$, $i \in \underline{n}^+$. Then $G_{\textcircled{H}} := \otimes_{i \in \underline{n}^+} G_i$ is a grammar for $L_{\textcircled{H}}$.

A regular language $L \subset L_{\textcircled{H}}$ is said to be well-posed if it is both partially invariant and partially blocking. Let F_G be the family of well-posed regular subsets of $L_{\textcircled{H}}$. Unfortunately F_G is not closed under the operation of set union.

5.4 Central Supervisor Problem

Let $\textcircled{H}$ be a DES and let $L_a \subset L_{\textcircled{H}}$ be an admissible behavior of $\textcircled{H}$.

Central Supervisor Problem (CSP):

Synthesize (if possible) an f-complete, f-trim, restricted supervisor for $\textcircled{H}$ such that $L_c \subset L_a$ and $(\textcircled{H},S)$ is deadlock free.

Let $L \subset L_{\textcircled{H}}$ be a partially invariant regular language and let A be an acceptor for Pre(L). For each $\sigma\in\Sigma$, let A_σ, $\bar{\bar{A}}_\sigma$ and $\bar{A}_\sigma$ be the subsets of the state set Q of A which were defined in the previous section. Clearly A_σ is the set of states at which σ must have the enabled status, $\bar{\bar{A}}_\sigma$ is the set of states from which σ can never occur and $\bar{A}_\sigma$ is the set of states at which σ must have the disabled status. We say σ is a blocked output of L if $\bar{A}_\sigma \neq \emptyset$. Let Δ_L be the set of blocked outputs of L.

For each $\sigma\in\Sigma$, define a partition π_σ of Q to be σ-amenable if π_σ has two cells C_σ and $\bar{C}_\sigma$ with $A_\sigma \subseteq C_\sigma$ and $\bar{A}_\sigma \cup (\bar{A}_\sigma)\Sigma^{-1} \subseteq \bar{C}_\sigma$. Let π^+_σ be the partition of Q with $C_\sigma = A_\sigma$ and $\bar{C}_\sigma = \bar{A}_\sigma \cup \bar{\bar{A}}_\sigma$. By Proposition 5.3, π^+_σ is a σ-amenable partition of Q. Thus there exists at least one σ-amenable partition for A. The family of σ-amenable partitions of Q is closed under neither the join nor meet operation of the lattice of partitions of Q.

We say a partition π of Q is amenable if π is the meet of a family of partitions $\{\pi_\sigma | \sigma\in\Sigma\}$, where for each $\sigma\in\Sigma$, π_σ is a σ-amenable partition of Q. There exists at least one amenable partition for A since $\pi^+ = \wedge\{\pi^+_\sigma | \sigma\in\Sigma\}$ is always amenable.

Theorem 5.1

CSP is solvable iff there exists an $L \in F_G$ such that $L \subset L_a$ and $\Delta_L \subset \Lambda$.

Proof. (If) Let A be a deterministic acceptor for Pre(L). Let π be an amenable partition for A. Since $\Delta_L \subset \Lambda$ then π is a control partition for A with $E_\sigma = C_\sigma$ and $\bar{D}_\sigma = \bar{A}_\sigma$,

$\sigma\epsilon\Sigma$. Then $(E_\sigma)\Sigma \cap \bar{D}_\sigma = (C_\sigma)\Sigma \cap \bar{A}_\sigma = \emptyset$ since $(\bar{A}_\sigma)\Sigma^{-1} \subset \bar{C}_\sigma$. Thus A is a control-consistent restricted supervisor with $L_c = |A| \cap L_{Ⓗ} = L$. If q is a state of A then there exists $u\epsilon Pre(L_c)$ with $q = (q_0)u$, and if $(q)\sigma$ is defined then $u\sigma \epsilon Pre(L_c)$. Hence A is f-trim. Clearly A is f-complete. Since $|A| \subset Pre(L_{Ⓗ})$ we have $Pre(L_c) = L_p$. Then since L is partially blocking, (Ⓗ,S) is deadlock free.

(Only if) Let $L = L_c$. Then $\Delta_L \subset \Lambda$ otherwise S would not be f-complete. By Proposition 5.5 and the definition of deadlock, L_c is partially invariant and partially blocking.

□

Let Sup$_{Ⓗ}$ be the category whose objects are pairs (S,β) with S a supervisor for Ⓗ and $\beta:\Lambda \to \{0,1\}$ a map specifying the initial status of each controlled output of Ⓗ. There is a morphism $f = (f_1,f_2,f_3):(A,\alpha) \to (S,\beta)$ in Sup$_{Ⓗ}$ provided $\alpha = \beta$, and (f_1,f_2,f_3) is a GSM morphism from A to S with $f_1 = id:\Sigma \to \Sigma$ and $f_3 = id:\Gamma \to \Gamma$.

If S is a control-consistent supervisor then we can replace the output map of S by its control partition. A control-consistent supervisor for Ⓗ is then a pair (S,π) with S a SM and π the control partition of S. The natural projection of π is the map $\pi:Q \to 2^\Lambda$ which gives for each state $q\epsilon Q$ of S the status of each controlled output of Ⓗ. A morphism of control-consistent supervisors from (A,λ) to (S,π) is a SM morphism $(f_1,f_2):A \to S$ with $f_1 = id:\Sigma \to \Sigma$ and $\lambda = \pi f_2$. Let Sup$^c_{Ⓗ}$ be the category of control-consistent supervisors and their morphisms.

We now present our two principal results.

Theorem 5.2

An f-trim control-consistent supervisor (S,π) solves CSP with the behavior of ((Ⓗ),S) = L iff there exists an acceptor A for Pre(L) with amenable partition λ, such that (A,λ) solves CSP with $L_c = L$, and there exists an epimorphism $f:(A,\lambda) \to (S,\pi)$ in Sup$^c_{Ⓗ}$.

Proof. (If) Let $L_p(A)(L_p(S))$ be the set of controlled output strings of ((Ⓗ),A) (((Ⓗ),S)). Let $w\epsilon L_p(A) = |A| \cap Pre(L_{Ⓗ})$. Then $w\epsilon |S| \cap Pre(L_{Ⓗ})$. Thus $L_p(A) \subset L_p(S)$. Let $\sigma\epsilon\Sigma$ be a string in $L_p(S)$. Then $\sigma\epsilon Pre(L_{Ⓗ})$ and σ is initially enabled in S. Hence σ is initially enabled in A and since A is f-complete, $\sigma\epsilon L_p(A)$. Assume each string in $L_p(S)$ of length k is an element of $L_p(A)$. Let $w = u\sigma\epsilon L_p(S)$ have length k+1 with $\sigma\epsilon\Sigma$. Let $x = (x_0)u$ $(q = (q_0)u)$ be the state of $S(A)$ after u. If $f_2:Q \to X$ is the state map of f, then $f_2(q) = x$. Hence σ is enabled at q and since A is f-complete, $u\sigma\epsilon L_p(A)$. Thus $L_p(A) = L_p(S)$. The other required properties follow similarly.

(Only if) Since S is control-consistent, $L_c = |S| \cap L_{Ⓗ}$. Hence $G_c = G_{Ⓗ} \cap S$ is a grammar for L_c. By the pullback construction of G_c, and since S is f-trim, there exists an epimorphism $h:G_c \to S$ in Aut with $f_1 = id:\Sigma \to \Sigma$. Let A be the supervisor for Ⓗ obtained from G_c by letting all states be halt states, and let $f:A \to S$ be the epimorphism corresponding to h. Define a partition λ on the state set Q of A by: $q \equiv q'(\lambda)$ iff $h_2(q) \equiv h_2(q')$ (π), where $h_2: Q \to X$ is the state map of h. If λ is not amenable then for some $q, q' \in Q$ and $\sigma \in \Sigma : q \in A_\sigma$, $q' \in \bar{A}_\sigma \cup (\bar{A}_\sigma)\Sigma^{-1}$ and $q \equiv q'(\lambda)$. Then $h_2(q) \in D_\sigma$, $h_2(q') \in \bar{D}_\sigma \cup (\bar{D}_\sigma)\Sigma^{-1}$ and $h_2(q') \in \Sigma_\sigma$. This contradicts the assumption that S solves CSP. Hence λ is amenable. By construction $\lambda = \pi h_2$. Thus $f:(A,\lambda) \to (S,\pi)$ is an epimorphism in $\underline{\mathrm{Sup}}^{c}_{Ⓗ}$. By the proof of Theorem 5.1 (A,λ) solves CSP.

□

Theorem 5.3 (Quotient Structure Theorem)

An f-trim supervisor (S,α) solves CSP with the behavior of $(Ⓗ,S) = L$ iff there exists an acceptor A for Pre(L) and an output map $\tau : \Sigma \times \phi \to \Sigma^*$ for A such that $(\mathcal{B} = (A,\tau),\alpha)$ solves CSP with $L_c = L$, and there exists an epimorphism $f:(\mathcal{B},\alpha) \to (S,\alpha)$ in $\underline{\mathrm{Sup}}_{Ⓗ}$.

6.0 CONCLUSIONS

The theorems of Section 5 show that the supervisor for a DES Ⓗ is essentially a quotient of a grammar for the resulting controlled behavior. The quotient structure is admissible since the supervisor does not need to precisely "track" Ⓗ but instead must only determine critical states where control action is required.

Further research is in progress to investigate more specific coordination tasks and the resultant algebraic structure of the supervisor.

REFERENCES

Eilenberg, S. (1974). Automata, Languages and Machines. Academic Press, New York.

MacLane, S. (1971). Categories for the Working Mathematician. Springer-Verlag, New York.

AUTOMATION AND SOCIETY

H.H. Rosenbrock

Control Systems Centre
The University of Manchester
Institute of Science and Technology

1. History

In science and technology, the history of a subject is usually considered as a separate study. The present state of say chemistry, or linear system theory, can be defined without reference to the way in which it came into being. Present knowledge incorporates all that was valid in past knowledge, and supersedes it.

In addressing an audience of system theorists, it may therefore be necessary to justify an excursion into past history. Some may be impatient of this, and suggest that it does not matter too much how we came to be where we are. What is important, is where we can go in the future.

In reply, I should like to suggest an analogy with delay-differential systems. These have as their initial condition, not just the present values of the variables, but their history over some previous interval. In a similar way, the initial condition for the future development of technology is not, I suggest, just our present condition. It is rather an interval of past history which stretches back at least a hundred years, and probably much more.

If you should press me upon the analogy, and ask what can possibly account for a delay of a hundred years or more, then I should answer: the formation of public opinion. This is not simply based upon past history; to a large extent it actually is that history. Urban conditions in the United States, for example, are not very different from those in most parts of Europe, yet the attitudes to gun control laws in the USA are often quite different from those in Europe. The American attitude is not just based upon a different historical experience. It is, in itself, often a composite historical picture: incorporating the hunter in a primitive land, the right to bear arms, the War of Independence, the breakdown of law and order in the development of the West; and the personal virtues that were appropriate in all of that development.

Northern Ireland illustrates the same point still more forcibly, and shows how one historical record, by a process of selection, can serve as the basis of two different and opposed views. Yet another illustration is the present state of industrial relations in Britain, which can only be understood in the light of two hundred years of history. Other examples will no doubt occur to each of you: as will the thought that such examples take on a quite different aspect when they are seen indifferently from outside, and when they are experienced from within.

The analogy, then, has perhaps served its purpose. Our starting-point is not a point in time. It is a long historical record which has been transmuted, perhaps in a refracted and selected form, into the views and opinions against which the present and the future are evaluated. And if it has served its purpose, we can admit that the analogy must not be pressed too far. The historical record does not change, but our interpretation of it can alter with time, so that we are not quite so much the prisoners of our past as the comparison might suggest.

The particular aspect of history which I wish to discuss is the development of automation since the beginning of the industrial revolution. It is usual to distinguish automation from mechanisation: for example mechanisation was the process of replacing human muscle power by mechanical power, while automation is the replacement of human mental activities by machines or instruments or computers. Alternatively, automation relies upon feedback, while mechanisation is open-loop. The distinction is not an easy one to maintain, and both developments are closely interwoven with a third: the division of labour.

It seems, indeed, to be better to think of one process with three aspects, any one of which may be more prominent in one case than in another, but all of which are usually present in some degree. For brevity, this single all-embracing process will be referred to as automation, thus making one aspect embrace the other two.

A full justification of this view would require an extensive development, but it is worth remarking that it is a return to an earlier opinion. The threefold separation, and the treatment of each part as existing independently, is relatively new. Babbage[1], in 1832, considers all three together: 'the possibility of performing arithmetical calculations by machinery... is connected with the subject of the <u>division of labour</u>', while the task of calculating numerical tables is similar to the operation of 'a cotton or silk-mill, or any similar establishment'.

The essential unity of the process of automation, with its three aspects, can be illustrated by the development of the 'self-acting mule' for cotton-spinning. In the mid-eighteenth century, spinning could not keep pace with weaving, so that weavers often suffered from a shortage of thread. The weaver James Hargreaves in 1764 overcame this difficulty as it affected himself by inventing the 'spinning jenny', by which eight or more threads could be spun simultaneously. Samuel Crompton, about 1779, invented the 'spinning mule', which operated on a different principle. Both of these were hand-operated, and both required a certain skill in operation. Both were intended for the inventor's own use. They are best regarded as highly-developed tools, extending the skill of the user and making it more productive, rather than examples of automation. Human muscle-power and human control were retained, and the spinner's task was not fragmented.

Automation came in 1830, when Richard Roberts invented the 'self-acting mule'. His motivation was different from that of Hargreaves or Crompton. He did not intend to operate the machine himself, and as described by Ure[2], his aim was to eliminate the spinner's skill: '...the only, or at any rate the principal benefit anticipated, was the saving of the high wages paid to the hand "spinner", and a release from the domination [through strikes] which he had for so long a period exercised over his employers and his fellow work-people...' A skilled spinner would be retained to oversee the operation of a group of self-acting mules, but the productive tasks which remained were fragmented and de-skilled: mending broken threads, cleaning the machine, removing spun thread, etc. These tasks were performed by 'young persons, or children'.

This result was achieved by what we should now call mechanisation. The spinner's skill consisted, among other things, in observing the way in which the thread wound itself onto the 'cop' and adjusting his actions to give the cop a suitable shape. This was closed-loop control, which the technology of the time could not duplicate. It was replaced by an open-loop system: the thread was led to the cop by a lever which followed a complicated and accurate path. Thus the greater absolute accuracy of the mechanical system was made to substitute for the corrective action of the spinner. 'The entire novelty and great ingenuity of which invention says Ure[3], 'was universally admitted, and proved the main step to the final accomplishment of that object which had so long been a desideratum'.

What is significant here is that open-loop control (which we should call mechanisation) replaced not only muscle-power but also human guidance and skill. The work that remained was fragmented, and required not skill but only dexterity, while its pace was largely controlled by that of the machine. This could no longer be considered a tool of the user, complementing his skill. Rather, the workers who remained were the servants of the machine. All of these were in fact the consequences desired by the inventor.

2. Tools and machines

A sharp contrast has been drawn above between the kind of machine invented by Hargreaves or Crompton, and that invented by Richard Roberts. The first one was an aid to the user, and a tool for his use. It accepted a previous level of skill, acquired with earlier machinery, and allowed it to develop to a higher and more productive level. As Ure[4] describes it, '...the skill and tact required in the operator deserve no little admiration... The spinner requires much skill and dexterity: first, to back off; secondly, to wind on the yarn without breaking; and thirdly, to give the cop such a shape as may facilitate the winding off, either in the shuttle, or upon the reel'.

The self-acting mule, on the other hand, was not intended to collaborate with the skill of the user, but to replace it. Its inventor's ideal would no doubt have been a machine which could operate with no human attention; but this he could not achieve. What he could achieve, and what he aimed at, was a situation in which the human aid required had as small an element of skill as possible. The jobs which remained were specialised fragments of the original skill of the spinner. Broken threads, for example, had to be mended, and this was made a separate job for the 'piecer', a job that was specially fitted for the nimble fingers of children.

These different types of machine, and the different aims of the designers, corresponded to their different motivations. Hargreaves and Crompton were seeking the benefit from an increased productivity of their own or their families' labour. They received something further as inventors, but not a great deal more. To patent their inventions, to defend them, and to enforce them against a multitude of small users or against a few large and powerful users, were not highly rewarding. Robinson, however, and the factory-owners who used his invention, were following in the steps of Arkwright, who was the first to set up a mechanical factory system for spinning, and who acquired from it a fortune and a knighthood. They sought, not the profit from an increase in their own productivity, but that from the increased productivity of the many workers they employed, and they wished this hired labour to be as cheap as possible. The point was not lost upon Ure[5]: 'What a warning voice does the fate of Hargreaves and Crompton send forth to inventors and improvers of the useful arts! how strongly does it justify the sound sense and self-respecting energy of Arkwright!'

3. Refinement and reaction

The subsequent history of these developments is one of refinement in their application, and reaction against their effects. The refinement is associated particularly with F.W. Taylor, the Gilbreths, and Henry Ford. The reaction can be seen in the development of Trade Unions, and of shop-floor working practices, and also in the development by social scientists of remedies for some of the worst kinds of fragmented work.

Taylor and the Gilbreths set out systematically to separate any mental component from manual work, so that the former could be done in a planning department. The physical tasks that remained were then to be studied to find the best way in which they could be done, the 'one best way'. The workers were to follow this way with no scope for initiative or control[6]: 'Under our system the workman is told minutely just what he is to do and how he is to do it; and any improvement which he makes upon the orders given to him is fatal to success'. Henry Ford[7] carried the fragmentation of work still further under the conditions of mass production: 'The man who puts in a bolt does not put on the nut; the man who puts on the nut does not tighten it'.

The determined effort which has been applied to these aims over many decades has led to results with which we are all familiar. They can be illustrated by a plant which in 1979 was producing electric light bulbs, with a metallised reflector, at the rate of 800 an hour. It was a small plant, and noisy, but physical conditions were otherwise not unpleasant. For the most part, its operation was automatic, with the lamp passing from one operation to the next on a chain conveyor.

Here and there, around the conveyor, were jobs which it was difficult or expensive to do automatically. For these, eight or ten middle-aged women were employed, whose work had to conform to the fixed rhythm of the plant. One was picking up each glass envelope as it arrived, inspecting it, and replacing it if it was sound. She was isolated from her companions by distance and noise, so that she could not talk with them, and all day long she lifted and replaced a light bulb every 4½ seconds. Other jobs on the plant were similar, except that some demanded a much greater degree of concentration.

The plant was in one of the socialist states, and similar working conditions can be seen in all industrialised countries, whatever the political system and whatever the type of firm. It is easy to recognise that such a plant makes excellent use of the physical machinery, but that it under-uses the human abilities of its workers to a gross and shameful degree. The under-use is damaging to the worker, because those abilities which are not used will in time decay.

It is also an economic loss: a point which we recognise much more readily if we substitute the under-use of a machine for the under-use of a man's or woman's abilities. Here, for example, is a technologist's comment on the scope for general-purpose robots in industry[8]: '...it is less obvious that robots will be needed to take the place of human beings in most everyday jobs in industry. To bring in a universal robot would mean using a machine with many abilities to do a single job that may require only one ability'. The reaction here against an economic waste by the under-use of a machine is sharp and almost instinctive. It is odd that we do not have this sharp and instinctive response to the under-use of human ability.

If the reaction to under-using human ability, regarded as an economic waste, has been muted, there has on the other hand been a growing recognition that it can lead to a resistance which undermines industrial relations. To be employed in work that uses the body as a machine, and would not justify the expense of a general-purpose robot, can induce a resentment that seeks to frustrate production. Walker and Guest[9] describe the car plant where 'The guys yell "hurrah" whenever the line breaks down ... you can hear it all over the plant'.

Within the social sciences, techniques have been developed to alleviate the worst of these consequences. There is job rotation, in which workers move from one fragmented job to another, in order to gain some variety and some relief from the stress of one or the boredom of another. Job enlargement puts together a number of fragmented tasks to make a more meaningful whole: say the assembly of a component. Job enrichment gives to the worker some elements of the supervisory function, say the responsibility for inspection. Autonomous groups of workers may be given responsibility for some moderately large section of the production process, and allowed to arrange among themselves how the work should be done. The Volvo Kalmar plant is perhaps the best-known example of this.

No detailed description of these techniques need be given, since accounts can readily be found elsewhere.[10] There is reasonably good evidence that they can lead to an improvement, and their intention is certainly benevolent. For both reasons they must be welcomed, even though it will be suggested later that there is a deeper problem which they fail to attack.

The same can be said of a remedy which often appeals to the technologist; that is, if a job is unsatisfactory for a man or woman, let it be done entirely by a machine. This in the recent past has commonly been a part of the justification for using robots. The direct economic gains have often been doubtful or marginal, but if the jobs which are eliminated are undesirable in themselves, this can be an added reason to proceed. Robots have in this way replaced dirty jobs such as grinding and polishing, dangerous jobs such as manipulating red-hot metal under a forging press, or laborious work done under high pressure of time such as the spot-welding of car bodies.

At the extreme, the whole justification for using a robot may lie in the unsatisfactory nature of the work. In a Japanese factory making large diesel engines, for example, the cylinder blocks in the past had been washed down after machining by a man using a pressure hose. He dressed from head to foot in protective clothing and used the hose to clean the cylinder block and to wash out any metal chips in bores and bolt holes. The system was recently changed, so that the hosing down was done by a robot. The man programmed the robot by leading it through all the required motions for the particular cylinder block, with no flow of liquid. Then he retired from the cubicle while the machine repeated his motions with the hose turned on.

Such developments are certainly welcome where the work replaced is entirely unsuited to human beings. Indeed, if a job can be done by robots as they exist at present, this is presumptive evidence that it makes no real use of human abilities. Present-day robots have no sense of sight or touch, and can only repeat a standard sequence of operations which they have once been taught. Yet, as with the remedies which have been developed in the social sciences, it will be suggested later that this technological remedy also fails to meet a more serious underlying difficulty.

4. New technology and the future

What has been said above is necessarily oversimplified and abbreviated because space does not permit a more extended account.[11] Nevertheless it will serve as a background against which we can consider the changes that the next twenty years are likely to bring. These, it is widely agreed, will be rapid and profound. They will be propelled by microelectronics, the microcomputer, and the associated advance of digital communications. These in turn will put in our hands new possibilities of automation. The beginnings of these changes are already evident in printing, in office automation, in robot technology and computer-aided design, and in a multitude of research and development projects which have not yet reached the stage of application.

The rate at which changes will take place should not be exaggerated. Studies of the 'diffusion of innovations' show that even the most profitable innovations will usually take 8 or 10 years to move from 10% to 90% of their final market penetration. The quartz crystal watch, for example, was introduced around 1970, and by 1979 its sales[12] amounted only to about 30% of world sales of wrist watches. Yet this appeared as a change of great rapidity and had a profound effect on the watch industry and on its distribution among the countries of the world.

Though great caution is necessary, it does seem that two predictions can be made with some confidence about the progress of new technology. The first is that it will influence a very wide range of activities, so that although each innovation may proceed on a time-scale of 10 years or more, the effect on society of a multitude of simultaneous innovations will be widespread and powerful. The second is that the effects will be much more profound on the way we make things than on the things that we make. There have been new products such as digital watches and pocket calculators and computer games, and no doubt there will be more. Communication and data services in the home will improve. Yet the range and scope of these developments seems likely to be small compared with the effects of new technology on industry and commerce.

In engineering production, large steps in automation will become possible, though their working out will not be rapid. About 30% of manufactured goods are mass-produced, and the rest are made in batches. The use of numerically-controlled machine tools and robots should permit the same kinds of gain in productivity for batch production as were obtained earlier in mass production. Design will be carried out through CAD systems, and these in time will be linked to CAM and to the computer-control of production.

Many kinds of clerical work will be affected by the development of the electronic office. Shops may be partly replaced by automated warehouses, with their catalogue on a Viewdata system and with automated ordering, fund transfer and stock control. Some activities which have been regarded as highly-skilled and professional may also be gradually affected through computer advisory services in law and medicine and banking and the like.

Predictions of these developments are likely to be wrong in detail, but the general trend seems moderately certain. The way in which work is carried out, in manufacture, in administration and banking and commerce, and in service industries, will most likely change out of all recognition during the next twenty years.

If so, we shall be presented with one of the rare opportunities to exert a strong influence on the kind of work which people will do, and the environment in which it will be done. Technology evolves through a multitude of decisions which are made by engineers and other technologists during its research and development and implementation. In taking these decisions, they embody in the technology the requirements and the beliefs of their society. When technology is changing rapidly, the frequency of these choices and their importance are both high, so that the opportunities for turning the development into a desired direction are correspondingly increased. When technological change is slow, on the other hand, the opportunities to intervene are much reduced, whatever the will to do so.

5. The direction of change

Granted that new technology will offer an opportunity, the question is whether we are able to take it, and how wide is the choice which is open to us. It is here, I suggest, that the influence of past history intervenes to block our view. We tend to see only one possibility, which is a continuation of the path we have followed so long.

If we follow this path, we shall take a new range of jobs, manual, clerical and professional, and we shall deal with them as we dealt in the past with so many jobs in manufacturing. That is, we shall fragment them, give some parts to robots or computers or other machines, and leave to the human being only those fragments which it is awkward or costly to do by machine. Then, perhaps, disliking the jobs that remain, we may apply the remedies of job enlargement or job enrichment or autonomous groups; or we may seek to press on to the ultimate goal where no human intervention is required.

At the best, if we follow this path, we shall face a long period during which a very large number of people (much larger than now) do fragmented, unskilled work over which they have no control, before we reach the stage where no human work remains to be done. And if you believe, as I do, that a world in which all work is done by machines is a phantasm which will never exist, then we could end by spreading and perpetuating the worst kind of fragmented work throughout the great majority of human activities.

Some objections will certainly be made to this conclusion. It will be said that new technology may fragment and de-skill some jobs, but it creates new jobs requiring high skill: maintenance, programming, systems analysis, and so on. This is true, though the number of jobs of this kind will be necessarily restricted if the investment in machines is to be justified. Moreover, the new jobs will themselves be susceptible to the same processes. Maintenance, for example, can be simplified by automatic diagnosis. More and more of the details of programming will be taken over by the computer. What will count is the intention, and if our intention, as in the past, is always to remove the need for skill wherever we find it, then there seems no reason why we should not succeed in doing so over an ever-wider range of activities.

It may also be said that the problem is a political one. It arose out of the particular self-interest of early industrialists, and can only be remedied by political change. Yet it has shown itself highly resistant to all such changes. The same outlook, and the same tendency to eliminate skill and control over work, can be seen in capitalist economies, whether authoritarian or democratic, in mixed economies, and in socialist economies and in firms whether privately-owned, or public corporations, or state-controlled or cooperatives. Whatever the political dimensions of the problem, it clearly has a technical dimension as well. If a solution is to be found it will require a direct attack on this technical aspect.

Finally, it may be suggested that the problem is an economic one: that technology has evolved in a competitive environment and is therefore more effective than any alternative. A number of answers are possible to this objection. The one which is most likely to appeal to system theorists is that the evolution of technology is a dynamical process. It is easy to construct models[13] in which the pursuit of a short-term optimum, enforced by market action, leads to a non-optimal technology. The pursuit of such a course is quite consistent with a belief that the final technology is optimal, because it is better than its predecessors, and we have no experience of a route which would lead to the still better technologies which are available. This, it will be understood, is a conceptual answer to a conceptual objection, and not an empirical verification of the suggested model.

6. Human skill and technology

The course which automation has followed so consistently - of eliminating skill, fragmenting jobs, machine pacing and the rest - requires more explanation. I suggest that this lies in a powerful complex of ideas to which we are still in thrall. The early replacement of machines like those of Hargreaves and Crompton, which cooperated with the skill of the user to make it more productive, by those such as Richard Roberts's which eliminated skill, clearly grew out of the commercial incentives acting on factory owners. That in itself, however, would not have provided a defensible long-term justification, particularly when public opinion and the power relationships in industry began to change.

At an early stage, a connection began to be drawn with science. Knowledge in the hands of the employer was regarded as scientific. Knowledge in the hands of the workman was unscientific. This was, in broad terms, a fair statement of the early situation. Industrialists such as Matthew Boulton and Josiah Wedgwood were involved in the scientific developments of their day, while craft knowledge was largely traditional and inexplicit. Ure[14] saw the work of Richard Roberts as the application of science to the purposes of the mill-owner: '...when capital enlists science in her service, the refractory hand of labour will always be taught docility'.

From these beginnings, the application of science to industry came to be associated with one particular line of development which has already been described. Jobs were fragmented, and made subservient to machines, while skill and control and initiative where possible were eliminated. Taylor called his system 'Scientific Management', and it has all these characteristics. It had the aim as he described it[15], of 'taking control of the machine shop out of the hands of the many workmen and placing it completely in the hands of management, thus superseding "rule of thumb" by scientific control'.

There are strong overtones here of a straightforward struggle for power, which were noted and resisted[16,17] in Taylor's own day. I suggest, however, that the force and the long-lasting influence of his ideas lies in something else. What Taylor did, following a line of development that was well-established before him, was to identify science in the service of industry with one special model. Only knowledge in the hands of management was 'scientific'. Knowledge held by the workmen was 'rule of thumb'. Progress must consist in the replacement of the latter by the former, and with knowledge should go control. Lacking control, and being 'told minutely just what he is to do and how he is to do it', the workman would have no opportunity to develop or make use of any skill.

There is indeed, though Taylor did not express it in this way, an antithesis set up between skill and science. Skill is typified by the handicraft, and belongs to the past. Science is the basis of modern industry, which is founded in theory and tested in practice. Skill is the limited embodiment of what was known; science is the unlimited potential for all knowledge that will develop in the future. Skill is finite, science is infinite, and the destiny of skill is to retreat before the advance of science.

If I question these ideas, as I wish to, I believe that I may raise a feeling of disquiet, because the ideas which I am describing are not something we view dispassionately from outside. They are still part of the framework of belief which we have inherited from the past and within which we operate and carry on our work. Any criticism is likely to be dismissed as a nostalgia for the past, and a proof that the critic has stepped aside from the mainstream of progress.

To close therefore, I wish to present an alternative view of the relation between skill and science. If it is valid, I suggest, it has important consequences for the development of technology.

7. The skilful use of science

The pattern for a different relation between skill and science as applied to industry already exists, though we have to go back before the industrial revolution to find it, in machines such as those of Hargreaves or Crompton. These did not fragment and destroy the skill of the spinner. On the contrary, they allowed his skill to develop in a new way in relation to themselves. The new skill, in relation to the new machine, was rewarded by a great increase in productivity.

In saying this, I am not advocating a return to the productive methods of the late eighteenth century. These are as much out of date as the wood and cast iron of the factory machinery by which they were replaced. What I am pointing to is a different tradition in the relationship between skill and science and technology. This alternative tradition was replaced by the industrial system as we know it, with its Tayloristic outlook, and therefore failed to develop. If it had developed, it would no more resemble Crompton's mule than a modern spinning machine resembles Arkwright's water-frame.

In this alternative tradition, skill was not regarded as finite, as simply a legacy of past technology. It was regarded as something which could grow and develop in collaboration with science. New and more productive machines would be developed, and these would demand new skills. Past skills would become obsolete, not by being fragmented and destroyed, but by developing into the new skills which were needed by the new machines: a process which need excite neither regret nor nostalgia.

This alternative view would have fundamental implications for our use of new technology. It is generally agreed that this technology will allow us, if we wish, to carry over the Tayloristic approach into a wide new range of jobs: in the office, in printing, in engineering design and draughting, and (in due course) into a range of managerial and professional tasks. Given the opportunity for such a development, we have never in the past resisted it, and there is a widespread foreboding about the use we shall make of new technology.

Yet it also seems clear that we have the opportunity, if we can take it, to adopt what I have called the alternative tradition. We need not develop CAD systems which refuse to use the special skills of the operator and the special properties of the human mind: we can instead develop systems which accept the skill of the user and collaborate with it to increase his productivity. We need not develop 'flexible manufacturing systems' which fragemnt and destroy the machinist's skill: we can allow that skill to develop into something new. The same can be said of office automation and (to the extent that it is not too late) to printing. Such professional areas as medical diagnosis[11] will offer us the same kind of alternative choice.

This I regard as the most important challenge facing engineers and technologists in the next twenty years. There are certainly difficulties in the way of developing a more human-centred technology - political, industrial, and managerial - and these are often put forward as reasons why such a course cannot succeed. None of these, however, seems nearly so important a barrier to me as our inability to break free from the Tayloristic tradition which we have inherited, and by which our thoughts are still moulded. If we could see this from the outside, as a system of beliefs handed down to us from a remote past and open to challenge by a quite different view of science and its relation to skill: then I believe that the other difficulties would be well on the way to being overcome.

8. References

1. Charles Babbage, The Economy of Machinery and Manufactures, first edition 1832; fourth edition 1835; reprinted 1963, pp. 196,195 (Kelly, New York)

2. Andrew Ure, The Cotton Manufacture of Great Britain, 1836; reprinted 1970, vol. 2, p. 199 (Johnson Reprint Corp.)

3. Andrew Ure, loc. cit., vol. 2, p. 197.

4. Andrew Ure, loc. cit., vol. 2, p. 154,164.

5. Andrew Ure, loc. cit., vol. 1, pp. 283-4.

6. Frederick Winslow Taylor, On the Art of Cutting Metals, Third Edition, revised, 1906, p. 55 (American Society of Mechanical Engineers)

7. Henry Ford, in collaboration with Samuel Crowther, My Life and Work, 1923, p. 83 (Heinemann)

8. F.H. George and J. D. Humphries (editors), The Robots are coming, 1974, p. 164 (NCC Publications)

9. Charles R. Walker and Robert H. Guest, The Man on the Assembly Line, 1952, p. 51 (Harvard Univ. Press)

10. Michael Argyle, The Social Psychology of Work, 1972; Pelican Edition, 1974 (Penguin Books)

11. Further discussion will be found in New Technology: Society, Employment and Skill, 1981 (Council for Science and Society)

12. Industrial Minerals, May 1979, p. 29.

13. Howard H. Rosenbrock, Human Resources and Technology, 1980, Paper given at Sixth World Congress of the International Economic Association, Mexico City.

14. Andrew Ure, The Philosophy of Manufactures, 1835, p. 368 (Charles Knight)

15. Frederick Winslow Taylor, loc. cit., p. 40.

16. Frederick Winslow Taylor, Testimony before the Special House Committee, 1912, in Scientific Management, 1947 (Harper and Row)

17. Robert Franklin Hoxie, Scientific Management and Labour, 1915 (D. Appleton Co.)

ABSTRACT REGULATION OF NONLINEAR SYSTEMS: STABILIZATION *

Eduardo D. Sontag
Department of Mathematics
Rutgers University
New Brunswick, NJ 08903, USA

1. INTRODUCTION

This paper addresses the following problem: given a control system S and an equilibrium state "0" of S, find natural internal ("state space") conditions on (S,0) for the existence of a controller which drives every state of S asymptotically to 0 while it applies inputs which themselves approach zero (internal stability). The controller is assumed to have access only to (partial) measurements of the state of S, and unknown and arbitrary (finite support) disturbances affect states and measurements.

For linear (time invariant, finite dimensional) systems, it is well known that such a controller exists if and only if S is stabilizable and detectable. The first property, to be called in this paper "asymptotic controllability", or just "asycontrollablity", means that each state can be (open-loop) driven asymptotically to the origin. The second, to be called, more precisely, "0-detectability", says that the subsystem defined by the "unobservable states", i.e. those states indistinguishable from 0, is asymptotically stable.

For general nonlinear systems, no such conditions have been given. In our view, part of the problem in the past has been the insistence in the synthesis of controllers having a "smooth", or even an algebraic, structure -for instance, bilinear controllers for bilinear systems. While such special controllers are of course to be desired if they exist, it may be impossible to derive a general theory under such artificial constraints. There are in fact many examples of simple control problems for which no "nice" synthesis is possible. In a recent paper (Sontag [1981]) we gave some results regarding conditions for the existence of constant-rate sampled piecewise linear controllers for a rather general class of nonlinear systems. Even when using such controllers, however, one does not have sufficient freedom to allow for natural characterizations. In the present paper, therefore, we have taken the most general approach that seems natural. Regulators will be just abstract control systems: a set with well defined transitions and output maps.

---

*This research was supported in part by Air Force Grant AFOSR-80-0196

A priori such an approach could lead to a mathematically trivial theory. In order for this not to happen, we shall restrict attention to the case were the original system has a certain amount of structure, at least topological, and progressing to analytic systems of differential equations and eventually to bilinear systems. For each such class the characterizations become progressively simpler, until in the last case one -rather surprisingly- recovers the full linear result. Although the proofs are rather abstract, the reader should realize that all constructions are in principle implementable numerically. In fact, and although such a question will not be studied here in detail, it seems to be true that one can develop the theory without leaving the category of "piecewise-C" systems, with C = say, analytic or continuous functions. The study of special structures for controllers associated to plants having themselves a given structure, can be seen as a subset of the general problem of finding abstract controllers; with the definitions and results given here one may pose questions like: "if an abstract controller exists, does there exist also one with the desired structure?". Such questions cannot be posed, much less answered, unless the abstract conditions have been first studied.

The organization of this paper is as follows. The next section states and discusses the precise definitions of the above concepts, including the notions of stability and regulability that we have chosen. The main results are also stated there. Section 3 includes sketches of the the proofs; details are omitted due to space limitations, and will be given in a forthcoming paper. The central theorem is that, for systems S whose state spaces admit a metric for which the system maps are continuous, regulability of (S,O) is equivalent to preregulability plus indy-detectability. The latter has a rather technical definition, but intuitively it boils down to the requirement that indistinguishability ("indy") classes can be estimated by appropiate "detectors" operating in closed loop. Preregulability means that S is O-detectable (identical definition to that in the linear case) and indy-asycontrollable (or, "indy classes are asycontrollable"), meaning basically that for each indy class there is some control sending all states in the class to the origin. The necessary part is straightforward. The intuitive idea of the sufficiency proof is easy to understand: alternate estimation of indy classes with appropiate control actions; technically the proof turns out to be rather delicate. One reason that the theorem is of interest is that, as it will turn out, indy-detectability is <u>automatic</u> for systems defined by analytic differential equations, so regulability will be in that case equivalent to just preregulability. In the bilinear (more generally, state-affine) case one can go even further and prove that preregulability is in fact equivalent to just asycontrollability and O-detectability.

2. DEFINITIONS AND STATEMENT OF RESULTS

We shall need a large number of definitions and notational conventions.

Systems and Signal Spaces.

A time-function will be any function defined on the nonnegative reals $\mathbb{R}(\geq 0)$. In any statement involving time functions, "for all t" will mean "for all $t \geq 0$" unless otherwise stated. Often, values will belong to a set having a distinguished element, to be called always "$\underline{0}$", and/or a set endowed with a metric d; some of the following notations assume this. Let w be a time function. Then $t(w) := \sup\{t \mid w(t) \neq 0\}$. The concatenation at $b \geq 0$ of v and w is denoted by $v|b|w$, and is equal to $v(t)$ for $t < b$ and to $w(t-b)$ otherwise. The time function 0 is the one having the constant value $\underline{0}$. The b-initial segment of v -restriction to $[0,b)$ followed by 0- is $v|b|0$. The right shift by b of v is just $0|b|v$; this suggests extending the above definition to deal with negative b, so that the left shift (with truncation) is $0|b|v$ with $b < 0$. Note the equality $u|a|(v|b|w) = (u|a|v)|(a+b)|w$. We adopt the convention that any function which we define on a subinterval I will be alternatively thought of as a time function, extended by $\underline{0}$ outside I.

For any metric space (X,d) and subset A of X, we denote by $B(A,r)$ [resp., $B^*(A,r)$] the open [resp., closed] ball $\{x \mid d(x,A) < r\}$ [resp., $\leq$]. When an element $\underline{0}$ has been distinguished in X, $\#(x) := d(x,\underline{0})$, and A is omitted in the above for $A = \{\underline{0}\}$. If v is a time function, $\#(v)$ will denote the sup-"norm", i.e. $\sup\{\#(v(t)),\ t \geq 0\}$, assuming values are on a metric space and there is a distinguished value. Consistently with this, $B(r)$ [$B^*(r)$] will be also used for time functions, i.e. for the set $\{v \mid \#(v) < r\}$ [$\leq r$]. For any T, the set $B(r;T)$ is the set of v with $\#(v) < r$ and with $t(v) \leq T$, and similarly for B^*. No confusion should arise from the fact that the same notations -$\#(.)$, $\underline{0}$, etc.- will be used in any given discussion to refer to objects associated to different sets; the meanings will be clear from the context.

One rather obvious terminology convention: most definitions and proofs will involve estimates depending on various (distinct) parameters -say T depending on k and e , b depending only on e, etc. A phrase like "for all $k,e > 0$ there exist positive $T(k,e)$ and $b(e)$ such that..." will be used sometimes instead of saying, more precisely: "there exist functions $T: \mathbb{R}(>0) \times \mathbb{R}(>0) \rightarrow \mathbb{R}(>0)$ and $b: \mathbb{R}(>0) \rightarrow \mathbb{R}(>0)$ such that, for all $k>0$ and $e>0$ the quantities $T(k,e)$ and $b(e)$ satisfy...".

Let $\underline{V}$ be a metric space with an element $\underline{0}$ and for which the balls $B^*(r)$ are all compact. A signal space V is a set of time functions with values in the (underlying signal-value set) $\underline{V}$ which (1) is closed under concatenations, (2) contains

0, (3) consists of locally bounded time functions, i.e., $\#(v|t|0)$ is finite for all v in V and all t, and (4) is extended, i.e. a time function v with all truncations $v|t|0$ in V is necessarily itself in V.

The cartesian product $U\times V$ of two signal spaces is itself a signal space with underlying set $\underline{U}\times\underline{V}$ and metric $d((u,v),(u',v')) = d(u,v) + d(u',v')$.

Let V,W be two fixed signal spaces (of "input" and "output" signals respectively). With respect to these, a system $S = (X,\phi,h)$ is given by a set X, a map ϕ: $\mathbb{R}(\geq 0)\times X\times V \to X$, and a map h: $X\times\underline{V} \to \underline{W}$, such that the following axioms hold for all x in X, all u,v in V, and all t,s:

(2.1) $\phi(0;x,v) = x$,

(2.2) $\phi(s;\phi(t;x,u),v) = \phi(t+s;x,u|t|v)$,

(2.3) if $u|t|0 = v|t|0$ then $\phi(t;x,u) = \phi(t;x,v)$,

(2.4) given $K,T>0$, there exists $K'>0$ so that, for all v in $B(K;T)$, $w(.) = h(\phi(.;x,v),v(.))$ is in W, and $w|T|0$ is in $B(K';T)$.

Note that of course K' depends on x. An initialized system $(S,0)$, or just S, is given by a system S and a state 0 in S which satisfies:

(2.5) $\phi(t;0,0) = 0$ for all t, and

(2.6) $h(0,\underline{0}) = \underline{0}$.

Note the use of "0" both to denote a state and elements of (input and output) signal spaces; there is a mild inconsistency in not denoting this state by "$\underline{0}$". The above definition is rather standard, except perhaps for (2.4), which must be added in the abstract setup but is automatically satisfied in the usual ("finite dimensional") cases. The causality axiom 2.3 follows from the consistency (2.1) and semigroup (2.2) axioms, but we include it for emphasis.

Some particular classes of systems will be of interest. A strictly causal system is one for which h is independent of v; more generally, if $V = C\times D$ and $W = E\times F$ then one says that (say) E is strictly causal on the C-coordinate if the first coordinate of $h(x,\underline{c},\underline{d})$ does not depend on $\underline{c}$. A discrete time (or a "sampled data" system) is one for which transitions occur at integer times only and depend only on samples of the input: $\phi(t;x,v) = x$ for $t<1$, $\phi(1;x,v)$ depends only on $v(0)$. This models the case of difference equations in the strictly causal case (with a 'one-second clock' added in each interval); for outputs depending on inputs one would modify the state space to store samples of past inputs. Note that the present definition allows for interconnections of discrete and other systems. A continuous-time system is one arising (in the obvious way) from equations

(2.7) $\dot{x}(t) = f(x(t),v(t))$, $w(t) = h(x(t),v(t))$,

where the state space X is a differentiable manifold, the map f: $X\times\underline{V} \to T(X)$ is continuous, and h is continuous. The space of input signals V consists of all piecewise continuous time functions to $\underline{V}$ (one could take locally bounded measurable inputs without changing any of the results), and it is assumed that solutions $x(t)$

exist and are unique for all $x(0)$ and all t. (Note that we are implicitely making the usual -and rather restrictive, in our view,- assumption that there are no finite escape times. This assumption simplifies considerably the exposition, but it would be interesting to have the general case treated in the future.) An analytic (continuous time) system is one for which X and $\underline{W}$ are real analytic manifolds, $\underline{V}$ is a subset of an Euclidean space with connected interior and no isolated points, and both f and h are real analytic (see Sussmann [1979]). Finally, a (continuous time) state-affine system is one for which $X, \underline{V}, \underline{W}$ are Euclidean, f is affine in x, and h is a constant linear function of x, i.e. one has equations

(2.8) $\dot{x}(t) = F(u(t))x(t) + G(u(t)), \quad y(t) = Hx(t),$

with $F(.)$ and $G(.)$ continuous. As an initialized system, 0 is the origin in X. (Bilinear systems have F, G linear on u.)

The most important class in what follows is that of metric systems. These are systems S for which X is a metric space with all $B^*(r)$ compact, and with h and ϕ jointly continuous in all their arguments (for the compact-open topology on V). The continuous systems defined before are all metric.

We need to introduce a few notions for a general system S. The output $out[x/v]$ is the signal in W defined by $w(.) := h(\phi(.;x,v),v(.))$, for any x in X and v in V. Two states x,x' are v-indistinguishable iff $out[x/v] = out[x'/v]$. They are indistinguishable if this happens for all v in V. An observable system is one for which no two states are indistinguishable. For a state x, input v, and subset A of X containing x, the indy class (resp., v-indy class) of x rel A is the set $[x//A]$ (resp., $[x/v/A]$) consisting of all states in A which are indistinguishable (resp, v-indistinguishable) from x. When $A = X$, we write just $[x]$ and $[x/v]$ respectively. This notation is consistent with the one for outputs, since the latter depend only on indy classes. Note that, for example, $[0/0]$ is for initialized systems the set of all states x giving identically zero outputs when the input signal 0 is fed into the system.

Let I be the equivalence relation defined by: xIx' iff $[x] = [x']$, and consider the quotient set X/I. When X is a topological space, X/I will be endowed with the (usual) finest topology for which the projection $[.]$ is continuous. The (A-) saturation $[B]$ (resp., $[B//A]$) of a subset B (resp., rel A) is the union of the sets $[x]$ for x in B (resp., $[x//A]$ for x in $A \cap B$). The set B is saturated (resp., rel A) iff $B = [B]$ (resp., $=[B//A]$). By continuity of the maps $out[./v]$, X/I is a Hausdorff space when X is a metric system.

B. Regulation concepts.

We shall say that a map $f: V \to W$ between signal spaces is stable iff the following two properties hold:

(2.9) for any $k,T,e>0$ there is a T' so that $\#(f(v)(t)) < e$ whenever v is in $B(k;T)$ and $t \geq T'$;

(2.10) for any $e>0$ there exist $d,T>0$ so that $\#(f(v)) < e$ whenever v is in $B(d;T)$.

An initialized system will be called <u>stable</u> iff its i/o map $f(v) := out[0/v]$ is stable.

In other words, outputs must converge to zero under any finite support inputs (to be thought as "disturbances" or "perturbations"), and this convergence is uniform on the "magnitude" of the disturbance; further, small disturbances should give rise to small outputs. This is just one of many possible definitions, and we use it because it is simple, mathematically convenient, and intuitively reasonable. The results to be given can be extended to cover stability under non-finite support but "sufficiently rapid decay" disturbances; the proofs are basically the same, but there seems to be no simple (elegant) way to make the corresponding statements precise. We leave as a suggestion for further research the search for similar results under other definitions -e.g., via extended spaces. From a purely mathematical standpoint, it would be highly desirable to have a definition of stability which is closed under cascades.

A (deterministic, general) <u>regulation problem</u> is specified by (i) an initialized system $(P,0)$ (the <u>plant</u>) whose input and output spaces split as $U\times V$ and $W\times Y$ respectively, with the Y-coordinate strictly causal on the U-coordinate, and (ii) a class OBJ of maps from V into $U\times W\times Y$. The signal spaces V,U,W,Y will be called the spaces of disturbances, controls, output-objectives, and measurements, respectively. A <u>solution</u> to such a problem is provided by an initialized system $(Q,0)$ (the <u>controller</u>) which satisfies the following properties:

(2.11) the input (resp., output) signal space of Q is Y (resp., U),

(2.12) the interconnection $P*Q$ (see below) is well-posed, and

(2.13) the i/o map $f(v) := out[(0,0)/v]$ of $P*Q$ is in OBJ.

In general, by the <u>interconnection</u> $P*Q$ of two systems $P = (X,\phi,h)$ and $Q = (X',\phi',h')$ with compatible signal spaces as above, we mean a system $(X\times X',\phi^*,h^*)$ such that (i) the input signal space of $P*Q$ is V, the output space is $U\times W\times Y$, and (ii) for any v in V, x in X and x' in X', let $(x(t),x'(t)) = \phi^*(t;(x,x'),v)$, $y(t) :=$ Y-coordinate of $h(x(t),.,v(t))$, $u(t) := h'(x'(t),y(t))$; then the following must hold:

(2.14) $out^*[(x,x')/v] = (out'[x'/y],out[x/(u,v)]) = (u,w,y)$, and

(2.15) $\phi^*(t;(x,x'),v) = (\phi(t;x,(v,u)),\phi'(t;x',y))$ for all t.

If both P and Q are initialized systems, one defines $P*Q$ to be initialized at $(0,0)$. We shall say that the interconnection $P*Q$ is <u>well-posed</u> if there is a unique such $P*Q$. (This can be equivalently expressed in terms of uniqueness of the signals u and y such that the above properties hold.) The $x(.)$, $x'(.)$, u, y will be refered to as the 'closed-loop' state trajectories and control and measurement signals.

We shall be interested here only in the state stabilization problem, but we feel that the above definition should be appropiate to the modeling of many other interesting regulation problems (decoupling when V,W are further split and OBJ consists of diagonal maps, etc.). One possible variation is to require the i/o maps associated to <u>every</u> initial state to be in OBJ, but this can be made equivalent to the above if one includes enough "disturbances" to set initial states. A rather interesting fact is that even some system theoretic problems not commonly thought of as "regulation" problems fit neatly in the above; for instance, if OBJ consists of a single map and P is the trivial system with $y:=v$ and $w:=u$ then a "controller" is just a <u>realization</u> of f; an inversion problem, on the other hand, can be modeled by letting $y:=f(v)$, $w:=u$, for given f, with OBJ = {delays} or {integrators}. Here we restrict ourselves to:

(2.16) DEFINITION. The plant P is <u>regulable</u> iff the regulator problem has a solution when OBJ = {stable maps}.

As explained in the introduction, we are going to treat only a particular case of this problem, namely that of state stabilization under any finite support perturbations. Specifically, <u>we make the following assumptions on P for the rest of the paper</u>:

(2.17) full state as output-objective: W-coordinate of $h(x,u,v)=x$;

(2.18) <u>independent</u> state and output disturbances: V is a product AxB, in such a way that ϕ is independent of B and also the second coordinate of h is independent of A ;

(2.19) the disturbances are <u>full</u> (see 2.21-2.22 below); and

(2.20) the system P is metric.

The argument A (resp., B) will be deleted from h (resp., ϕ).

The notion of full disturbance corresponds to requiring that arbitrary effects can be achieved by the perturbations. The typical example, and the standard case in the regulation literature, is that of additive disturbances -e.g., $\dot{x}=f(x,u)+a$, $y=h(x)+b$, for continuous time systems on an Euclidean space. The axioms are as follows:

(2.21) For each T,u the map $\phi[T,u](a):=\phi(T;0,u,a)$ (i) is open at $a=0$ with respect to the compact-open topology on A, and (ii) for each k,u there is a k' such that the image of B(k';T) under $\phi[T,U]$ contains B(k).

(2.22) Let T>0. Then, (i) for each e>0 there is a d>0 such that, for each admissible trajectory x(.) with $\#(x(t))<d$ for t<T and for each y in B(d;T), there is a b in B with $\#(b|T|0)<e$ and y(t) =

Y-coordinate of $h(x(t),\not{b}(t))$ for $t<T$, and (ii) for each $k>0$ there is a $k'>0$ such that, if $x(.)$ is a trajectory with $x(t)$ in $B(k)$ for all $t<T$, and if y is in $B(k;T)$, then there is a $\not{b}$ with $\#(\not{b}|T|0)<k'$ and with $y(t) = h(x(t),\not{b}(t))$ for $t<T$.

We introduce also the underlying system $S:= P/0$ of the plant P; this is the system (X,ϕ,h) obtained when the maps ϕ,h are restricted to $v = 0$ and the first coordinate of h is ignored. This system, with input and output spaces U and Y respectively, will be very important in what follows, since most properties will depend only on S. (Suggestion for further research: if 2.19 does not hold, one may derive the results using for S the set of states reached from $x=0$, and if 2.17 doesn't hold the results can be derived with some variation relating to controllablity to the set of states giving $w=0$.)

We now introduce the notions which will be used to characterize regulability. Let P be a fixed plant, and take $S = P/0$. All notions of indistinguishability will be with respect to the underlying system S, not to P. The same notations ϕ,h will be used for the system maps of P and S; each has one less argument ($\not{a}$, or $\not{b}$, respectively) in the case of S. Definitions 2.23 and 2.24 depend only on S, and 2.26 only involves P in the well-posedness of the feedback system for arbitrary disturbances. Recall that X has a metric $\not{d}$.

(2.23) DEFINITION. The system S is 0-detectable iff $[0/0]$ is asymptotically stable, i.e., (i) $\phi(t;x,0)$ converges to $x=0$ (as $t \to \infty$) for any x in $[0/0]$, and (ii) for each $e>0$ there is a $d>0$ such that $\#(\phi(.;x,0))<e$ for $\#(x)<d$.

Note that $[0/0]$ is positively invariant under $u=0$, so the above could have been also defined as stability (in the sense of this paper) for the "subsystem" $[0/0]$ of S, assuming one introduces appropiate "disturbances" to set the initial states in $[0/0]$. Note also that a standard stability argument can be used to prove that the convergence is uniform on compacts, i.e. that for each $k>0$ and $r>0$ there is a T such that $\phi[T,0]$ maps $B(k)$ into $B(r)$.

The next definition requires, intuitively, that each state x be open-loop asymptotically controllable to the origin, given only the knowledge of the indy class $[x]$; the technical conditions ask for for uniform convergence on compacts, and for small excursions and control values for small states:

(2.24) DEFINITION. The system S is indy-asycontrollable iff there exists a function $d: \mathbb{R}(>0) \to \mathbb{R}(>0)$ and for each indy class L there is a control signal u such that:

(i) if $\#(x)<d(e)$ for some x in L and some e, then both $\#(u)<e$ and $\#(\phi(.;x,u))<e$; and

(ii) for each k,r there is a T such that, whenever there is an x in L with $\#(x)<k$, then $\#(\phi(T;x,u))<r$.

(2.25) DEFINITION. The system S is <u>asycontrollable</u> iff S is indy-asycontrollable with respect to the <u>identity</u> measurement function.

The most important of 2.24 and 2.25 will be the former. Note that in particular this condition implies that $L = [0]$ is asymptotically stable when $u = 0$, but this is in general weaker than 2.23 because $[0]$ may be a proper subset of $[0/0]$. Note that 2.24 and 2.25 coincide for observable systems. In the "classical" linear case, 2.24 is equivalent to 2.23 plus 2.25. No definition like 2.24 appears to have been given before. Some definitions of asycontrollability do not require controls to converge to zero. Since we are studying here "regulation with internal stability", the present definition is the appropiate one. The reference Sontag [1982] deals with a Lyapunov ("direct method") characterization of asycontrollability for continuous time systems. We conjecture that a similar necessary and sufficient condition can be given for indy-asycontrollability (and for 0-detectability). Roughly, 2.24 should be equivalent to the existence of a continuous, positive definite, proper functional V on states with the property: for each indy class L there exists a (relaxed) control u with $DV(x,u)<0$ for any x in L, where D denotes the Dini (lim sup) derivative along the trajectory $\phi(.;x,u)$, evaluated at $t=0$, plus the requirement that u can be chosen small if inf V(L) is small. Along these lines we wish to also suggest the study of Lyapunov-like characterizations of other (not just stabilization) regulator problems. Note that we are suggesting here "internal" characterizations, <u>not</u> the search for conditions under which a given regulator configuration gives a solution to a regulator problem (which is an interesting but different problem).

(2.26) DEFINITION. The system S is <u>preregulable</u> iff it is 0-detectable and indy-asycontrollable.

While the intuitive content of the next definition is very natural -indy classes can be estimated on compacts by a suitable "detector",-we do not wish to imply that the particular formalization given here is in any sense best possible. In fact, we would like to suggest for further study the search for a simpler definition compatible with the results to be obtained. In particular, an "open-loop" definition would be intuitively desirable. Note the interpretation of the Q(k,e) below as "detectors" for states in the large compact $B^*(k)$ which do not "disturb" the states in a small ngbd B(e).

(2.27) DEFINITION. The plant P is <u>indy-detectable</u> iff (i) for each $e>0$ there is a $d(e)>0$, (ii) for each $k,k',e > 0$ there is an $m(k,e,k')>0$, (iii) for each $k,r,e > 0$ there is a $T(k,e,r)>0$, (iv) for each $k,e > 0$ there is a system Q(k,e) with

$P*Q(k,e)$ well-posed, a state $q(k,e)$ in each such system, and a $b(k,e)>0$, and (v) there is given for each such k,e a function $\ell: X' \to X/I$ (primes indicate objects associated to $Q(k,e)$), such that the following properties hold for arbitrary k,e,r. Consider first the closed loop system $P*Q(k,e)$ and, for each x in X the trajectory with $x'(0) = q(k,e)$, $x(0) = 0$, and $v = 0$, and let $T := T(k,e,r)$; then:

(a) if $\#(x)<d(e')$ then $\#(x(t))<e'$ and $\#(u(t))<e'$ for all $t\leq T$ and all e' in the interval $[e,1]$,

(b) if $\#(x)<k$ then $\#(x(t))<b(k,e)$ for all $t\leq T$, and $x(T)$ is in $B(\ell(x'(T)),r)$,

(c) if $\#(x)<k'$ for some k', then $\#(x(t))<m$ and $\#(u(t))<m$ for all $t\leq T$ and $m=m(k,e,k')$.

Consider now the system P by itself, with input signals $u=v=0$.
It is then also required that:

(d) if $\#(x)<d(e)$ and if $y|t'|0 = 0$ for some t', then either $[\phi(t';x,0)/0] = [0/0]$ or $\#(x(t))<e$ for $0\leq t\leq t'$, and

(e) if $\#(x)<k'$ and $y|t'|0 = 0$ for some k',t', then either $[\phi(t';x,0)/0] = [0/0]$ or $\#(x(t))<m(1,1,k')$ for $t\leq t'$.

The definition could be generalized to allow for the estimation function to depend on the present output $y(t)$; this would give an equivalent concept since one may always enlarge the state space of Q to allow for the memory of y. The 1 in (e) is chosen arbitrarily; it is only used in order to simplify notations by not adding yet another estimate depending only on k'. We shall see later that:

(2.28) THEOREM. Every plant P with S analytic is indy-detectable.

But many -even smooth- systems are not indy-detectable. For instance, consider the system -plant- with $X = A = Y = U = \mathbb{R}$, ($W$ is irrelevant and B can be taken trivial,) and equations $\dot{x} = u + \not{s}$, $y = h(x)$, where h is any function which is bijective on the interval $[-2,-1]$ and is zero outside $[-3,-1]$. The underlying system S ($\not{s} = 0$) is in fact observable. but the above definition cannot hold for, say, $r=.1$, $e=.5$. Indeed, compare two trajectories remaining in $B(.5)$. Both result in $y=0$, independently of $x(0)$, so the controller state trajectory $x'(.)$ is independent of the initial state of the plant, say $x(0) =x$, or $=z$. It follows that the control u is also the same, and thus the distance $d(x(t),z(t))$ is constant in t (where $z(.)$ is the trajectory starting at z). But (b) requires that for some $t=T$ this distance be less than (.2), because the set (indy class) $\ell(x'(T))$ consists of a single state (observability). Taking $x=0$, $z=.4$ gives a contradiction.

These are the main results to be proved:

(2.29) THEOREM. Regulability = indy-detectablity + preregulability.

(2.30) COROLLARY. For S analytic, regulability = preregulability.

(2.31) THEOREM. For S state-affine, preregulability = 0-detectability + asycontrollability.

3. SUMMARY OF PROOFS

A. Necessary part.

Let P be a regulable plant, S = P/0, and Q a regulator for P. Recall that 2.17-2.20 are supposed to hold, and that OBJ = all stable maps. We wish to show that P is preregulable and indy-detectable. A state x' in X' is reachable (from 0) iff there is some input signal y for Q and some t such that x' = ϕ'(t;0,y). The following lemma gives consequences of regulability for S.

(3.1) LEMMA. Consider P*Q, and assume that v=0. Then, for each k,e > 0, and for each x' in X' , there exist positive numbers d(e), m(k,x'), and T(k,e,x') such that: (a) if #(x)<d(e) and x'=0, then the closed loop trajectories have #(x(.))<e and #(u)<e, and (b) if #(x)<k and x' is reachable, then (i) #(x(t))<e and #(u(t))<e for all t $\geq$ T=T(k,e,x'), and (ii) x(.), y, u are in B(m(k,x')).

IDEA OF PROOF: use independence of disturbances to "set" reachable states in the plant while not affecting measurements. Regulability then gives the result. */

(3.2) PROPOSITION. S is 0-detectable.

PROOF. Let [x/0] = [0/0]. Consider in P*Q the initial state (x,0), and assume that v=0. Consider the signals u=0 in U and y=0 in Y. Let x(t) = ϕ(t;0,0), x'(t) = 0 for all t. By well-posedness, it follows that ϕ*(t;(x,0),0) = (x(t),0). The conclusion follows from (3.1). */

(3.3) PROPOSITION. S is indy-asycontrollable.

PROOF. Let d(.) be as in 3.1. Let L be an indy class. Pick any x in L, and consider ϕ*(t;(x,0),0). We claim that the ensuing u and y are independent of the particular x chosen. But this follows from well-posedness, since out[x/u] = out[z/u] whenever [x] = [z] = L. Again the result follows from 3.1. */

(3.4) PROPOSITION. P is indy-detectable.

PROOF. Again the proof uses 3.1. We let d(.) be as there, and define Q(k,e):= Q, and q(k,e):= 0, for all k,e, T' via T'(k,e,r):= the T(k,r,0) from 3.1, b(k',e) = m(k,e,k'):= m(k',0), and the functions ℓ all constant and equal to [0].

Given e,k,r, axioms (a),(b), and (c) in 2.27 then follow from 3.1. In particular, note that (a) holds for <u>any</u> e' (not just those in $[e,1]$), because Q is independent of e, and for (b) note that $d(x(T),\bar{l}(x'(T))) = d(x(t),[0]) \le d(x(t),0)$. To prove (d), assume that $x(0) = x$ is in $B(d(e))$, $x'(0)=0$, $u = v = 0$, and $y|t'|0 = 0$ for some t'. Let $z := x(t')$. Consider the evolution of P^*Q with $v:=0$ and starting at (z,0). Let u', y', z(.), z'(.) be the trajectories obtained. The signals $0|t'|u'$ and $0|t'|y'$ are consistent with the trajectory equal to $x(t)$ for $t \le t'$ and to $z(t'-t)$ otherwise, and similarly for x'. By well-posedness, these are the trajectories corresponding to the initial state (x,0) and $v=0$. Thus $\#(x(t))<e$ for all t, and in particular for $t \le t'$, as wanted. The proof of (e) is analogous. */

B. <u>Sufficiency</u>.

Throughout this section, P is an indy-detectable, preregulable plant. We shall need a couple of easy technical lemmas. Only indy-asycontrollability is needed in 3.5-3.7.

(3.5) LEMMA. For each $k',r',e > 0$ with $r'<k'$, $e<k'$ there are positive $d'(e)$, $c(k')$, and $T(k',r')>1$ such that, for each indy class L', there is an input u' with: (i) if x is in $L' \cap B^*(d'(e))$ then both $\phi(t;x,u')$ and $u'(t)$ are in $B(e)$ for all t, and (ii) if x is in $L' \cap B^*(k')$ then $x(t) = \phi(t;x,u')$ is in $B(r')$ for all $t \ge T'(k',r')$, and $\#(x(t))<c(k')$, $\#(u)<c(k')$.

(3.6) LEMMA. Let T(.,.) and d(.) be the functions in the statement of 3.5. For each $k,r,e > 0$ with $r<k$ and $e \le 1 \le k$ there exists a $c(k)$, and for each indy class L intersecting $B^*(k)$ there exist an input u and an open set N , such that (i) $L \cap B^*(k) \subseteq N$, (ii) N is saturated rel $B^*(k)$, (iii) for any x in $N \cap B^*(k)$, $\#(\phi(T;x,u))<r$ and $\#(x(t))<c(k)$ for all $t \le T$, (iv) if $e \le e' \le 1$ and x is in $N \cap B^*(d(e'))$ then $\#(x(t))<e'$ for $t \le T$, and (v) $\#(u)<c(k)$.

(3.7) LEMMA. For each $k',k'',r,e,e' > 0$, $e \le 1$, there are $T=T'(k',r,e)>1$, positive $c'=c'(k')$, $m'=m'(k',e,k'')$, $g=g(e')$, $b'=b'(k',e)$, and systems $Q=Q(k',r,e)$ with states $x'=x'(k',r,e)$ such that the following properties hold: (i) the interconnection P^*Q is well posed, and for $v=0$, $x(0)=x$, and $x'(0)=x'$ the following hold for the ensuing closed loop trajectories: (ii) if $\#(x)<g(e')$ for an e' in the interval $[e,1]$, then $\#(x(t))<e'$ and $\#(u(t))<e'$ for $t \le T$, (iii) if $\#(x)<k'$ then $\#(x(T))<r$, and $\#(x(t))<c'$ and $\#(u(t))<b'$ for $t \le T$, and (iv) if $\#(x)<k''$ then $\#(u(t))<m'$ for $t \le T$.

IDEA OF PROOF: by the above lemma, control into B(r) is possible if the indy class is estimated sufficiently well; the systems Q are then obtained by the sequence "first apply a suitable detector, then apply an appropiate control". */

(3.8) LEMMA. Consider the system $S = P/0$. For each $k,e > 0$ there exist positive $\delta(e)$, $\mu(k)$ such that, for any t, x for which $out[x/0]|t|0 = 0$, the following properties hold: (a) if $\#(x)<\delta(e)$ for some e, then $\#(\phi(t';x,0))<e$ for all $t'<t+1$, and (b) if $\#(x)<k$ for some k, then $\#(\phi(t';x,0))<\mu(k)$ for all $t'<t+1$.

The proof that a given system description involving different kinds of states indeed defines a system Q, and that an interconnection P*Q is well posed, is in general very tedious, since it involves careful checking of all axioms for each kind of state. The following technical lemma can be used, at least in the case that we shall need to consider, to establish directly these properties.

(3.9) LEMMA. Assume given a family of systems $Q(i)$, $i\geq 0$, for which the interconnections $P*Q(i)$ are well posed for each i. States $q(i)$ and a subset $Z(i)$ are specified for each state set $X(i)$. Further, a function f_i: $Z(i) \to \underline{N}$ (=non-negative integers) is given for each i. Let $\beta(x,y) := \inf\{t \mid \phi_i(t;x,y)$ is in $Z(i)\}$, for each input y and each x in $X(i)$. It is assumed that there is an $\beta(x)$ (possibly infinite) such that $\beta(x,y) \geq \beta(x) > 0$ for all y, for each x not in $Z(i)$, and that $\beta(q(i)) = T(i) \geq 1$ for all i. The claim is that there is a unique system Q such that P*Q is well posed, and such that for each x in $X(i)$:

(3.10)
$$\begin{aligned}\phi(t;x,y) &= \phi_i(t;x,y) \quad \text{if } t<\beta(x,y), \text{ and}\\ &= \phi_i(t-\beta(x,y);q(j),0|(\beta(x,y)-t)|y) \quad \text{otherwise,}\end{aligned}$$

where $j := f_i(\phi_i(\beta(x,y);x,u))$, and where the state space of Q is the disjoint union of the $X(i)$ mod the identifications $x = q(f_i(x))$ for each x in $Z(i)$. The output map is assumed to be $h(x,\underline{y}) := h_i(x,\underline{y})$ for x in $X(i)$ but not in $Z(i)$.

PROOF (sketch). The definition of ϕ is by induction on n, defining $\phi(t;x,y)$ for each (t,x,y) such that, either $\beta(x)>1$ and $t\leq n$, or $\beta(x)\leq 1$ and $t\leq n-1$. Formulas 3.10 then define ϕ for all values. The axioms for Q being a system follow from those for the subsystems $Q(i)$. To prove well-posedness, consider a state x of P and a state x' of Q, say in $X(i)$. By the above identification, one may assume that x' is not in $Z(i)$. Thus a unique pair of closed-loop signals (u,y) exists, by well posedness of $P*Q(i)$, for $t<\beta(x)$. Using this y one obtains the next j as in 3.10, and the result follows by induction. Note the use of the assumption that $T(i)>1$ in the induction step, insuring that there are finitely many transitions, from one type of state to another type of state, in any finite time interval. */

We now start constructing the controller P. The definition will use the construction in 3.9. Define, using 3.7, $e(1) := 1$ and inductively $e(i+1) := g(e(i))$. For each $a>0$, define $f(a)$ to be any integer j such that $e(j)<a$.

A system $Q(0)$ is introduced as follows. Its state set is $X(0):=$ set of initial segments of observations, i.e. the set of pairs (y,t), y in Y, with $y|t|0 = 0$. We denote by 0 the pair $(0,0)$, and define $\phi(t;0,y):= 0$ if $T(y):= \inf\{t| y(t)\neq 0\}$ is greater than t, $\phi(t;0,y):= (y|(t-T(y))|0,\ t-T(y))$ otherwise. For a state (y,t) with $t>0$, define $\phi(t';(y,t),y'):= (y|t|y',\ t+t')$. The output h is constantly $= 0$. Let $Z(0):= \{(y,t)|\ t=1\}$, and $f(y,1):= f(\#(y|.5|0))$. The hypothesis of 3.9 hold then for $Q(0)$, when $q(0):= 0$.

Define now functions $K(i,j)$ and $m(i,j)$ as follows, by induction on i. Let $K(1,j):= j$ for all $j\geq 1$. Assume that $K(i,j)$ has been defined for all j. Let $k'(i):= K(i,i)$. Consider using 3.7 the system $Q'(i):= Q(k'(i),\ e(i+2),\ e(i+1))$, and obtain $T(i):= T(k'(i),\ e(i+2),\ e(i+1))$. Let $m(i,j):= m'(k'(i),\ e(i+1),\ K(i,j))$, for all j. The induction is then completed by letting $K(i+1,j)$ be any number bounding

(3.11) $\{\#(\phi(t;x,u))\ |\ t\leq T(i),\ \#(x)\leq K(i,j),\ \#(u)\leq m(i,j)\}$.

Note that $K(i+1,j) > K(i,j) > j$ for each i. Now introduce the systems $Q(i)$ as follows. The state space $X(i)$ of $Q(i)$ is $X'(i)\times\mathbb{R}(\geq 0)$, where $X'(i)$ is the state space of $Q'(i)$, and let $q(i):= (x'(i),0)$, using the states $x'(i)$ given in 3.7. The set $Z(i)$ consists of all the (x,t) with $t=T(i)$. If ϕ' is the transition map of $Q'(i)$, then let for $Q(i)$:

(3.12) $\phi(t';(x,t),y):= (\phi'(t';x,y),t'+t)$,

and output $h((x,t),\underline{y}):= h'(x,\underline{y})$. Let f be constantly equal to $i+1$. The hypothesis in 3.9 hold by well posedness of the original $P^*Q'(i)$. Note that $\emptyset(x,t):= T(i)-t \leq T(i)$ for all x. And $\emptyset(q(i)) = T(i)$, so the notations $T(.)$ in 3.9 and 3.7 are consistent.

Let Q be constructed from these $Q(i)$ via 3.9. It must be proved that P^*Q is stable. We shall use primes (') to indicate objects associated to Q. For simplicity, conditions 2.9-2.10 will be only proved explicitely for u and x: by the continuity of the output measurement y on x, the full result for $out[(0,0)/v] = (u,x,y)$ can be obtained trivially from this. Note the following two properties that hold by construction. Let $v=0$, and consider the states $x(0)=0$ in P and $x'(0)=q(i)$ in X'. Then, for each $i\geq 1$:

(3.13) assume $\#(x)<e(i+1-i')$, $i'\geq 0$, and take any $j\geq i$; let $T:= 0$ if $j=i$, $=T(i+1)+\ldots+T(j)$ otherwise; then $\#(x(t))$ and $\#(u(t))$ are necessarily $< e(j-i')$ for $t\geq T$;

(3.14) assume $\#(x)<K(i,j)$, $j\geq i$, and let $T:= T(i)+\ldots+T(j)$; then $\#(x(t))$ and $\#(u(t))$ are $< e(j)$ for $t\geq T$.

Let k,T be given, and consider a disturbance v in $B(k;T)$. Let $j>0$ be chosen such that $T \leq T(0)+\ldots+T(j)$. Then $x'(T)$ must be in one of the sets $X(i)$ with $0\leq i\leq j$. Assume first that $i>0$. Let $t':= \emptyset(x'(T))$. Then $x'(T+t') = q(i+1)$. Let K be a number large enough so that $x(T+t')$ is in $B(K)$. Note that K can be chosen independently of the particular v, since $x(T)$ is bounded for v in, say,

$B(k;T+1)$, and since the input disturbance is identically zero in the interval $(T,T+T')$. Let $j \geq i$ be such that $K < K(i,j)$ -recall always $K(i,j)>j$. Then property 2.9 in the definition of stability follows from 3.14. Assume now that $i=0$, and suppose that $x(T)$ is in $[0/0]$ and $x'(T)=0$. Then 2.9 follows by 0-detectability. If $x=x(T)$ is not in $[0/0]$, consider $t':=$ inf of those t for which the output y of S has $y(t)\neq 0$, assuming $x(0)=x$ and control $u=0$. Then $x(T+t')$ is again bounded as a function only of k and T, by the second part of 3.8. Since $x'(T+t') = q(1)$, one may again apply 3.14. If $x'(T) \neq 0$, then there is a $t'<1$ as above, so we again apply 3.14. Thus 2.9 is true in every case.

To prove 2.10, let e be given. Define $T:=.5$. Pick a j' such that $e(j')<e$. Let d be such that, with $u=0$ and $\#(v)<d$, necessarily $\#(x(t)) < \delta(e(j'+1))$ -c.f. 3.8- and $\#(y(t)) < e(j')$ for $t\leq .6$ (any number larger than .5 is suitable here). Let $t'':= \inf\{t \mid y(t)\neq 0$ if $u=0$ is applied to $P\}$. If $y = 0$, then in the closed loop operation always $x'(t)=0$. Thus $u=0$ is indeed applied (note the delay which prevents this reasoning from being circular). In that case, 3.8 gives the desired result. The other possibility is that t'' is finite. Then $x'(t''+1)=$ some $q(i)$, where $i = f(a)$, a = magnitude of the output y restrited to $[t'',t''+.5)$. By the choice of d, $a < e(j')$. By definition of f, then, $i > j'$. Again by 3.8, one knows that $\#(x(t''+1)) < e(j'+1)$. Let $i':= i-j'$. By 3.13, then, $\#(x(t)) < e(i-i') < e(j') < e$ for all t, and the same happens for the control values $\#(u(t))$.

This completes the proof of regulability. Close inspection of the above proof suggests an area for further research: develop all the theory in the category of 'piecewise continuous' systems, where one requires all maps appearing to be continuous on each of the elements of a covering; the sets in this covering could be e.g. intersections of open and closed sets, and in order for the setup to be of applied interest one should require this covering to be locally finite around each nonzero state. Similarly for piecewise analytic, etc. Note that these constructions result a priori in very large state spaces; as remarked in the introduction we are here interested only in abstract regulability. But various simplifications are immediate. In most situations one is only interested in appropiate behavior for disturbances which are bounded in magnitude and/or in controlling to a sufficiently close tolerance around the origin, not necesarily asycontrollability. All that is required is then to only consider those $Q(i)$ controlling states of a given bounded magnitude, to a ngbd $B(j)$ small enough.

C. Some particular classes of systems.

(3.15) LEMMA. Let P be a plant, and assume that the following properties hold for $S=P/0$, for some $T>0$: (i) for each e there is a u with $\#(u)<e$ and such that u determines final states, i.e., for any x,z in X, if $out[x/u]|T|0 = out[z/u]|T|0$ then $[\phi(T;x,u)] = [\phi(T;z,u)]$, and (ii) for any x in X, either

out$[x/0]|T|0 \neq 0$ or $[\phi(T;x,0)] = [0/0]$. Then P is indy-detectable.

For analytic systems, the hypothesis of the lemma hold; this is an immediate consequence of the main result in Sussmann [1979]; part (ii) is just an exercise in analiticity. This establishes 2.28. We remark also that one has the same result for (at least) polynomial discrete time systems; see Sontag [1979].

Turning now to preregulability, we ask when is this property equal to just asycontrollability plus 0-detectability. Call a system weakly preregulable if these two latter properties hold. (Note that this is unrelated to "weak regulation" in the sense of Sontag [1981].) It can be proved that the analytic system defined below is weakly preregulable but not preregulable. The state space is the plane, $U = \mathbb{R}$, and its equations are:

$$\dot{x} = 4x \arctan(xz-1)(u+1) + 2x(1-z^2)(u^2+1)$$
$$\dot{z} = -z(u+1)$$

and output $y=z$. It is interesting that this system is pretty well behaved near (0,0), since its linearization there is asymptotically stable. Study of this example suggests the definitions below.

Call the system S locally indy-asycontrollable if there exists a ngbd N of 0 and a function d such that the properties in 2.4 hold for all indy classes rel N. Call S strongly locally indy-a.c. if there is some saturated such ngbd. In the 'hyperbolic' case when the linearization of S at zero makes sense and is regulable, one concludes only (non-strong) local indy-a.c. The following is easily proved by controlling first into such a saturated ngbd, and then sending this ngbd into a small ngbd of the origin.

(3.16) LEMMA. A weakly regulable and strongly locally indy-a.c. system S is necessarily preregulable.

There are particular cases of interest in the above, e.g. relating to local indy-a.c. with [0] compact, or under 'exact' local controllability. We have no space to treat these here. Instead, we give a proof of 2.31.

Let S be a weakly preregulable state affine system. Consider [0], which is a subset of [0/0]. This set is a subspace for which we may decompose the equations for S as

(3.17) $\dot{x} = F(u)x + G(u)$,

(3.18) $\dot{z} = A(u)z + E(u)x + B(u)$,

with $y = Hx$, and where [0] is the space of vectors of the form $\{(0,z)\}$. Since [0/0] is asymptotically stable, [0] also is (under input = 0). Thus A(0) is a Hurwitz matrix. By Bellman [1969, Theorem 2.2.2], and by continuity of A on u, there is an e>0 such that $\dot{x}(t) = A(u(t))x(t)$ is asystable for any u in B(e). In the definition of asycontrollability, let d be such that, for any (x,z) in B(d), there is a u in D(e) controlling (x,z) asymptotically to zero. Consider now N:= saturated nbhd of $B(d)$=space $\{(x,z) \mid \#(x)<d\}$. Pick any indy class $L(x)$ which contains an $(x,0)$ in $B(d)$. Let u control $(x,0)$ as above. In general,

(3.19) $\dot{z}(t) = \underline{A}(t)z + g(u,x)(t)$,

where $\underline{A}$ is the fundamental matrix of $\dot{z} = A(u(t))z$. For $z=0$, then, $g(u,x)(t)$ converges to zero. Thus for any other z this is still true, for the same fixed u, because the first term in the right converges by the choice of e. Thus $L(x)$ is controlled, and the estimates $d(.)$, $T(.,.)$ follow from those for asycontrollability applied to $(x,0)$ plus the linear theory. Thus S is strongly locally indy-a.c., and the result follows from the previous lemma. */

4. REFERENCES

Bellman, R. [1969] Stability Theory of Differential Equations, Dover, NY.

Sontag, E.D. [1979] "On the observability of polynomial systems,I: Finite-time problems", Siam J.Control and Opt. 17:139-151.

Sontag, E.D. [1981] "Nonlinear regulation: The piecewise linear approach", IEEE Trans. Autom Control AC-26: 346-358.

Sontag, E.D. [1982] "A Lyapunov-like characterization of asymptotic controllability", to appear.

Sussmann, H.J. [1979] "Single-input observability of continuous time systems", Math. System Theory 12:31-52.

TIME-OPTIMAL CONTROL IN THE PLANE

H. J. Sussmann
Mathematics Department
Rutgers University
New Brunswick, NJ 08903 /USA

Summary: We outline a research program in control theory, whose goal is to characterize local properties of systems in terms of Lie bracket configurations, and we illustrate it by carrying out part of the analysis for time-optimal control in the plane, for systems with a scalar control that enters linearly. □

The purpose of this lecture is to illustrate, for the rather simple case of time-optimal control in the plane, a general research program on some aspects of nonlinear control theory. The fundamental principles of this program are:

a) the systematic use of Lie bracket conditions, i.e. of assumptions about the Lie brackets of the vector fields $x \to f(x,u)$ that correspond to a control system

(1) $$\dot{x} = f(x,u)$$

b) the use of the concept of "codimension" in order to organize Lie bracket conditions into groups of increasing "degree of degeneracy",

c) the search for coordinate-free, control-theoretic conclusions. Specifically, one wants to obtain information about

(c.1) structure of reachable sets,

(c.2) sufficient classes of trajectories and trajectory reduction,

(c.3) regular synthesis.

To explain what all this means, let us begin with a remark made by Lobry in his pathbreaking paper [1]. Consider a system in $\mathbb{R}^3$, of the form

(2) $$\dot{x} = (1-u)X(x) + uY(x), \quad 0 \leq u \leq 1,$$

where X and Y are smooth vector fields. Assume that

(I) $X(p)$, $Y(p)$ and $[X,Y](p)$ are linearly independent.

Lobry pointed out that, if (I) holds, then one can get significant control-theoretic information about the system (2) near p. Specifically, let us use F^X, F^Y to denote the flows of X and Y (so that $t \to F_t^X(x)$ is the X-trajectory which goes through x at time $t = 0$). Let $A_U(p)$ denote the set of all points $q \in U$ that can be reached from p by means of a trajectory of (2) that does not leave U. Lobry showed that, if U is a suitably small neighborhood of p, then $A_U(p)$ has the shape shown in Figure 1, i.e. it is bounded by two smooth surfaces S_{XY} and S_{YX} (and by $\mathrm{Clos}\, A_U(p) \cap \partial U$). Here S_{XY} is the XY-surface, i.e. the set of points of the form $F_t^Y F_s^X(p)$, $t \geq 0$, $s \geq 0$, and S_{YX} is defined similarly. Lobry's observation actually tells us more. If q is an arbitrary point in $A_U(p)$, then we can follow the X-trajectory through q backwards in time until we hit the boundary of $A_U(p)$ at a point q', which will necessarily lie in S_{XY}. So q is of the form $F_r^X F_t^Y F_s^X(p)$ for some $s \geq 0$, $t \geq 0$, $r \geq 0$. In general, if

$\xi_1,\ldots,\xi_m$ are symbols for types of trajectories, let us refer to any trajectory which is a concatenation of at most m pieces, of types $\xi_1,\ldots,\xi_m$ (in that order, and with the possibility that some of the pieces may be missing) as a $\xi_1\ldots\xi_m$-trajectory. Then we have shown that every point $q \in U$ that can be reached at all from p by a trajectory in U can actually be reached by an XYX-trajectory. This says that XYX-trajectories are sufficient for reachability from p within U. (In general, let Γ, Γ' be collections of trajectories, and let P be a map from $\Gamma \cup \Gamma'$ to some set. Think of $P(\gamma)$ as "the effect of γ". We call Γ' sufficient for (Γ,P) if $P(\Gamma') = P(\Gamma)$, i.e. if "anything that can be done with a $\gamma \in \Gamma$ can also be done with a $\gamma' \in \Gamma'$". If Γ is the set of all trajectories in U that start at p, and $P(\gamma)$ is the terminal point of Γ, then Γ' is sufficient for reachability from p within U if every point that can be reached from p within U can actually be reached by a trajectory in Γ.) We have achieved a sizable trajectory reduction of the reachability problem from p, since we now know that, if we need a control that will steer p to some point q, then we can limit our search to controls of a very special type, knowing that, if q can be reached at all, then it can be reached by a control in the smaller class. Our reduction is in fact, finite-dimensional, in the sense that our sufficient class of trajectories is smoothly parametrized by a finite-dimensional parameter (actually three-dimensional, in our case). Finally, we can actually find a regular synthesis. (A synthesis, or feedback control law, for the problem of reaching a point p, is a selection, for each point q from which p can be reached, of a trajectory γ_q that goes from q to p, in a way which is compatible (i.e. "memoryless": if γ_q gets to a point q' at a time t then, from that time on, γ_q must agree with $\gamma_{q'}$).) Consider a system (2) which satisfies (I), and apply our remarks to this system with X,Y replaced by $-X,-Y$ (i.e. draw Figure 1 with the arrows reversed). Now $A_U(p)$ is the set of points that can be steered to p, and the synthesis consists of selecting the XYX-trajectory for q in the interior of $A_U(p)$, the XY- or YX-trajectory if q is in one of the boundary surfaces, and the X- or Y-trajectory if q is in the frontier of one of the boundary surfaces. A regular synthesis is, roughly, a synthesis which is "piecewise smooth" in some sense. (At present, no unanimous consensus exists as to what the definitive definition of "regularity" should be, but the synthesis we have obtained in our case is clearly "piecewise smooth" in any reasonable sense.)

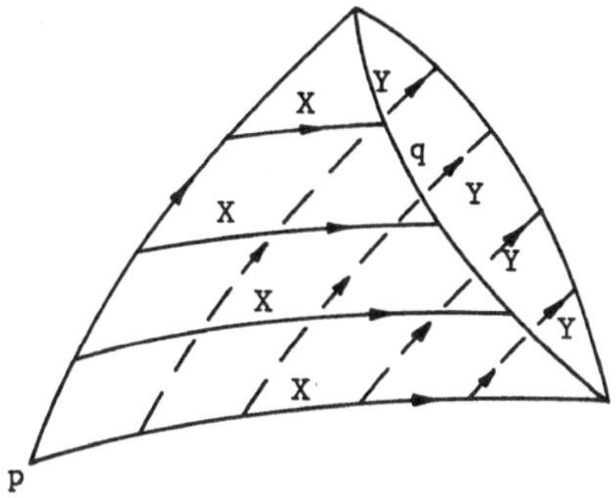

Figure 1

Notice that all the consequences derived so far from condition (I) (i.e. that the reachable set near p is a finite union of smooth manifolds, that everything reachable can be reached with a bang-bang control with no more than two switchings, that there is a regular synthesis) are control-theoretic properties of (2) that remain unaltered if we make an arbitrary smooth nonlinear change of coordinates near p. We refer to any such property as an invariant property. That only invariant properties appeared in our conclusion (but not, say, properties such as convexity of the reachable set) is far from surprising, in view of the fact that condition (I) is itself invariant.

The goal of the program of research referred to above is to carry out this same type of analysis for more general systems, and under more general assumptions on Lie brackets. Ideally, one would like to have a complete "dictionary" that, given the Lie bracket configuration (LBC) of a system at a point p, would give us the structure of the reachable set from p, information about good sufficient families of trajectories, and a regular synthesis (if it exists). This would include, in particular, the solution of the similar problem for optimal control. (The definitions of "sufficient families", "finite-dimensional reductions" and "regular synthesis" are the same for the optimal control case as for the general reachability problem, except that the word "trajectory" is replaced throughout by "optimal trajectory". The optimal control problem is really contained in the reachability problem, since every optimal control problem can be recast as a reachability problem by adding an extra variable. The analysis of optimal trajectories then becomes, simply, the analysis of trajectories that stay in the lower part of the boundary of the reachable set for the extended problem.)

Whether or not such a "dictionary" can actually be built remains to be seen, but there is a remarkable theorem, due to Nagano, which says that the "dictionary" exists "in principle", i.e. that all invariant properties having to do with reachability from p are determined by the LBC at p, if the system is analytic. Precisely, suppose that $f = \{f_a : a \in A\}$ is a family of real analytic vector fields on the analytic manifold M. Let $p \in M$. Let L_A denote the free Lie algebra generated by a family of indeterminates $\{X_a : a \in A\}$, and let $Ev(f,p)$ be the map which takes an element of L_A, substitutes the f_a for the X_a, and evaluates the result at p. The kernel of this map is called the set of Lie bracket relations between the f_a at p, or the LBC of the f_a at p. (Example: $X_1+3X_2+[X_1,X_2]-5[X_1,[X_1,X_3]]$ is a Lie bracket relation between f_1, f_2, f_3 at p iff $(f_1+3f_2+[f_1,f_2]-5[f_1,[f_1,f_3]])(p)=0$.) Suppose we have two analytic systems $\{f_a : a \in A\}$, $\{f'_a : a \in A\}$ on M, M', and let $p \in M$, $p' \in M'$. Nagano's theorem says that, if a certain extra condition holds, then there is an analytic diffeomorphism D from a neighborhood U of p onto a neighborhood U' of p', that transforms the f_a to the f'_a, iff the LBC of the f_a at p is the same as that of the f'_a at p'. (The extra condition is that both systems be "transitive" at p, p'. A system $\{f_a\}$ is transitive at p if the Lie algebra generated by the f_a has full rank at p. If the systems are not transitive, then the conclusion of Nagano's Theorem remains true, except that we must now take U, U' to be neighborhoods of p, p' in the integral manifolds S, S' through p, p' of the Lie algebras generated by the f_a and by the f'_a. Since all trajectories from p lie in S, and similarly for those from p', we see that, if the LBC's of the f_a at p and of the f'_a at p' agree, then the reachable sets from p and p' are diffeomorphic, even if the transitivity condition fails.)

Nagano's Theorem implies that, for analytic systems with the accessibility property, the truth-value of any statement about local properties of a system near a point p should be determined by the LBC of the system at p. The research program referred to above consists of looking, first of all, for a collection of inequalities involving Lie brackets at p, under which the question of interest to us can be answered completely. (In the Lobry example, the question is: "What does the reachable set from p look like near p?" The inequality under which we can answer it is $D \neq 0$, where D is the determinant of X, Y, and [X,Y].) After this "nondegenerate" case is taken care of, we try to determine what happens when some of the inequalities actually become equalities. This may be possible provided that some other inequalities hold, i.e. that the "degeneracy" occurs in a "least degenerate" way. The codimension of a particular type τ of LBC's is, roughly, the number of independent equalities characterizing τ. The following analogy may be useful. Let f be a smooth function of n variables, defined near 0. Suppose we want to describe the zero set Z(f) of f, near 0. The least degenerate (zero-codimensional) case is when $f(0) \neq 0$. In this case, Z(f) is empty. If $f(0) = 0$, we can still describe Z(f) if $(\text{grad } f)(0) \neq 0$. This situation has codimension one (and Z(f) turns out to be a smooth hypersurface). If $f(0) = 0$ and $(\text{grad } f)(0) = 0$ (i.e. n+1 equalities), then we can still describe Z(f) if the Hessian matrix Hf(0) of f at 0 is nonsingular (i.e. one more inequality, $\det Hf(0) \neq 0$). In this case (of codimension n+1), Morse's Lemma says that, after a change of coordinates, f equals $x_1^2+\ldots+x_k^2-x_{k+1}^2-\ldots-x_n^2$, and this determines what Z(f) looks like. If $f(0) = 0$, $(\text{grad } f)(0) = 0$, $\det Hf(0) = 0$, we have a situation of codimension $\geq n+2$, whose analysis is much more difficult and requires more inequalities.

We now show how all this works for the case of time-optimal control in the plane.

Consider a system of the form (2), where now x takes values in some open subset M of the plane. We want to analyze in detail how various properties of the LBC of X and Y at a point p entail conclusions about the structure of the time-optimal trajectories near p. Moreover, we will also try, if possible, to prove some results of a more global nature.

It will be useful cometimes to write the system in the equivalent form

$$\dot{x} = f(x) + vg(x), \quad -1 \leq v \leq 1 \tag{3}$$

where

(4.a) $$f = \frac{1}{2}(X + Y),$$

(4.b) $$g = \frac{1}{2}(Y - X),$$

and where the controls u, v are related by

(4.c) $$v = 2u - 1.$$

One useful tool for the analysis of time-optimal trajectories is the Maximum Principle (MP). Let $u(\cdot)\colon [a,b] \to [0,1]$ be a control, and $x(\cdot)\colon [a,b] \to M$ a trajectory corresponding to $u(\cdot)$. The MP says that, if $(u(\cdot), x(\cdot))$ is time-optimal, then there exist

a) a constant $\lambda_o \geq 0$, and

b) an $\mathbb{R}^2$-valued map $\lambda(\cdot)$ on $[a,b]$, which is not identically zero, and is a solution of the adjoint equation

(5.a) $$\dot{\lambda}(t) = -\frac{\partial H}{\partial x}(\lambda(t), x(t), u(t)),$$

which satisfy

(5.b) $$H(\lambda(t), x(t), u(t)) \leq H(\lambda(t), x(t), u)$$

for almost all $t \in [a,b]$, and all $u \in [0,1]$, and

(5.c) $$H(\lambda(t), x(t), u(t)) + \lambda_o = 0$$

for almost all t.

Here H is the Hamiltonian:

(6) $$H(\lambda,x,u) = \langle\lambda, (1-u)X(x) + uY(x)\rangle.$$

If $\gamma = (u(\cdot), x(\cdot))$ is a trajectory, and $\lambda(\cdot)$, λ_o are such that $\lambda_o \geq 0$, $\lambda(\cdot)$ is not identically zero, and (5.a,b,c) hold, then we will call $\gamma^* = (u(\cdot), x(\cdot), \lambda(\cdot), \lambda_o)$ an <u>extremal lift of</u> γ. If γ^* is an extremal lift of γ, then we define the <u>switching function</u> $\phi_{\gamma^*}\colon [a,b] \to \mathbb{R}$ by

(7) $$\phi_{\gamma^*}(t) = \langle\lambda, Y(x)-X(x)\rangle.$$

Then condition (5.b) shows that $u(t) = 1$ if $\phi_{\gamma^*}(t) < 0$, and that $u(t) = 0$ if $\phi_{\gamma^*}(t) > 0$. Hence, if the zeros of ϕ_{γ^*} on $[a,b]$ are isolated, it follows that γ is bang-bang. Moreover, if ϕ_{γ^*} has N zeros then γ has at most N switchings.

The derivative of ϕ_{γ^*} is easily seen to be

(8) $$\dot{\phi}_{\gamma^*}(t) = \langle\lambda(t), [X,Y](x(t))\rangle.$$

So, if $Y(x(t))-X(x(t))$ and $[X,Y](x(t))$ are linearly independent for each t, see that ϕ_{γ^*} and $\dot{\phi}_{\gamma^*}$ can never vanish simultaneously (because $_*(t) = \dot{\phi}_{\gamma^*}(t) = 0$ would imply $\lambda(t) = 0$, and so $\lambda(s) = 0$ for all s, because .a) is a linear homogeneous O.D.E. in λ). Moreover, if $\phi_{\gamma^*}(t_o) = 0$, then the lue of the Hamiltonian H at t_o is $2\langle\lambda(t_o), f(x(t_o))\rangle + \lambda_o$. Since $H+\lambda_o$ nishes along γ^*, and $\lambda_o \geq 0$, we conclude that

(9) $$\langle\lambda(t_o),\ f(x(t_o))\rangle \leq 0.$$

If $X(x(t_o))$ and $Y(x(t_o))$ are linearly independent (i.e. if $f(x(t_o))$ and $g(x(t_o))$ are linearly independent), then (9) cannot be an equality (since $\langle\lambda(t_o),\ g(x(t_o))\rangle = 0$, but $\lambda(t_o) \neq 0$). So we have

(10) $$\langle\lambda(t_o),\ f(x(t_o))\rangle < 0.$$

Let us say that a point p is an *A-point* if $X(p)$ and $Y(p)$ are linearly independent, and that it is a *B-point* if $Y(p)-X(p)$ and $[X,Y](p)$ are independent. Let $\Omega(A)$, $\Omega(B)$ denote, respectively, the sets of A points and of B points. Let $p \in \Omega(A) \cap \Omega(B)$. Then we can define $\omega(p)$ to be 1 if the bases $(f(p), g(p))$ and $(g(p), [f,g](p))$ have the same orientation, and we let $\omega(p) = -1$ if they have opposite orientation. (Since $[X,Y] = 2[f,g]$, the definition of $\omega(p)$ could also be stated, in an obvious way, in terms of X and Y.) Now suppose again that $\gamma = (u(\cdot), x(\cdot))$ is a trajectory, and that $\gamma^* = (u(\cdot), x(\cdot), \lambda(\cdot), \lambda_o)$ is an extremal lift of γ. Let $p = x(t_o)$ be a point such that $\phi_{\gamma^*}(t_o) = 0$, and suppose p is an A-point and a B-point. Then (8) and (10) imply that $\dot{\phi}_{\gamma^*}(t_o)$ has the same sign as $\omega(p)$. If γ is entirely contained in $\Omega(A) \cap \Omega(B)$, then $\omega(x(t))$ is the same for all t. Therefore, either $\dot{\phi}_{\gamma^*}(t_o) > 0$ at every point where $\phi_{\gamma^*}(t_o) = 0$ (if $\omega(x(t)) = 1$ for all t), or $\dot{\phi}_{\gamma^*}(t_o) < 0$ for all t_o such that $\phi_{\gamma^*}(t_o) = 0$ (if $\omega(x(t)) = -1$ for all t). In either case, γ is necessarily bang-bang with at most one switching. If $\omega = 1$, then the switching is from $u = 1$ to $u = 0$ (so that γ is a YX-trajectory) whereas, if $\omega = -1$, then γ is an XY-trajectory.

If $U \subseteq \Omega(A) \cap \Omega(B)$ is connected, define $\omega(U)$ to be the common value of the $\omega(p)$ for all $p \in U$. We have proved:

PROPOSITION 1. *Let* U *be a connected subset of* $\Omega(A) \cap \Omega(B)$.
(a) *If* $\omega(U) = +1$, *then every time-optimal trajectory which is contained in* U *is a* YX *trajectory, and*
(b) *if* $\omega(U) = -1$, *then every time-optimal trajectory in* U *is an* XY *trajectory.*

Proposition 1 completely characterizes the optimal trajectories in $\Omega(A) \cap \Omega(B)$. Let us refer to the points in $\Omega(A) \cap \Omega(B)$ as *ordinary points*. A point which is not ordinary will be called a *singular point*, or a *singularity*. Our problem is to study the optimal trajectories near singular points. From now on, we will use letters such as A, B, C, ... to denote conditions that a point may satisfy, and which are defined by a set of inequalities. Then the same letter with a prime will denote the negation of the condition. The conjunction will be denoted by juxtaposition. For instance, an AB point is what we have called an ordinary point. An A B point is a point which is in $\Omega(B)$ but not in $\Omega(A)$. If σ is any symbol such as A, B, AB, AB', AC, etc., we will use $S(\sigma)$ to denote the set of all σ-points. When $S(\sigma)$ is open, we will also write $\Omega(\sigma)$, instead of $S(\sigma)$. (E.g.: M is the disjoint union of $\Omega(AB)$, $S(A'B)$, $S(AB')$ and $S(A'B')$.)

The conditions that $p \in \Omega(A)$, $p \in \Omega(B)$ are equivalent to the nonvanishing of certain determinants. Precisely, let us write vectors as columns, and let $\det(w_1,w_2)$ denote the determinant of the 2×2 matrix whose columns are w_1, w_2. Let

(11.a) $$\Delta_A(p) = \det(f(p), g(p)),$$

(11.b) $$\Delta_B(p) = \det(g(p), [f,g](p)).$$

Then $\Omega(A)$, $\Omega(B)$ are the sets of points where $\Delta_A \neq 0$, $\Delta_B \neq 0$, respectively. So, the singularities are the points where one of the functions Δ_A, Δ_B vanishes. In order to study what happens there, it is natural to look first at the least degenerate cases, i.e. to require that some other inequalities be satisfied. For instance, let us define a C <u>point</u> to be a point p such that $\Delta_B(p)$ and grad $\Delta_B(p)$ do not both vanish. Then $\Omega(C)$ is an open subset of M, and $\Omega(B) \subseteq \Omega(C)$. The complement of $\Omega(B)$ in $\Omega(C)$ is $S(B'C)$, which is the zero set of a smooth function on $\Omega(C)$ which never vanishes together with its gradient. So $S(B'C)$ is a smooth one-dimensional submanifold of $\Omega(C)$, i.e. a locally finite union of smooth arcs.

Suppose p is an AB'C point. Then $\Delta_B(p) = 0$ but $(\text{grad}\,\Delta_B)(p) \neq 0$. Since X(p) and Y(p) are linearly independent, at least one of the inequalities

(12.a) $$X\Delta_B(p) \neq 0$$

(12.b) $$Y\Delta_B(p) \neq 0$$

holds. The case when both inequalities (12.a,b) hold is the least degenerate one, and so we give it a special name. We say that p is a D <u>point</u> if a) $\Delta_B \neq 0$ or b) $\Delta_B = 0$ but both inequalities (12.a,b) hold.

Suppose $p \in S(AB'D)$. Then X(p) and Y(p) are both nontangent to the smooth arc S(AB'D). If they point to opposite sides of this arc, then it is possible to find, at each point $q \in S(AB'D)$, close to p, a convex combination

(13) $$Z(q) = [1-u(q)]X(q) + u(q)Y(q), \quad 0 \leq u(q) \leq 1,$$

which is tangent to S(AB'D) at q. The solution of

(14) $$\dot{x}(t) = Z(x(t))$$

is a (parametrized) arc which is contained in S(AB'D) and is a trajectory of our system (2). Such an arc is referred to as a <u>singular arc</u>, or a <u>type</u> S arc. We will want to consider trajectories that are concatenations of bang-bang and singular arcs, and we use the notational convention described earlier (so that, for instance, an XYSY arc is a concatenation of at most four pieces, of which the first one is an X trajectory, the second and fourth are Y trajectories, and the third one is a singular arc).

If $p \in S(AB'D)$, then S(AB'D) divides a suitably small neighborhood U of p into two parts U_+, U_-. The function Δ_B has different signs on U_+ and U_-, but Δ_A does not vanish. Hence $\omega(U_+)$ and $\omega(U_-)$ are opposite. After relabelling

U_+ and U_-, if necessary, we may assume that

(15) $$\omega(U_+) = 1, \quad \omega(U_-) = -1.$$

Then we must distinguish two cases:

(I.1) $X(p)$ points towards U_+, and $Y(p)$ points towards U_-,

(I.2) $X(p)$ points towards U_- and $Y(p)$ points towards U_+.

A point $p \in S(AB'D)$ where (I.1) holds will be called a D_1 point, or a <u>turnpike point</u>. If (I.2) holds then p will be called a D_2 point, or an <u>antiturnpike point</u>. There is, finally, a third type of D point, namely, a point $p \in S(AB'D)$ such that

(I.3) $X(p)$ and $Y(p)$ point to the same side of $S(AB'D)$.

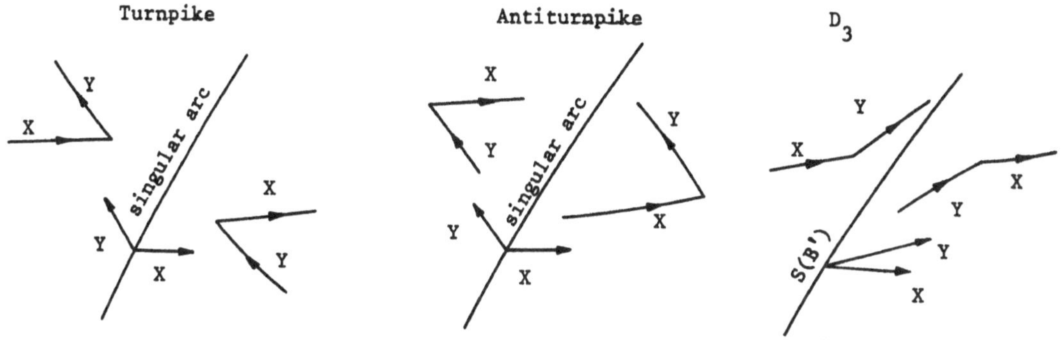

Figure 2 (In each case, some optimal trajectories are shown)

It is easy to see that, if $q \in \Omega(AB)$, then $\omega(q)$ has the same sign as $(\Delta_B\Delta_A)(q)$. Therefore, if $p \in S(AB'D)$, then U_+ is the side of $S(AB'D)$ where Δ_B has the same sign as Δ_A. So the three cases considered above are characterized by

(D_1) $$\Delta_A(p)(X\Delta_B)(p) > 0 > \Delta_A(p)(Y\Delta_B)(p),$$

(D_2) $$\Delta_A(p)(X\Delta_B)(p) < 0 < \Delta_A(p)(Y\Delta_B)(p),$$

(D_3) $$(X\Delta_B \cdot Y\Delta_B)(p) > 0.$$

Time-optimal trajectories are easiest to analyze in the D_1 and D_3 cases. Actually, we can prove a somewhat more global result:

PROPOSITION 2. <u>Suppose U is a connected open set, and Σ is a smooth connected arc which is contained in U and divides U into two parts. Suppose that $U-\Sigma \subseteq \Omega(AB)$, and that $\Sigma \subseteq S(AB'D)$. Then</u>

a) <u>Σ is a subset of one of the $S(AB'D_i)$, $i = 1,2,3$.</u>

b) <u>If $\Sigma \subseteq S(AB'D_1)$, then every time-optimal trajectory in U is of one of the following types</u>: XSX, XSY, YSX, YSY.

c) <u>If $\Sigma \subseteq S(AB'D_3)$ then every time-optimal trajectory in U is</u> XYX <u>or</u> YXY.

Proof. a) is trivial, because Σ is connected. To prove b), suppose $\gamma = (u(\cdot), x(\cdot))$ is time-optimal. If γ is entirely contained in Σ, then γ is clearly singular. If not, let $[a,b]$ be the interval where γ is defined, and let $[t_1,t_2]$ be a maximal closed subinterval of $[a,b]$ such that $x(t) \notin \Sigma$ for $t_1 < t < t_2$. Then the points $x(t)$, $t_1 < t < t_2$, all lie in one of the two components of $U-\Sigma$. Let $\tilde{U}$ denote this component. Suppose that $t_1 > a$. Then $x(t_1) \in \Sigma$. Moreover, for $t > t_1$ but sufficiently close to t_1, γ is either an X-trajectory or a Y-trajectory. In the former case, we have $\omega(\tilde{U}) = 1$. But this implies that, for $t_1 < t < t_2$, γ may switch from Y to X, but never the other way around. Hence γ remains an X-trajectory for ever and, in particular, γ never returns to Σ (because X points from Σ into $\tilde{U}$). So $t_2 = b$, and γ is an X trajectory for $t \geq t_1$. Similarly, if γ is a Y-trajectory for $t > t_1$ but close to t_1, then $t_2 = b$ and γ is a Y-trajectory for $t \geq t_1$. Also, if $t_2 < b$, it follows that $t_1 = a$ and that γ, restricted to $[t_1,t_2]$, is an X-trajectory or a Y-trajectory. So there are at most two maximal intervals $[t_1,t_2]$ such that $x(t) \notin \Sigma$ for $t_1 < t < t_2$, and any such interval contains an endpoint of $[a,b]$. Statement b) now follows easily.

The proof of c) is even easier. Every trajectory in U (optimal or not) crosses Σ at most once. The structure of optimal trajectories is completely described by Proposition 1, as long as they are contained in a component of $U-\Sigma$. So, if γ is a time-optimal trajectory in U, then γ must be bang-bang with at most three switchings (one for each of the pieces in $U-\Sigma$, and one as γ crosses Σ). The possibility of three switchings is easily excluded. Suppose, for instance, that γ is an XYXY trajectory. Then the second switching must occur as γ crosses Σ. Therefore the first and third switching occur in different components of $U-\Sigma$. But both are switchings from X to Y. If U_1, U_2 are the components of $U-\Sigma$, then $\omega(U_1) \neq \omega(U_2)$. So, by Proposition 1, a switching from X to Y can only occur in one of the components, but not in the other. So we have reached a contradiction.

Q.E.D.

Antiturnpike points are harder to analyze. The reason is that, contrary to the turnpike case, the permitted switchings always have the effect of leading back to Σ. So it is not completely obvious how to exclude the possibility of trajectories such as the one in Figure 3, with an arbitrarily large number of switchings. To exclude such trajectories, one needs a more sophisticated analysis, based on

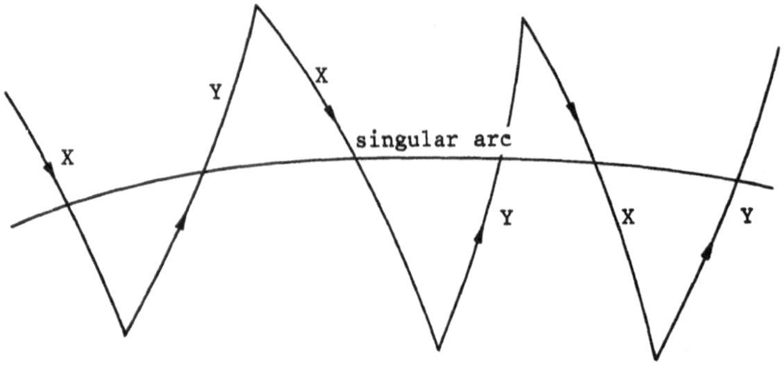

Figure 3

a study of <u>conjugate points</u>.

Suppose $\gamma = (u(\cdot), x(\cdot))$ is a trajectory. A <u>variational vector field</u> (vvf) <u>along</u> γ is a map $w(\cdot)$ from the domain of γ to $\mathbb{R}^2$, which is a solution of the variational equation

(16) $$\dot{w}(t) = ((1-u(t))D_xX(x(t))+u(t)D_xY(x(t)))\cdot w(t).$$

If $\gamma^* = (u(\cdot), x(\cdot), \lambda(\cdot), \lambda_o)$ is an extremal lift of γ, then it is easy to see that, if $w(\cdot)$ is a vvf along γ, then the function $t \to \langle\lambda(t), w(t)\rangle$ is a constant. Suppose now that $t_o < t_1$ satisfy $\phi_{\gamma^*}(t_o) = \phi_{\gamma^*}(t_1) = 0$. Let $w(\cdot)$ be the vvf such that $w(t_o) = g(x(t_o))$. Then $\langle\lambda(t_o), w(t_o)\rangle = 0$ and so $\langle\lambda(t_1), w(t_1)\rangle = 0$. Also $\langle\lambda(t_1), g(x(t_1))\rangle = 0$. Since $\lambda(t_1) \neq 0$, we conclude that $w(t_1)$ and $g(x(t_1))$ are linearly independent. This latter fact is therefore a necessary condition for γ to admit an extremal lift and, in particular, for γ to be optimal.

If $\gamma = (u(\cdot), x(\cdot))$, and if $t_1 < t_2$, we say that the points $x(t_1)$, $x(t_2)$ are <u>conjugate along</u> γ if $w(t_2)$ and $g(x(t_2))$ are linearly dependent, where $w(\cdot)$ is the vvf along γ such that $w(t_1) = g(x(t_1))$. Then we have proved:

PROPOSITION 3. <u>If</u> $\gamma = (u(\cdot), x(\cdot))$ <u>is a trajectory which has switchings at times</u> t_1, t_2, <u>then a necessary condition for</u> γ <u>to be time-optimal is that</u> $x(t_1)$ <u>and</u> $x(t_2)$ <u>be conjugate along</u> γ.

In order to prepare the ground for the analysis of antiturnpikes, we first consider the more general case of a point p such that

(II.a) $p \in S(AB')$,

(II.b) $(\Delta_A \cdot X\Delta_B)(p) < 0$.

If (II.a,b) hold, then we may choose coordinates (x_1,x_2) on an open set U, containing p, such that:

(III.i) p has coordinates $(0,0)$,

(III.ii) $S(AB') \cap U$ is the subset of U defined by $x_1 = 0$,

(III.iii) X has components $(1,0)$,

(III.iv) U corresponds, via the coordinates (x_1,x_2), to the square

(17) $$C(\varepsilon) = \{(x_1,x_2): |x_1| < \varepsilon, |x_2| < \varepsilon\}.$$

By means of the coordinates x_1, x_2, we shall identify U with $C(\varepsilon)$.

Let Y have components α, β. Then

(18.a,b) $$\Delta_A = \frac{1}{2}\beta, \quad \Delta_B = -\frac{1}{4}\xi,$$

where

(19) $$\xi = \alpha_1\beta + (1-\alpha)\beta_1.$$

(Here, if ρ is any function on $C(\varepsilon)$, ρ_i stands for $\frac{\partial\rho}{\partial x_i}$.)

The zero set of ξ on U is the x_2 axis. So ξ has the factorization

(20) $$\xi(x_1,x_2) = x_1\eta(x_1,x_2)$$

where η is a smooth function such that $\eta(p) = \xi_1(p) = -4(X\Delta_B)(p) \neq 0$. Hence we may assume that

(III.v) η never vanishes on U.

Since $\beta(p) \neq 0$, we may assume that $\beta(p) > 0$ (otherwise, change x_2 to $-x_2$). Then we may also assume that

(III.vi) $\beta > 0$ throughout U.

Clearly, (II.b) implies that $\eta(p) > 0$ and so $\eta > 0$ throughout U.

If q, q' are in U, let us write $q \sim q'$ if q and q' lie on an X-trajectory γ (i.e. both have the same x_2 coordinate) and are conjugate along γ.

LEMMA A. <u>If</u> ε <u>is small enough, then:</u>

(III.vii) <u>there exists a smooth function</u> ψ, <u>from</u> $C(\varepsilon)$ <u>to</u> $\mathbb{R}$, <u>such that</u> $\psi(0,x_2) = 0$, $\psi_1(0,x_2) = -1$, <u>for</u> $|x_2| < \varepsilon$, <u>and that, if</u> $q = (x_1,x_2)$, $q' = (x_1',x_2')$, $q \neq q'$, <u>then</u> $q \sim q'$ <u>if and only if</u> $x_2' = x_2$ <u>and</u> $x_1' = \psi(x_1,x_2)$.

<u>Proof.</u> Let $q = (x_1,x_2)$, $q' = (x_1',x_2)$. Then $q \sim q'$ iff

(21) $$\zeta(x_1, x_1', x_2) = 0$$

where

(22) $$\zeta(x_1, x_1', x_2) = \theta(x_1,x_2) - \theta(x_1',x_2),$$

(23) $$\theta(x_1,x_2) = \frac{\alpha(x_1,x_2)-1}{\beta(x_1,x_2)}$$

We have $\theta_1 = \beta^{-2}\xi$ and so, in particular

(24) $$\theta_1(0,x_2) = 0.$$

Clearly, $\zeta(x_1, x_1', x_2)$ vanishes when $x_1 = x_1'$, so it has a factorization

(25) $$\zeta(x_1, x_1', x_2) = (x_1'-x_1)\tilde{\zeta}(x_1, x_1', x_2)$$

where $\tilde{\zeta}$ is smooth. Hence $q \sim q'$ iff

(26) $$\tilde{\zeta}(x_1, x_1', x_2) = 0.$$

On the other hand, (24) implies that

(27) $$\theta(x_1,x_2) = \theta(0,x_2) + x_1^2\tilde{\theta}(x_1,x_2)$$

for some smooth $\tilde{\theta}$. If we write

(28) $$\tilde{\theta}(x_1',x_2) - \tilde{\theta}(x_1,x_2) = (x_1'-x_1)\theta^*(x_1,x_1',x_2),$$

we get

(29) $$\zeta(x_1,x_1',x_2) = (x_1'-x_1)[(x_1+x_1')\tilde{\theta}(x_1',x_2) + x_1^2\theta^*(x_1,x_1',x_2)]$$

so that

(30) $$\tilde{\zeta}(x_1, x_1', x_2) = (x_1+x_1')\tilde{\theta}(x_1',x_2) + x_1^2\theta^*(x_1, x_1', x_2).$$

Clearly

(31) $$\tilde{\theta}(0,x_2) = 2\theta_{11}(0,x_2) = (2\beta^{-2}\xi_1)(0,x_2).$$

So $\tilde{\theta}(0,0) \neq 0$, and $\tilde{\zeta}(0,0,0) = 0$, but $\frac{\partial\zeta}{\partial x_1'}(0,0,0) \neq 0$. Therefore, by the Implicit Function Theorem, there exist $\delta > 0$, $\delta' > 0$, such that $\delta < \varepsilon$, $\delta' < \varepsilon$, and a smooth $\psi: C(\delta) \to (-\delta',\delta')$ such that, if $|x_1| < \delta$, $|x_1'| < \delta'$, $|x_2| < \delta$, then (21) holds if and only if $x_1' = \psi(x_1,x_2)$. If we take our new ε to be $\min(\delta,\delta')$, then ψ has all the desired properties. Q.E.D.

Now, let U, ε, and the chart (x_1,x_2) be such that (III.i,...,vii) hold. We consider the situation depicted in Figure 4. That is, we assume that γ_1 and γ_2 are trajectories in U, with γ_1 to the left of $x_1 = 0$, and γ_2 to the right, and that γ_1 steers q_1 to $\hat{q}_1$, and γ_2 steers q_2 to $\hat{q}_2$, with each pair (q_1,q_2), $(\hat{q}_1,\hat{q}_2)$ lying on a horizontal line. Assume, moreover, that the x_2 coordinate is strictly increasing along both γ_1 and γ_2. Let $\tilde{\gamma}_1$ be the trajectory from q_1 to $\hat{q}_2$ obtained by concatenating γ_1 and the X-trajectory from $\hat{q}_1$ to $\hat{q}_2$, and let $\tilde{\gamma}_2$ be the concatenation of the X-trajectory from q_1 to q_2 and of γ_2. Let T_1, T_2 denote the corresponding times.

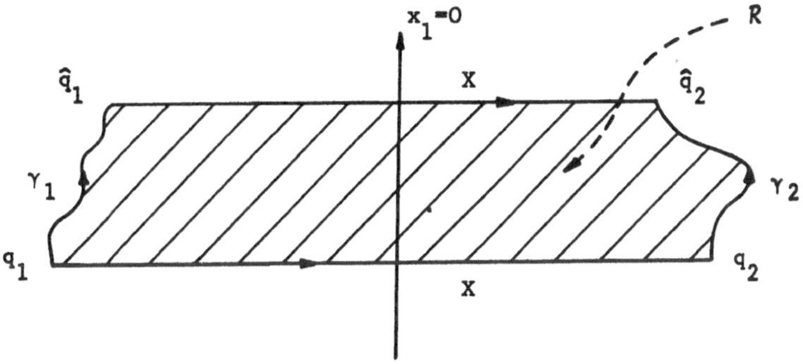

Figure 4

LEMMA B. <u>Suppose that, for each point</u> r_1 <u>on</u> γ_1, <u>the point</u> r_2 <u>where the X-trajectory</u> $\tau(r_1)$ <u>through</u> r_1 <u>meets</u> γ_2 <u>is conjugate to</u> r_1 <u>along</u> $\tau(r_1)$. <u>Then</u> $T_1 = T_2$.

<u>Proof</u>. Let σ be the 1-form such that $\langle\sigma,f\rangle = 1$, $\langle\sigma,g\rangle = 0$. Then

(32) $$\sigma = dx_1 + \frac{1-\alpha}{\beta}dx_2.$$

The time along any trajectory γ is, simply, $\int_\gamma \sigma$. So our conclusion will be proved if we show that

(33) $$\int_{\tilde{\gamma}_1^{-1}\tilde{\gamma}_2} \sigma = 0$$

i.e. that the integral of $d\sigma$ along the shaded region R of Figure 4 is zero.

But

(34) $$d\sigma = \left(\frac{1-\alpha}{\beta}\right)_1 dx_1 dx_2 .$$

If we integrate $\left(\frac{1-\alpha}{\beta}\right)_1$ along a horizontal segment going from r_1 on γ_1 to r_2 on γ_2, we get

$$\frac{1-\alpha}{\beta}(r_2) - \frac{1-\alpha}{\beta}(r_1),$$

which is equal to zero, because r_1 and r_2 are conjugate. So $\iint_R d\sigma = 0.$ Q.E.D.

Now, let p be such that (IIa,b) hold, and let U, ε, (x_1,x_2) be such that (III.i,...,vii) hold. Let U_o be the set of those $(x_1,x_2) \in U$ for which $|\psi(x_1,x_2)| < \varepsilon$. Let

$$K(x_1,x_2) = (\psi(x_1,x_2),\ x_2).$$

Clearly, K maps U_o to U_o diffeomorphically, and K^2 = identity. Let $K_*(q)$ denote the Jacobian matrix of K at q. Then $K_*(0,x_2)$ has rows $(-1,0)$, $(0,1)$. For $q \in C_o$, let

$$Y_*(q) = K_*(K(q))\cdot Y(K(q)).$$

Then it is clear that, if γ is a Y-trajectory which is contained in U_o, then $K\circ\gamma$ is a Y_*-trajectory.

LEMMA C. Suppose p satisfies (II.a,b), and let U, ε, (x_1,x_2) be such that (III.i,...,vii) hold. Let γ be a time-optimal YXY trajectory in U, which switches from Y to X at a point q_1 and then from X to Y at q_2. Then: a) $q_1 \in U_o$ and $q_2 \in U_o$. b) $Y_*(q_2)$ is not a linear combination of $Y(q_2)$ and $X(q_2)$ with positive coefficients.

Proof. Clearly, q_1 and q_2 are conjugate along γ. Since X-trajectories go from left to right, and the conjugate of a point to the left of $x_2 = 0$ must be to the right of $x_2 = 0$ (because $\theta_1(x_1,x_2) < 0$ for $x_1 < 0$, and $\theta_1(x_1,x_2) > 0$ for $x_1 > 0$), it follows that q_1 is to the left, and q_2 to the right, of the x_2 axis. Let γ_1 be the restriction of γ to an interval $[t_o-\delta,\ t_o]$, with t_o such that $\gamma(t_o) = q_1$. Let $\gamma_2 = K\circ\gamma_1$. Then γ_2 is a Y_*-trajectory. If $Y_*(q_2)$ is a positive combination of $X(q_2)$ and $Y(q_2)$, it follows that, if δ is small enough, then γ_2 is a trajectory of (2), after a suitable reparametrization. Then we can apply Lemma B (cf. Figure 5), and conclude that the times to go from $\gamma(t_o-\delta)$ to q by the two trajectories shown in Fig. 5 are equal. Since γ is time-optimal, it would follow that γ_2 is time-optimal, which is a contradiction, since γ_2 is contained in $\Omega(AB)$ but is not bang-bang.

Q.E.D.

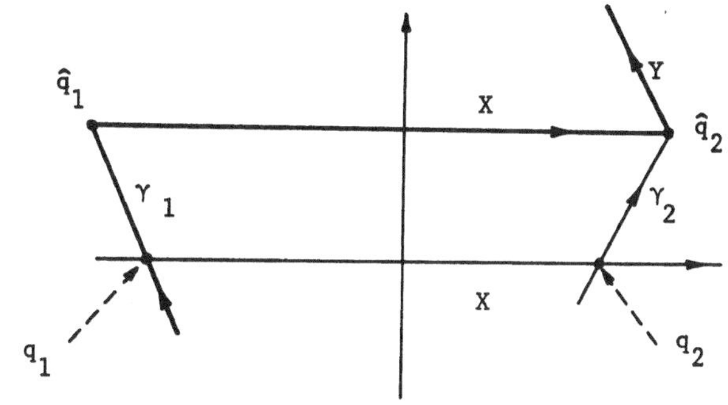

Figure 5

Now suppose that $p \in S(AB'D_2)$. Then we can select U, ε, and a chart (x_1, x_2) such that (III.i,...,vii) hold. Moreover, Y(p) points to the left of the x_2 axis, and so we may assume, by making ε even smaller that

(III.viii) $\alpha < 0$ throughout U.

Clearly, $Y_*(0,x_2) = (-\alpha(0,x_2), \beta(0,x_2))$, and therefore, by making ε even smaller, we may also assume:

(III.ix) both components of Y_* are positive on U_o.

Then it follows easily that Y_* is a positive combination of X and Y, and so Lemma C applies. Therefore no YXY trajectory in U, with two switchings, can be optimal. Since the roles of X and Y are interchangeable, we may assume, by making ε smaller, that the same conclusion holds for XYX trajectories, i.e. that

(III.x) no bang-bang trajectory in U which has two switchings is optimal.

To complete our analysis of antiturnpike points, we must exclude trajectories that are not bang-bang. First notice that, if (III.i,...,x) hold, then no point in $S(B') \cap U$ can be conjugate to a point not in $S(B') \cap U$ along an X-trajectory. By making ε smaller, we can assume that the same conclusion holds for Y-trajectories as well. So, if an optimal trajectory $\gamma = (u(\cdot), x(\cdot))$ in U is not bang-bang, and if $[t_o, t_1]$ is a maximal subinterval of the domain of γ such that $x(t) \notin S(B')$ for $t_o < t < t_1$, then $b_o = a$ or $t_1 = b$, and the restriction of γ to $[t_o, t_1]$ is an X- or Y-trajectory. So any time-optimal γ in U is XSX, or YSX, or XSY, or YSY. To conclude, we must prove that no singular arc in U can be optimal. Let $\gamma = (u(\cdot), x(\cdot))$ be singular. We can pick q_1, q_2 in $S(B') \cap U$ such that there is an XY-trajectory $\hat{\gamma}$ which goes from q_1 to q_2 in time $\hat{T}$, and is such that $q_i = x(t_i)$, $t_1 < t_2$. Let $T = t_2 - t_1$. Let R be the region bounded by γ and $\hat{\gamma}$. To prove that $\hat{T} < T$, we must show that $\iint_R d\sigma < 0$, where σ is the 1-form that was introduced in the proof of Lemma B. Using (34) and (19), we see that

$$d\sigma = -\xi\beta^{-2} dx_1 dx_2 .$$

Since R is contained in $\{(x_1,x_2): x_1 > 0\}$, whereas $\xi > 0$ when $x_1 > 0$, we see that $\iint_R d\sigma < 0$ and that $\hat{T} < T$. Therefore γ is not optimal.

We have proved:

PROPOSITION 4. Let $p \in S(AB'D_2)$. Then p has a neighborhood U such that every time-optimal trajectory in U is bang-bang with at most one switching.

It is now easy to analyze a slightly more degenerate situation, namely, when p is a point where an antiturnpike starts or ends in a nondegenerate way. Precisely, suppose p is an $AB'CD'$ point. Then p satisfies $\Delta_B(p) = 0$ and grad $\Delta_B(p) \neq 0$. Since $X(p)$ and $Y(p)$ are independent, at least one of the inequalities $X\Delta_B \neq 0$, $Y\Delta_B \neq 0$ holds at p. Since $p \in S(D')$, one of those inequalities fails to hold, i.e. we have $X\Delta_B = 0$ or $Y\Delta_B = 0$. Let us say that p is an E point if, when the equality $X\Delta_B = 0$ or $Y\Delta_B = 0$ holds, then the inequality $X^2\Delta_B \neq 0$, or $Y^2\Delta_B \neq 0$, is satisfied. If $p \in S(AB'CD'E)$, suppose that $X\Delta_B(p) \neq 0$ but $Y\Delta_B(p) = 0$. Then Y is tangent to $S(B')$ at p, but the inequality $Y^2\Delta_B(p) \neq 0$ implies that p is an isolated point of tangency of Y to $S(B')$, and that the component of Y in a direction transversal to that of $S(B')$ changes sign at p. So, for $q \in S(B')$, q on one side of p, the vectors $X(q)$, $Y(q)$ point to opposite sides of $S(B')$, while they point to the same side of $S(B')$ for q on the other side of p. So p is a "transition point", which separates an arc of $AB'D_3$ points from an arc of $AB'D_1$ or $AB'D_2$ points. That is, p is a "turnpike beginning or end" (notation: E_1) or an "antiturnpike beginning or end" (notation: E_2).

Let us study the optimal trajectories near a $p \in S(AB'CD'E_2)$. Suppose that (II.b) holds, and that $Y\Delta_B(p) = 0$, $Y^2\Delta_B(p) \neq 0$. (The case when $(\Delta_A \cdot Y\Delta_B)(p) < 0$, $X\Delta_B(p) = 0$, $X^2\Delta_B(p) \neq 0$, is similar.) Then we can choose U, ε, and coordinates x_1, x_2, so that (III.i,...,vii) hold. Since $Y\Delta_B(p) = 0$, $Y^2\Delta_B(p) \neq 0$, we can conclude that $\alpha(0,0) = 0$, $\alpha_2(0,0) \neq 0$. Let us assume that $\alpha_2(0,0) < 0$. (The case when $\alpha_2(0,0) > 0$ is similar.) Then $Y(q)$ points left for $q \in S(B') \cap U$ above p, and it points right for q below p. The vector field Y_* is defined near p. Let $Y_* = (\alpha_*,\beta_*)$. Then $\alpha_*(0,x_2) = -\alpha(0,x_2)$. Therefore $\alpha_*(0,0) = 0$, $\alpha_{*2}(0,0) \neq 0$. By the Implicit Function theorem, there are functions σ, σ_* on an interval $(-\delta,\delta)$, with values in $(-\varepsilon,\varepsilon)$, such that $\alpha(x_1,x_2) = 0$ iff $x_2 = \sigma(x_1)$, and $\alpha_*(x_1,x_2) = 0$ iff $x_2 = \sigma_*(x_1)$, for $|x_1| < \delta$, $|x_2| < \varepsilon$. Since the Y and Y_* trajectories ρ, ρ_* through p are tangent to the x_2 axis at p we may assume, by making ε smaller, that $\alpha(x_1,x_2) < 0 < \alpha^*(x_1,x_2)$ for all (x_1,x_2) such that $x_2 > 0$ and that (x_1,x_2) is in the region R that lies between ρ and ρ^* (cf. Figure 6). Then, on R, Y^* is a positive combination of X and Y. So Lemma C applies, and we can conclude that no time-optimal trajectory in R can contain a YXY piece with two switchings.

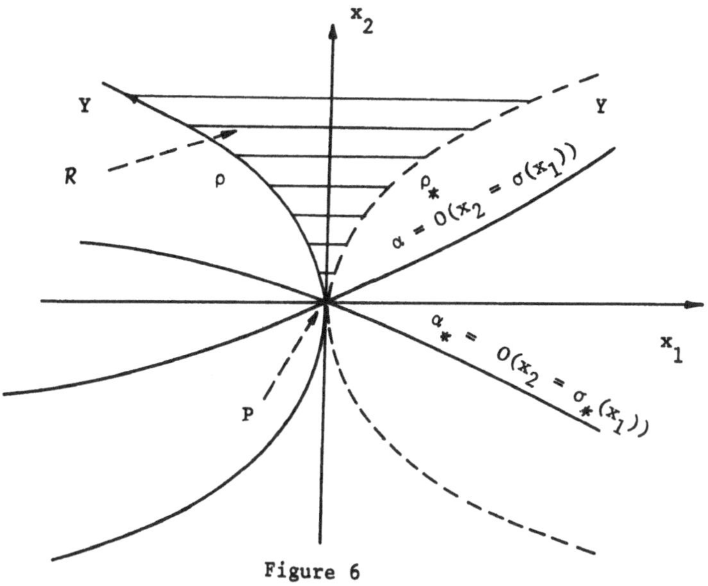

Figure 6

If γ is a time-optimal trajectory in U, then γ is bang-bang, because every point of U (with the possible exception of p), has a neighborhood V such that every optimal trajectory in V is bang-bang. Let P be the region above the x_1 axis and to the right of ρ. Then γ consists of at most two pieces γ_1, γ_2, one in U-P, and the other one in P. The part γ_1 in U-P can be studied using Proposition 2. We conclude that γ_1 is bang-bang with at most two switchings. As for γ_2, it is clear that it must be bang-bang.

Suppose γ_2 contains a YXY piece with two switchings. If the switchings occur at q_1, q_2 (in that order), then q_1 is to the left of the x_2 axis, and q_2 to the right. But q_1 is between ρ and the x_2 axis, and so q_2 is between the x_2 axis and ρ_*. So γ_2 actually contains a YXY piece in R, with two switchings. As pointed out above, this implies that γ_2 is not time-optimal, which is a contradiction. So γ_2 is an XYX trajectory. Therefore γ is a YXYXYX trajectory. We have proved:

PROPOSITION 5. <u>If</u> $p \in S(AB'CD'E_2)$, <u>then</u> p <u>has a neighborhood</u> U <u>such that every time-optimal trajectory in</u> U <u>is bang-bang with at most five switchings</u>.

A similar (in fact, easier) proof applies if $p \in S(AB'CD'E_1)$. In this case, we can split γ into γ_1 and γ_2 as above, and we can apply Proposition 2.c to γ_1, and Proposition 2.b to γ_2, to conclude:

PROPOSITION 6. <u>If</u> $p \in S(AB'CD'E_1)$, <u>then</u> p <u>has a neighborhood</u> U <u>such that every time-optimal trajectory in</u> U <u>is a concatenation of at most six pieces, each of which is either an</u> X, <u>or a</u> Y, <u>or an</u> S <u>trajectory (but no more than one piece is</u> S).

So far, we have analyzed several types of LBC's of X and Y at a point p, and shown how to characterize the time-optimal trajectories near p. We now list the inequalities and equalities characterizing each case, as well as the codimension c:

AB: $\Delta_A \neq 0$, $\Delta_B \neq 0$ (c = 0).

AB'CD: $\Delta_A \neq 0$, $\Delta_B = 0$, $X\Delta_B \neq 0$, $Y\Delta_B \neq 0$ (c = 1).

AB'CD'E: $\Delta_A \neq 0$, $\Delta_B = 0$, $X\Delta_B \neq 0$, $Y\Delta_B = 0$, $Y^2\Delta_B \neq 0$

or $\Delta_A \neq 0$, $\Delta_B = 0$, $X\Delta_B = 0$, $Y\Delta_B \neq 0$, $X^2\Delta_B \neq 0$ (c = 2).

The preceding list does not contain all possible cases with $c \leq 2$. The other possibilities arise when $\Delta_A = 0$ but $\Delta_B \neq 0$ (A'B points) or when both Δ_A and Δ_B vanish. It turns out that a complete analysis of all the cases of codimension ≤ 2 can be given, although the analysis becomes rather complicated, especially for the A'B' case. The details will be given elsewhere (cf. [2]).

For completeness, we remark that all the systems of equalities and inequalities listed above can be reformulated as statements about LBC's. (For instance, the condition that $\Delta_B = 0$ but $X\Delta_B \neq 0$ is equivalent to the statement that Y-X and [X,Y] are dependent, but Y-X and [X,[X,Y]] are independent.) This can be done easily, and we omit the details.

Finally, let us point out that, by Thom's Transversality Theorem, systems of the form (2) in the plane will, generically, only have singularities of codimension ≤ 2. Hence our results provide, in particular, a complete classification of the time-optimal trajectories for generic problems in the plane.

REFERENCES

1. Lobry, C., Controlabilité des Systèmes Non Linéaires, SIAM J. Control 8 (1970), pp. 573-605.

2. Sussmann, H. J., Lie brackets and time-optimal control in the plane, to appear.

CONTINUOUS-TIME QUANTUM MECHANICAL FILTER

T. J. Tarn, John W. Clark, C. K. Ong, G. M. Huang
Washington University
St. Louis, Missouri 63130

ABSTRACT

Attempts in recent years to detect gravitational radiation have led to the introduction of a new concept called Quantum Nondemolition Measurement. We apply this concept to the problem of demodulating optical signals and obtain a continuous time quantum mechanical filter. The advantage of our approach over previous approaches is that no optimization is necessary and we can construct the filter systematically.

1. INTRODUCTION

Recent interest in optical communication has stimulated the development of new methods of detection and estimation which incorporate the axioms of quantum mechanics. These methods come under the name of Quantum Detection and Estimation. Helstrom, Kennedy, Holevo, and others [1-4] have developed minimum mean-square estimators for estimation of parameters of electromagnetic fields. Baras [5, 6] and our group [7, 8] have generalized the problem to estimation of a member of a random sequence by considering quantum-mechanical measurements at discrete time intervals. Attempts have also been made by Davies and Baras [9-11] toward a formulation of a continuous-time quantum-mechanical filter. The latter formulation involves the analysis of operator differential equations satisfied by the density operator. The essential problem in either case is to choose the quantum measurement process and construct the estimator as a functional of the measurement results, in such a manner that the error is minimized. The above investigations have culminated in a beautiful mathematical theory of optimal quantum mechanical filters. Unfortunately, this theory assumes one can make a precise measurement of one observable or another; it is incomplete in the important practical sense that it does not tell us how such precise measurements can be realized.

Attempts in recent years to detect gravitational radiation have led to the introduction of a new concept called Quantum Nondemolition Measurement. Ordinarily, if we carry out successive precise measurements of an observable quantity of a quantum-mechanical system (e.g., position), the contamination of the quantum state due to free evolution under the system Hamiltonian will rule out the possibility of definite and complete prediction of the results of the second and further measurements on the basis of the result of the first. However, it turns out that one will, in general,

This research was supported in part by the National Science Foundation under Grant Nos. ECS-8017184 and INT-7902976.

be able to identify certain observables (called quantum nondemolition (QND) observables), such that this contamination does not occur -- i.e. such that the result of each measurement after the first is completely predictable, in the absence of external agents (e.g., gravitational waves) acting on the system. A QND measurement, of course, consists of a sequence of precise measurements of a QND observable.

We shall apply the quantum nondemolition idea to the problem of demodulating optical signals. The advantage of our approach over previous approaches [9-11] is that no optimization is necessary and we can construct the quantum nondemolition filter systematically.

Because of lack of space, we shall only present the results and omit proofs, which may be found in [12].

2. PROBLEM FORMULATION

We consider systems described by the following Hamiltonian:

$$H = H_0 + u(t)H_1,$$

where H_0 is the time-independent Hamiltonian of the unperturbed radiation field, the operator H_1 depends on the modulation scheme used, and $u(t)$ is a bounded, real analytic signal. For optical communication systems, such a Hamiltonian includes the common modulation schemes such as AM and PM. The dynamics of the state $\psi(t)$ is given by the Schrodinger equation

$$i\hbar \frac{d\psi}{dt} = (H_0 + u(t)H_1)\psi, \quad \psi \in S_H, \quad \psi(0) = \psi_0. \tag{1}$$

Here H_0 and H_1 are self-adjoint operators and S_H, denotes the unit sphere of some underlying Hilbert space H. Without loss of generality we take $\hbar = 1$ and divide (1) by i to obtain

$$\frac{d\psi}{dt} = (H_0 + u(t)H_1)\psi, \quad \psi \in S_H, \quad \psi(0) = \psi_0. \tag{2}$$

The new H_0 and H_1 are skew-adjoint operators and (2) is an infinite-dimensional bilinear system.

Suppose we make measurements of a physical quantity C with corresponding observable C, a self-adjoint operator acting in H. With the system in state ψ, the expected value of the measurement result is [13]

$$y(t) = (C(t)\psi, \psi), \tag{3}$$

where $(\cdot,\cdot)$ denotes the Hilbert-space inner product. Our problem is to recover u(t) from the appropriate derivatives of the measurement outcomes, i.e. we are faced with an invertibility problem.

All measurements considered herein are assumed to be "of the first kind" [14]. The properties of such a measurement are the following: (i) Suppose the physical system is in an eigenstate of C, with eigenvalue c_i, at the time of observation; then the outcome of the measurement is precisely equal to that eigenvalue. (ii) Let the system be in an arbitrary state ψ at the time of measurement, and suppose the outcome c_n(necessarily one of the eigenvalues of C) is obtained; then the measurement leaves the system in an eigenstate of C corresponding to that outcome ("collapse of the wave pocket").

In general the outcome in situation (ii) is uncertain and can be predicted only in a statistical sense. Thus, by a basic postulate of quantum mechanics, the hypothesized result c_n is realized with probability $|(P_n \psi, \psi)|^2$, where P_n is the projection operator onto the subspace of H of eigenvalue c_n of C. There is evidently a fundamental acausality (an "uncontrollable disturbance") endemic to quantum observation. Consequently, we must content ourselves with a nondeterministic formulation of the problem posed by (2), (3) -- except in the special circumstance that ψ is an eigenstate of C and remains one through subsequent evolution and observation. This "escape clause" leads directly to the idea of quantum nondemolition (QND) measurement [15-17]. QND measurement consists of a time sequence of precise, instantaneous measurements of a special sort of observable, conforming to one or both of the following two definitions.

Definition 1. In the absence of any signal ($u(t) \equiv 0$), C qualifies as a quantum nondemolition observable (QNDO) iff the result of each measurement of C after the first is uniquely determined by the outcome of the first measurement. (For simplicity of presentation we assume C has non-degenerate spectrum. If not the initial measurement must in general include the determination of some other quantities in addition to C.)

Definition 2. C qualifies as a quantum nondemolition filter (QNDF) iff in the presence of an arbitrary signal of the given class, a sequence of measurements of C can reveal with arbitrary accuracy the time dependence of the signal u(t).

If C is a QNDF observable, then y is truly the measurement outcome and not merely the expected value, and therefore (2) and (3) completely describe the system and the measurement device. In addition to our earlier delineations of the operators H_0, H_1, and C and the signal u(t), these assumptions will be made:

(a) The Lie algebra $A \equiv \mathcal{L}(H_0, H_1)$ generated by the skew-adjoint operators H_0, H_1 is finite dimensional.

(b) The observable C has the structure $C(t) = \sum_{p=1}^{N} \gamma_p(t)\, iQ_p$, where the functions $\gamma_p(t)$ are real analytic in t and the Q_p are time independent skew-adjoint operators.

3. MAIN RESULTS

It is presumed throughout that the quantum system (2) and (3) admits an analytic domain $\mathcal{D}_\omega$. Sufficient conditions for the existence of an analytic domain are given in refs. [18, 19]. On such a domain, standard techniques, developed for the analysis of finite-dimensional bilinear systems, are applicable [20].

If (2) and (3) is invertible on $\mathcal{D}_\omega$, the system will be called analytically invertible. Since our system is nonlinear, invertibility is contingent on the initial state; we take the lead of Hirschorn [21] in formulating suitable definitions.

Definition 3. Let M be some submanifold on S_H. Then:

(i) System (2) and (3) is analytically invertible at $\psi_0 \in M \cap \mathcal{D}_\omega$ if distinct inputs $u_1(t)$, $u_2(t)$ give rise to distinct outputs, i.e., $y(t, u_1, \psi_0) \neq y(t, u_2, \psi_0)$.

(ii) System (2) and (3) is strongly analytically invertible at ψ_0 if there exists an open neighborhood N of ψ_0 such that analytic invertibility holds at ϕ for all $\phi \in N \cap \mathcal{D}_\omega$.

(iii) System (2) and (3) is strongly analytically invertible if there exists an open submanifold M_0 of M, dense in M, such that strong analytic invertibility holds at ψ, for all $\psi \in M_0 \cap \mathcal{D}_\omega$.

Definition 4. The relative order μ of system (2) and (3) is the smallest positive integer k such that $[C_{k-1}(t), H_1] \neq 0$ for almost all t, where for almost all t means that the Lebesgue measure of $\{t \mid [C_{k-1}(t), H_1] = 0\}$ is zero. The operator $C_k(t)$ for arbitrary positive integer k is given by the recursive relation

$$C_k(t) = [C_{k-1}(t), H_0] + \frac{\partial}{\partial t} C_{k-1}(t),$$

with

$$C_{k=0}(t) = C_0(t) = C(t).$$

Definition 5. The inverse submanifold for the system (2) and (3), having relative order μ, is defined as

$$M = \{\phi \varepsilon M \cap \mathcal{D}_\omega \mid ([C_{\mu-1}(t), H_1]\phi, \phi) \neq 0 \text{ for almost all } t\}.$$

Theorem 1. Given that system (2) and (3) admits an analytic domain $\mathcal{D}_\omega$, the system is strongly analytically invertible if its relative order μ is finite. If indeed μ is finite, and if the initial state ψ_0 belongs to the inverse submanifold M_μ, the system specified by

$$\frac{d}{dt}\hat{\psi}(t) = a(\hat{\psi}(t)) + \hat{u}(t)\, b(\hat{\psi}(t)), \quad \psi(0) = \psi_0,$$

$$\hat{y}(t) = d(\hat{\psi}(t)) + \hat{u}(t)\, e(\hat{\psi}(t)).$$

provides a (left) inverse for the quantum control system (2) and (3), with

$$a(\hat{\psi}\,(t)) = H_0\hat{\psi} - ([C_{\mu-1}, H_1]\hat{\psi}, \hat{\psi})^{-1}([C_\mu, H_0]\hat{\psi}, \hat{\psi})H_1\hat{\psi},$$

$$b(\hat{\psi}(t)) = ([C_{\mu-1}, H_1]\hat{\psi}, \hat{\psi})^{-1}\, H_1\hat{\psi},$$

$$d(\hat{\psi}(t)) = -([C_{\mu-1}, H_1]\hat{\psi}, \hat{\psi})^{-1}([C_\mu, H_0]\hat{\psi}, \hat{\psi})H_1\hat{\psi},$$

$$e(\hat{\psi}(t)) = ([C_{\mu-1}, H_1]\hat{\psi}, \hat{\psi})^{-1}\, \hat{\psi}.$$

Corollary. Suppose (2) and (3) admits an analytic domain and also assume that $d^\rho C/dt^\rho$ vanishes for some positive integer ρ. Then $\mu < \infty$ provides a necessary as well as sufficient condition for the system to be invertible.

We define a sequence of sets of operators for any positive integer k by the following recursive relation

$$\Lambda^{(k)} = \begin{cases} \{C_k\}, & k < \mu \\ \{\dot{L} + [L, H_0], [L, H_1] \,|\, L \in \Lambda^{(k-1)}\}, & k \geq \mu . \end{cases}$$

Theorem 2. Observable C in (3) is a QNDF (Continuous-Time Quantum Mechanical Filter) iff system (2) and (3) is invertible and $[C, L] = 0$, $\forall\, L \in \Lambda^{(k)}$ $\forall k$.

Example (Electrooptic Amplitude Modulation). Take

$$H = -i\,\omega\, a^+ a + u(t)\,(a^+ - a),$$

where a and a^+ are the annihilation and creation operators, respectively, for the mode in question. It is easy to check that $C = ae^{i\omega t} + a^+e^{-i\omega t}$ is a QNDF.

4. EPILOGUE

The underlying idea of quantum filtering pursued in this paper is drawn from Helstrom [1]. We have adapted the concept of "quantum nondemolition measurement" [15-17] to the investigation of the continuous-time quantum mechanical filtering problem. We showed that the well-known property of invertibility from systems theory together with the property that the output observable C is a QNDO furnish a necessary and sufficient condition for the existence of a QNDF for continuous-time quantum mechanical systems.

The result presented in this paper is for a scalar signal. The case of multiple signals requires a more elaborate treatment along the same line. This problem and the problem of construction of a quantum mechanical filter will be addressed in another paper.

5. REFERENCES

1. C. W. Helstrom, Quantum Detection and Estimation Theory, Academic Press, 1976.
2. C. W. Helstrom, R. S. Kennedy, "Noncommuting Observables in Quantum Detection and Estimation Theory", IEEE Trans. Inform. Theory, Jan. 1974.
3. C. W. Helstrom, et. al., "Quantum Communication Theory", Proc. IEEE, Oct. 1970.
4. A. S. Holevo, "Statistical Decision Theory for Quantum Systems", J. of Multivariate Analysis, 1973.
5. J. Baras, et. al., "Quantum-Mechanical Linear Filtering of Random Signal Sequences", IEEE Trans. Inform. Theory, Jan. 1976.
6. J. Baras, R. O. Hargar, "Quantum-Mechanical Filtering of Vector-Signal Processes", IEEE Trans. Inform. Theory, Nov. 1977.
7. D. Ilic, Simultaneous Quantum Mechanical Observations and Nonlinear Quantum Stochastic Filter, D.Sc. Dissertation, Washington Univ. (St. Louis), May 1978.
8. D. Ilic, T. J. Tarn, "Nonlinear Quantum Stochastic Filter for Discrete Time Systems", Pro. of the International Symposium on Mathematical Theory of Networks and Systems, Delft University of Technology, July 3-6, 1979.
9. E. B. Davies, "Quantum Communication Systems", IEEE Trans. Inform. Theory, July 1977.
10. J. Baras, "Stochastic Partial Differental Equations in Optical Communication Problems", Proc. 3rd US-Italy Seminar on Variable Structure Systems, Taormina 1977.
11. J. Baras, "Continuous Quantum Filtering", Proc. 15th Allerton Conf., 1977.
12. C. K. Ong, Quantum Nondemolition Filters, D.Sc. Dissertation, Washington University (St. Louis), February, 1982.
13. A. Messiah, Quantum Mechanics, John Wiley, 1958.
14. J. M. Jauch, Foundations of Quantum Mechanics, Addison-Wesley, 1968.

15. C. M. Caves, K. S. Thorne, R. W. P. Drever, V. D. Sandberg, M. Zimmermann, "On the Measurement of a Weak Classical Force Coupled to a Quantum-Mechanical Oscillator. 1. Issues of Principle", Reviews of Modern Physics, Vol. 52, No. 2, Part 1, pp. 341-392, April 1980.

16. W. G. Unruh, "Analysis of Quantum-Nondemolition Measurement", Physical Review D, Vol. 18, No. 6, pp. 1764-1772, September 1978.

17. V. B. Braginskii and Yu. I. Vorontsov, "Quantum-Mechanical Limitations in Microscopic Experiments and Modern Experimental Technique", Soviet Physics-Uspekhi, Vol. 17, No. 5, pp. 644-650, March-April 1975.

18. E. Nelson, "Analytic Vectors", Annals of Mathematics, Vol. 70, pp. 572-615, 1959.

19. A. O. Barut and R. Raczka, Theory of Group Representations and Applications, Polish-Scientific Publishers, Warszawa, 1977.

20. G. M. Huang, T. J. Tarn, John W. Clark, "On the Controllability of Quantum-Mechanical Systems", submitted for publication.

21. R. N. Hirschorn, "Invertibility of Nonlinear Control Systems", SIAM J. Control and Optimization, Vol. 17, No. 2, pp. 289-297, March 1979.

APPROXIMATE DISTURBANCE DECOUPLING
BY MEASUREMENT FEEDBACK

JAN C. WILLEMS
Mathematics Institute
P.O. Box 800
9700 AV GRONINGEN
The Netherlands

ABSTRACT

The purpose of this paper is to outline, in a self-contained style, but without giving details of the proofs, the necessary and sufficient condition for the solvability of the approximate disturbance problem by measurement feedback (ADDPM) for linear finite-dimensional time-invariant systems.

PROBLEM STATEMENT

0. Notation: $\mathbb{R}$ denotes the real line, $\mathbb{R}^+ = [0,\infty)$, $\mathbb{R}^n$ n-dimensional vectorspace, and $\mathbb{R}^{p\times q}$ the (p×q) real matrices. $L_p(\mathbb{R}^+;\mathbb{R}^n)$ denotes the L_p-space with domain $\mathbb{R}^+$ and codomain $\mathbb{R}^n$, while $L^{loc}(\mathbb{R}^+;\mathbb{R}^n)$ denotes the space of maps from $\mathbb{R}^+$ into $\mathbb{R}^n$ whose restriction to a compact set is integrable. $\mathbb{C}$ denotes the complex plane and σ the spectrum (i.e. the set of eigenvalues) of a square matrix . Finally, $\mathbb{Z}$ ($\mathbb{Z}^+$) denotes the set of (nonnegative) integers.

1. Consider the linear finite-dimensional time-invariant plant

$$\boxed{\Sigma_p : \dot{x} = Ax + Bu + Gd, \quad y = Cx, \; z = Hx}$$

with $x \in \mathbb{R}^n$, the state, $u \in \mathbb{R}^m$, the control input, $d \in \mathbb{R}^q$, the disturbance input, $y \in \mathbb{R}^p$, the measured output, and $z \in \mathbb{R}^\ell$ the (to-be-) controlled output. The plant is thus specified by the parameter matrices $\Sigma_p \approx (A,B,G,C,H)$. We will assume throughout that the input signals $\underline{u}$, $\underline{d}$ belong to $L^{loc}(\mathbb{R}^+, \mathbb{R}^m)$ and $L^{loc}(\mathbb{R}^+; \mathbb{R}^q)$, respectively. This implies that for any initial condition $\underline{x}(0) = x_o$, there are unique well-defined output signals $\underline{y}$, $\underline{z}$ which belong to $L^{loc}(\mathbb{R}^+; \mathbb{R}^p)$ and $L^{loc}(\mathbb{R}^+; \mathbb{R}^\ell)$, respectively.

2. One of the most easily motivated control synthesis problems is that of (approximate) disturbance decoupling. There the question is to come up with a feedback compensator Σ_{fb} such that in the closed loop system $\Sigma_{c\ell}$ the disturbances have (almost) no influence on the controlled output (see figure). It may be shown that

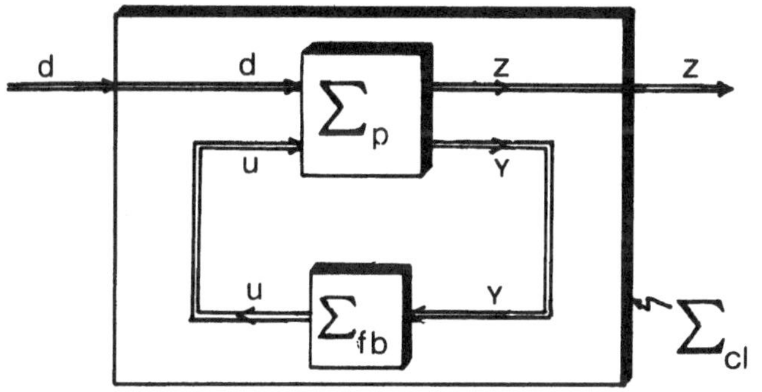

for the problems which we will consider there is no loss of generality in assuming that the feedback compensator Σ_{fb} is selected from the same category as the plant Σ_p, i.e., we will take Σ_{fb} to be linear finite-dimensional and time-invariant as well:

$$\Sigma_{fb} : \dot{w} = Kw + Ly, \quad u = Mw + Fy$$

with $w \in \mathbb{R}^r$, the state of the feedback compensator. The nonnegative integer r and the compensator matrices $\Sigma_{fb} \approx (K,L,M,F)$ are to be chosen. The closed loop system $\Sigma_{c\ell}$ may be obtained by substituting the equations of Σ_{fb} in Σ_p which yields

$$\Sigma_{c\ell}: \begin{bmatrix} \dot{x} \\ \dot{w} \end{bmatrix} = \begin{bmatrix} A+BFC & BM \\ LC & K \end{bmatrix} \begin{bmatrix} x \\ w \end{bmatrix} + \begin{bmatrix} G \\ 0 \end{bmatrix} d \;, \quad z = \begin{bmatrix} H & 0 \end{bmatrix} \begin{bmatrix} z \\ w \end{bmatrix}$$

which, when written compactly in the obvious notation with $x_e : = \begin{bmatrix} x \\ w \end{bmatrix}$, becomes

$$\Sigma_{c\ell} : \dot{x}_e = A_e x_e + G_e d, \quad z = H_e x_e$$

3. The response $\underline{z}$ is given in terms of $\underline{x}_e(0)$ and $\underline{d}$ by

$$\underline{z}(t) = H_e e^{A_e t} \underline{x}_e(0) + (\underline{W}_{c\ell} * \underline{d})(t)$$

where * denotes convolution and $\underline{W}_{c\ell} : \mathbb{R}^+ \to \mathbb{R}^{\ell \times q}$ denotes the closed loop impulse response d → z, i.e. $\underline{W}_{c\ell}(t) : = H_e e^{A_e t} G_e$. Let us now compute the closed loop tranfer function. The plant is described by the transfer function $\begin{bmatrix} u \\ d \end{bmatrix} \to \begin{bmatrix} y \\ z \end{bmatrix}$ given by

$$G(s) = \begin{bmatrix} G_{11}(s) & G_{12}(s) \\ G_{21}(s) & G_{22}(s) \end{bmatrix} = \begin{bmatrix} C \\ H \end{bmatrix} (Is-A)^{-1} [B \;\; G].$$

The feedback compensator has the transfer function $X(s) = F + M(Is-K)^{-1}L$ (we use X because in synthesis problems it is the unknown). Eliminating u from the equations yields as transfer function d → z in $\Sigma_{c\ell}$:

$$G_{c\ell}(s) = G_{22}(s) + G_{21}(s)[I - X(s)G_{11}(s)]^{-1} X(s)G_{12}(s)$$

$G_{c\ell}$ is a $(\ell \times q)$-matrix of strictly proper rational functions.

4. In the (exact) disturbance decoupling problem by measurement feedback (DDPM) one wants z to be independent of d, i.e., the problem is to choose Σ_{fb} such that $\underline{W}_{c\ell} = \underline{0}$, or equivalently, $G_{c\ell} = 0$. One can combine this requirement with conditions on the transient behavior of $\Sigma_{c\ell}$, for example by requiring that the spectrum of A_e should be contained in a (non-empty, symmetric w.r.t. the real axis, and containing at least one point of the real axis) set $\mathbb{C}_g \subset \mathbb{C}$. This problem is called that (exact) disturbance decoupling problem with measurement feedback and stability (DDPMS). If one asks that this should be possible for any $\mathbb{C}_g$ then we speak of disturbance decoupling by measreument feedback with pole placement (DDPMPP). These problems have recently been solved [1-4]. We will mention their solution later on.

5. Our main interest in the present paper is in an approximate version of this problem. The influence of the disturbance d on the controlled output z is given by $\underline{z} = \underline{W}_{c\ell} * \underline{d}$, where * denotes convolution. In disturbance decoupling it makes a great deal of sense to measure the degree of disturbance decoupling by means of the induced norm of this convolution operator. We would then say that the approximate disturbance decoupling problem is solvable if this induced norm can be made arbitrarily small by choosing Σ_{fb}. We denote this induced norm by $\|\Sigma_{c\ell}\| := \sup\limits_{\underline{d} \neq 0} \frac{\|\underline{W}_{c\ell} * \underline{d}\|_Z}{\|\underline{d}\|_D}$, where $\|\cdot\|_Z$ and $\|\cdot\|_D$ denote norms on the space of disturbance inputs and controlled outputs respectively. Employing L_p-norms, for example, yields

$$\|\Sigma_{c\ell}\| = \sup_{\underline{d} \neq 0} \frac{\|\underline{W}_{c\ell} * \underline{d}\|_{L_p(\mathbb{R}^+;\mathbb{R}^\ell)}}{\|\underline{d}\|_{L_p(\mathbb{R}^+;\mathbb{R}^q)}}$$

For $p = 1, 2, \infty$ explicit conditions for approximate disturbance decoupling may thus be derived. For $p = 1, \infty$ this requires that for all $\varepsilon > 0$ there should exist Σ_{fb} such that $\| \underline{W}_{c\ell} \|_{L_1(\mathbb{R}^+;\, \mathbb{R}^{\ell \times q})} \leq \varepsilon$ while for p = 2 this requires that $\underline{W}_{c\ell} \in L_1(\mathbb{R}^+;\, \mathbb{R}^{\ell \times q})$ and that $\sup_{\omega \in \mathbb{R}} \sigma_{max} [G_{c\ell}(i\omega)] \leq \varepsilon$, where σ_{max} denotes the maximal singular value, i.e. the maximum eigenvalue of $G^T_{c\ell}(-i\omega)\, G_{c\ell}(i\omega)$. Other (semi-)norms which one could use are for example $\|\underline{d}\|_D := \lim_{T\to\infty} \sup \frac{1}{T} (\int_0^T \|\underline{d}(\tau)\|^p d\tau)^{1/p}$, with $\|\cdot\|_Z$ similarly defined. The resulting induced norms however yields precisely the L_p-induced norm, and other than allowing a more general and perhaps more realistic class of disturbances, this introduces no new difficulties. Another possibility is to assume the disturbances to be stochastic processes. For example, we could assume that the disturbance input is white noise and require that the mean square of the controlled output can be made arbitrarily small. This requires

$$\sigma_{max} [\int_0^\infty \underline{W}_{c\ell}(t)\, \underline{W}^T_{c\ell}(t) dt] \leq \varepsilon .$$

Alternatively, one could require that it be possible to obtain, for any stationary disturbance input with finite mean square, that the mean square of the controlled output should be arbitrarily small. This requires again

$$\sup_{\omega \in \mathbb{R}} \sigma_{max} [G_{c\ell}(i\omega)] \leq \varepsilon .$$

6. It would be unpleasant if all of the above slight variations would lead to a different condition for solvability. Fortunately that is not so and (because of linearity, finite dimensionality, and time-invariance) we can reduce the problem of approximate quenching of the disturbances in any of the above senses to the following formulation of the *approximate disturbance decoupling problem by measurement feedback* (ADDPM):

> *Given* $\Sigma_p \approx (A,B,G,C,H)$ *when does there exist, for any* $\varepsilon > 0$, *a feedback compensator* $\Sigma_{fb} \approx (K,L,M,F)$ *such that the impulse response of* $\Sigma_{c\ell}$, $\underline{W}_{c\ell}$, *satisfies* $\| \underline{W}_{c\ell} \|_{L_1(\mathbb{R}^+;\mathbb{R}^{\ell\times q})} \leq \varepsilon$

Of course, it is possible to combine (ADDPM) with stability or pole placement requirements. Some results in this direction have been given in [5 , 6]. However the full problem of approximate disturbance decoupling with measurement feedback and stability is at this time still unsolved. In the present paper we will consequently not consider the stability issue and concentrate on approximate disturbance decoupling sec.

PRELIMINARY BACKGROUND RESULTS

7. In this section we will introduce the main ideas and results from [5, 6] which we need. Let us take a look at the linear system $\Sigma : \dot{x} = Ax + Bu,\ z = Hz$ and consider the following notions:

* $x_0 \in \mathbb{R}^n$ belongs to the *output nulling set* if there exists $\underline{u}$ such that the solution of $\dot{\underline{x}} = A\underline{x} + B\underline{u}$, $\underline{x}(0) = x_0$, satisfies $H\underline{x} = \underline{0}$;
* $x_0 \in \mathbb{R}^n$ belongs to the *L_p-approximate output nulling set* if $\forall \varepsilon > 0$, $\exists \underline{u}$ such that $\| H\underline{x} \|_{L_p(\mathbb{R}^+;\mathbb{R}^\ell)} \leq \varepsilon$
* $x_0 \in \mathbb{R}^n$ belongs to the *distributional output nulling set* if $\exists$ a distribution $\underline{u}$ with support on $\mathbb{R}^+$ such that the solution $H\underline{x}$, in the sense of distributions, satisfies $H\underline{x} = \underline{0}$. Here $\underline{x}$ is the sum of $e^{At}x_0$ $(t \in \mathbb{R}^+)$ and $\underline{W} * \underline{u}$ where $\underline{W}: t \in \mathbb{R}^+ \to e^{At}B$ and $*$ denotes convolution.

These output nulling subspaces may be computed easily in terms of the notions of almost controlled invariant subspaces. In fact:

The output nulling set equals $V^*_{\ker H}$

The L_∞-approximate output nulling set equals $V^*_{a,\ker H}$

The L_p-approximate output nulling set for $1 \leq p < \infty$ equals $V^*_{b,\ker H}$ and is also equal to the distributional output nulling set

Here $V^*_{\ker H}$ denotes the supremal controlled invariant ('(A,B)-invariant') subspace contained in $\ker H$, $V^*_{a,\ker H}$ denotes the supremal almost controlled invariant (' almost (A,B)-invariant') subspace contained in $\ker H$, and $V^*_{b,\ker H} = AV^*_{a,\ker H} + V^*_{\ker H} + \operatorname{im} B$.

These subspaces have been studied in full detail in [5, 6] and explicit finite recursive linear algorithms for computing these subspaces are given there.

8. In [5] it is also shown that if approximate output nulling is possible then it is possible by a state feedback control law. This implies that the inputs required in approximate output nulling have rational Laplace tranforms. Similarly if distributional output nulling is possible, then it is possible with a distribution which has a rational Laplace transform. This makes clear that the output nulling results are very much related to the solvability of matrix equations in rational functions. However, there is more: by considering almost controllability subspaces it is possible to obtain an even more complete theory. We will now outline these results. First, however, more subspaces:

$R^*_{\ker H}$ = the supremal controllability subspace contained in ker H

$R^*_{a, \ker H}$ = the supremal almost controllability subspace contained in ker H

$R^*_{b, \ker H} = AR^*_{a, \ker H} + \text{im } B$ = the supremal distributional controllability subspace 'contained' in ker H

9. Notation: Let $\mathbb{R}[s]$ and $\mathbb{R}(s)$ denote respectively the ring of polynomials and the field of rational functions with coefficients in $\mathbb{R}$. The ring of strictly proper rational functions will be denoted by $\mathbb{R}_+(s)$. More generally, let $\mathbb{R}_n(s)$ denote the set of rational functions with the degree of the denominator $\geq$ n + the degree of the numerator. In this notation $\mathbb{R}_+(s) = \mathbb{R}_{+1}(s)$ and $\mathbb{R}_o(s)$ is the ring of proper rational functions. We will use the notation $\mathbb{R}^n[s]$, $\mathbb{R}^n(s)$, $\mathbb{R}^{p\times q}[s]$, $\mathbb{R}^{p\times q}(s)$, etc. to denote the n-vectors, respectively the (p×q)-matrices, with coefficients in $\mathbb{R}[s]$, $\mathbb{R}(s)$, etc.

10. Consider now the matrix equation

$$(L):\quad H(Is - A)^{-1} G + H(is - A)^{-1} B U(s) = 0$$

which is an equation of the form $R(s) + M(s) U(s) = 0$ with $M(s) \in \mathbb{R}_+^{\ell\times m}(s)$ and $R(s) \in \mathbb{R}_+^{\ell\times q}(s)$ given and $U(s)$ the unknown. We are interested in solving (L) with $X(s) \in \mathbb{R}_+^{m\times q}(s)$, $\mathbb{R}^{m\times q}(s)$, $\mathbb{R}^{m\times q}[s]$, or $\mathbb{R}_n^{m\times q}(s)$, and we will refer to this as the solvability over $\mathbb{R}_+(s)$, etc. Furthermore we will say that (L) is L_p-approximately solvable over $\mathbb{R}_+(s)$ (or $\mathbb{R}_n(s)$) if $\forall \varepsilon > 0$, $\exists U(s) \in \mathbb{R}_+^{m\times q}(s)$ such that the $L_p(0,\infty)$-norm of the inverse Laplace transform of $R(s) + M(s)U(s)$ is $\leq \varepsilon$. Note that equation (L) is exactly equally general as the equation $R(s) + M(s)U(s) = 0$. This may be seen by realizing, if necessary, $[R(s)\ M(s)]$ in state space form, which yields $[R(s)\ M(s)] = H(Is - A)^{-1}\ [G\quad B]$. The following gives a reasonably complete picture on the solvability conditions for (L) in terms of geometric concepts:

(i) $\{(L)$ is solvable over $\mathbb{R}_+(s)\} \Leftrightarrow \{\mathrm{im}\, G \subset \mathcal{V}^*_{\ker H}\}$

(ii) $\{(L)$ is solvable over $\mathbb{R}(s)\} \Leftrightarrow \{\mathrm{im}\, G \subset \mathcal{V}^*_{b,\ker H}\} \Leftrightarrow$
{for a given $1 \leq p < \infty$, (L) is L_p-approximately solvable over $\mathbb{R}_+(s)\} \Leftrightarrow \{(L)$ is L_p-approximately solvable over $\mathbb{R}_n(s)$ for some (or all) $n \in \mathbb{Z}\} \Leftrightarrow \{(L)$ is solvable over $\mathbb{R}_n(s)$ for some $n \in \mathbb{Z}\}$

(iii) $\{(L)$ is solvable over $\mathbb{R}[s]\} \Leftrightarrow \{\mathrm{im}\, G \subset \mathcal{R}^*_{b,\ker H}\}$

(iv) $\{(L)$ is solvable over $\mathbb{R}[s]$ and $\mathbb{R}_+(s)$ simultaneously$\} \Leftrightarrow$
$\{\mathrm{im}\, G \subset \mathcal{R}^*_{\ker H}\}$

(v) $\{(L)$ is L_∞-approximately solvable over $\mathbb{R}_+(s)\} \Leftrightarrow \{\mathrm{im}\, G \subset \mathcal{V}^*_{a,\ker H}\}$

The above solvability results are intriguing in the following sense: $\mathbb{R}(s)$ is a field with two complementary subrings: $\mathbb{R}[s]$ and $\mathbb{R}_+(s)$ ($\mathbb{R}[s] \cap \mathbb{R}_+(s) = \{0\}$, and every element $f \in \mathbb{R}(s)$ allows a unique decomposition into $f = [f] + f_+$ with $[f] \in \mathbb{R}[s]$ and $f_+ \in \mathbb{R}_+(s)$). Solvability of (L) is of interest in $\mathbb{R}(s)$, $\mathbb{R}[s]$, $\mathbb{R}_+(s)$ and in $\mathbb{R}[s]$ and $\mathbb{R}_+(s)$ simultaneously and all these situations lead to conditions which have a very natural geometric interpretation.

11. In the present paper however, we are mainly interested in the connection between approximate solvability over $\mathbb{R}_+(s)$ and exact solvability in $\mathbb{R}[s]$, as given in 10(ii). We will lift the statement which we will need later on and reformulate it slightly.
Let L be a linear operator from $\mathbb{R}^q(s)$ into $\mathbb{R}^p(s)$ and assume that $L\,\mathbb{R}_o^q(s) \subset \mathbb{R}_+^p(s)$ (i.e. L is representable by a (p × q)-matrix with elements in $\mathbb{R}_+{}^{p\times q}(s)$). Let $y \in \mathbb{R}_+^p(s)$ be a given and consider the equation Lx = y. In the obvious nomenclature, we have as an immediate corollary of 10(ii):

<u>Proposition 1</u>: *The equation* Lx = y *is approximately solvable over* $\mathbb{R}_o(s)$ *if and only if it is exactly solvable over* $\mathbb{R}(s)$.

The above proposition has the nice feature that it allows us to conclude the hard analysis question of approximate solvability over a ring by the soft algebraic question of solvability over a field. We will exploit this to our advantage later on. Note finally that (because of the fact that the elements of the matrix L need not have denominators with roots in the left half of the complex plane) there is, to our knowledge, no way of proving the above proposition by simply approximating distributions (the inverse Laplace transforms of the solution over $\mathbb{R}(s)$) by smooth functions (the inverse Laplace transforms of approximate solutions over $\mathbb{R}_+(s)$).The theory of almost invariant subspaces appears in fact to be a key element in deriving Proposition 1.

A LINEARIZATION LEMMA

12. In this we will reformulate (ADDPM) in terms of a linear question about convolution operators. Let $A(p,q)$ be defined as follows: $A(p,q) := \{(F_0,\underline{F}_1) \mid F_0 \in \mathbb{R}^{p\times q},\ \underline{F}_1 \in L^{loc}(\mathbb{R}^+;\mathbb{R}^{p\times q})\}$ and denote by $A^+(p,q) := \{(F_0,\underline{F}_1) \in A(p,q) \mid F_0 = 0\}$. Elements of $A(p,q)$ define via $\underline{u} \mapsto F_0\underline{u} + \underline{F}_1 * \underline{u}$ convolution operators from $L^{loc}(\mathbb{R}^+;\mathbb{R}^q)$ into $L^{loc}(\mathbb{R}^+;\mathbb{R}^p)$. We define the multiplication A_1A_2 of an element $A_1 = (F_0^1, \underline{F}_1^{2}) \in A(p,q)$ with an element $A_2 = (F_0^{2}, \underline{F}_1^{2}) \in A(q,r)$ by $A_1A_2 = (F_0^{1}F_0^{2}, \underline{F}_1^{1}F_0^{2} + \underline{F}_0^{1}\underline{F}_1^{2} + \underline{F}_1^{1} * \underline{F}_1^{2})$. This makes $A(n,n)$ into an algebra with unit (I,0). Clearly $A^+(n,n)$ is an ideal. It is easy to see that an element $(F_0,\underline{F}_1) \in A(n,n)$ is invertible if and only if F_0 is a nonsingular matrix.

Elements of $A(p,q)$ are called *bounded* if $\|\underline{F}_1\|_{L_1(\mathbb{R};\,\mathbb{R}^{p\times q})} < \infty$. Let $B(p,q)$ denote the bounded elements. Define the norm of $(F_0, \underline{F}_1) \in B(p,q)$ by $\|(F_0,\underline{F}_1)\| := \|F_0\|_{p\times q} + \int_0^\infty \|\underline{F}_1(t)\|_{p\times q}dt$, where $\|\cdot\|_{p\times q}$ denotes an induced norm on $\mathbb{R}^{p\times q}$. $B(n,m)$ is a Banach algebra. If in $(F_0,\underline{F}_1) \in A(p,q)$, $\underline{F}_1$ is of exponential growth, then we may define its Laplace transform by $\hat{F}(s) := F_0 + \int_0^\infty \underline{F}_1(t)e^{-st}dt$. We will say that an element $(F_0,\underline{F}_1)$ is *Bohl* if $\underline{F}_1$ may be written as $\underline{F}_1(t) = R\,e^{St}T$ for suitable matrices R,S,T, or, equivalently, if and only if its Laplace transform is a matrix of (proper) rational functions. The Bohl elements form a subalgebra of $A(n,n)$ which is closed under inversion, but the bounded Bohl elements donot form a closed subalgebra of $B(n,n)$.

13. It seems most appropriate to state our linear lemma in the language of 12. The following bijection will turn out to be very useful to us:

Lemma 1: Let $g \in A^+(p,q)$ be given. Then the map $x \mapsto (I - xg)^{-1}x$ defines a bijection on $A(q,p)$. If g is Bohl, then this bijection maps the set of Bohl elements of $A(q,p)$ onto itself.

Proof: It is obvious that $I - xg$ is invertible. Furthermore, it is easy to verify that $y \mapsto y(I + gy)^{-1}$ is the inverse of the map displayed in the lemma. This expression for the inverse also puts in evidence that the Bohl elements are mapped onto. □

14. In 3 we have seen that the closed loop transfer function was given by $G_{c\ell} = G_{22} + G_{21}(I - XG_{11})^{-1}XG_{12}$ and the problem was to choose X such that the inverse Laplace transform of $G_{c\ell}$ has arbitrarily small L_1-norm. By the above lemma we see that instead of treating X as the unknown we may as well treat the whole expression $Y = (I - XG_{11})^{-1}X$ as an arbitrary unknown. This leads to:

Proposition 2: (ADDPM) *is solvable if and only if the equation* $G_{22}(s) + G_{21}(s)\,Y(s)\,G_{12}(s) = 0$ *is approximately solvable over* $\mathbb{R}(s)$, (that is to say for any $\varepsilon > 0$ there should exist a $Y(s) \in \mathbb{R}_0^{m\times p}(s)$ such that the inverse Laplace transform of $G_{22}(s) + G_{21}(s)\,Y(s)\,G_{12}(s)$ has L_1-norm $\leq \varepsilon$)

A SOLVABILITY CONDITION IN TERMS OF RATIONAL FUNCTIONS

15. Note that the equation $G_{22} + G_{21} YG_{12}$ is linear in Y. It is an equation of the type studied in 11, and Proposition 1 applied in Proposition 2 immediately yields:

Proposition 3: (ADDPM) *is solvable if and only if the equation*

$$(L_*^*) : H(Is - A)^{-1}G + H(Is - A)^{-1}B\, Y(s)\, C(Is - A)^{-1}G = 0$$

is solvable over $\mathbb{R}(s)$, *i.e., if and only if there exists* $Y(s) \in \mathbb{R}^{m\times p}(s)$ *such that* (L_*^*) *is satisfied.*

We will now derive necessary and sufficient conditions for the solvability of (L_*^*)

A LEMMA ON MATRIX EQUATIONS OVER A FIELD

16. If we look at equation (L_*^*) of 15 and compare it with equation (L) from 10 then it is obvious that (L_*^*) is a special case of (L). It turns out, in fact, that solvability of (L_*^*) may be deduced from the solvability of (L) and its dual. Very important in this is that (L_*^*) is being solved over the field $\mathbb{R}(s)$. Consider an arbitrary field $\mathbb{F}$ and let $\mathbb{F}^{p\times q}$ denote the (p×q)-matrices with coefficients in $\mathbb{F}$. Let $R \in \mathbb{F}^{\ell\times q}$, $M \in \mathbb{F}^{\ell\times m}$, and $N \in \mathbb{F}^{p\times q}$ be given and consider the equation $R + MXN = 0$ in the unknown matrix $X \in \mathbb{F}^{m\times p}$. We have:

Lemma 2: The matrix equation $R + MXN = 0$ has a solution $X \in \mathbb{F}^{m\times p}$ if and only if both:

(i) $R + MU = 0$ has a solution $U \in \mathbb{F}^{m\times q}$

and (ii) $R + LN = 0$ has a solution $L \in \mathbb{F}^{\ell\times p}$

Proof: (only if): immediate

(if) : (i) is solvable if and only if $\operatorname{im} R \subset \operatorname{im} M$, while (ii) is solvable if and only if $\ker R \supset \ker N$.

Let $U \in \mathbb{F}^{m\times q}$ be a solution of $R + MU = 0$. Define now X as follows: $X|\ker N = 0$ and $X|Z = V|Z$ where Z is any complement of ker N in $\mathbb{F}^p$.

Then $(R + MXN)|\ker N = 0$ since $\ker R \supset \ker N$ and $(R+MXN)|Z = (R+MU)|Z = 0$ since $R + MU = 0$. This yields $R + MXN = 0$ and X is indeed the solution which we were after . □

17. With the above lemma and the fact that $\mathbb{R}(s)$ is a field we obtain immediately the following

Proposition 4: (ADDPM) *is solvable if and only if both*

$$\text{(i)}\quad (L):\quad H(Is-A)^{-1}G + H(Is-A)^{-1}BU(s) = 0$$

and $$\text{(ii)}\quad (L^*):\quad H(Is-A)^{-1}G + L(s)\,C(Is-A)^{-1}G = 0$$

are solvable over $\mathbb{R}(s)$.

THE SOLVABILITY OF (L) AND (L^*) OVER $\mathbb{R}(s)$

18. The solvability of (L) over $\mathbb{R}(s)$ was studied in [5] and some results have already been discussed in 10. We will now connect this up with the solvability of the approximate disturbance decoupling problem by state feedback. Consider therefore the plant Σ_p introduced in 1 but with $y = x$, i.e., the measured output is the whole state x. It may be shown that the approximate disturbance decoupling problem may be solved in this case if and only if it can be solved by memoryless state feedback, i.e. by a feedback compensator of the type Σ_{fb} : $u = Fx$. This problem is called (ADDP) and may be stated as follows: *Given* $\Sigma_p \approx (A,B,G,H)$ *when does there exist, for any* $\varepsilon > 0$ *a feedback gain matrix* F *such that the impulse response of* $\Sigma_{c\ell}$ *satisfies* $\int_0^\infty \|He^{(A+BF)t}G\|\,dt \leq \varepsilon$?

19. The solution of (ADDP) and its connection with the solvability of (L) are given in

Proposition 5: {(ADDP) *is solvable*} $\Leftrightarrow$ {(L) *is solvable over* $\mathbb{R}(s)$} $\Leftrightarrow$ {$\mathrm{im}\, G \subset V^*_{b,\ker H}$}

20. We will consider the equation:

$$(L^*) : H(Is - A)^{-1}G + L(s)\,C(Is - A)^{-1}G = 0$$

Obviously (L^*) is in a sense dual to (L) since if we transpose (L^*) we obtain an equation the type (L) with the unknown matrix $L^T(s)$ at the right of the second term. In fact, this duality is exactly in the sense that controllability and observability are dual, pole placement and observers are dual, and LQ optimal control and Kalman filtering are dual.

Let us take a look at the system $\Sigma : \dot{x} = Ax$, $y = Cx$. We will also consider an observer for this system defined by $\dot{w} = Kw + Ly$, $\underline{w}(0) = 0$. Consider now the following notions:

* a subspace $S \subset \mathbb{R}^n$ is said to be *conditionally invariant* if $\exists$ K,L such that $\underline{x}(0) \in S$ implies $\underline{x}(t) \pmod{S} = \underline{w}(t)$
* a subspace $S_a \subset \mathbb{R}^n$ is said to be L_p*-almost conditionally invariant* if $\forall \varepsilon > 0$ $\exists$ K, L such that $\underline{x}(0) \in S_a$ implies $\|\underline{x}(\cdot) \pmod{S_a} - \underline{w}(\cdot)\|_{L_p(\mathbb{R}^+;\, \mathbb{R}^k)} \leq \varepsilon$
* a subspace $S_D \subset \mathbb{R}^n$ is said to be *distributionally conditionally invariant* is $\exists$ K, L, $L_0, \ldots, L_N$ such that $\underline{x}(0) \in S_D$ implies $\underline{x}(\text{mod } S_D) = \underline{w} + L_0\underline{y} + \ldots + L_N\underline{y}^{(N)}$ on $\mathbb{R}^+$ (for the precise sense of this equality, see [6]).

These clases of subspaces are closed under subspace intersection and hence there exists an infimal element containing a given subspace of $\mathbb{R}^n$. We denote these by:

S^*_{imG} = the infimal conditionally invariant subspace containing im G

$S^*_{\text{a,im G}}$ = the infimal L_∞-almost conditionally invariant subspace containing im G

$S^*_{\text{b,im G}}$ = $(A^{-1} S^*_{\text{a,imG}}) \cap S^*_{\text{im G}} \cap \ker C$ = the infimal L_p-almost conditionally invariant subspace containing im G for $1 \leq p < \infty$ = the infimal distributionally conditionally invariant subspace containing im G

21. The relevance of conditonally invariant subspaces may best be seen by considering the disturbance decoupled estimation problem (see figure). In this problem we are given the plant Σ_p: $\dot{x} = Ax + Gd$, $y = Cx$, $z = Hx$, and

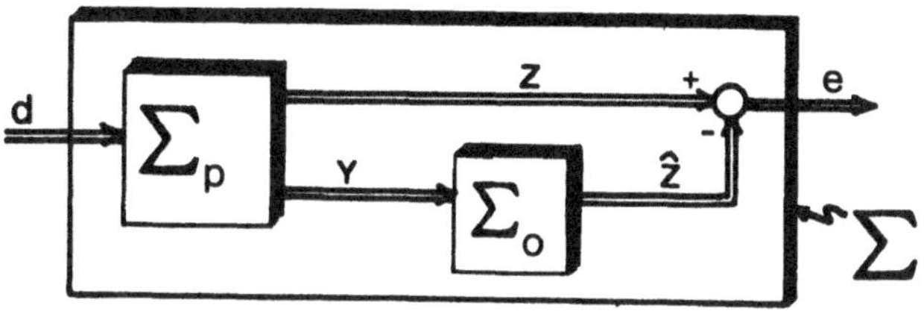

the problem is to synthesize an observer Σ_o: $\dot{w} = Kw + Ly$, $\hat{z} = Mw + Fy$ such that in the cascaded system Σ_s the estimation error e: $= z - \hat{z}$ is independent, or almost independent, of the disturbance d. As argued in 5 it is natural to formulate the approximate version of this problem (ADDEP) as follows: *Given the plant* $\Sigma_p \approx (A,G,C,H)$ *when will there exist, for any* $\varepsilon > 0$, *an observer* $\Sigma_o \approx (K,L,M,F)$ *such that the cascaded system* Σ *has an impulse response* $d \to e = z - \hat{z}$ *which has an* L_1*-norm* $\leq \varepsilon$? If we denote the transfer function of the unknown observer by L(s) then it is easily calculated that the transfer function $d \to e$ is given by

$$H(Is - A)^{-1}G - L(s)\, C(Is - A)^{-1}C$$

i.e., up to a minus sign, it is the left hand side of (L^*) and it may be seen from this and from the duality of (almost) controlled and (almost) conditionally invariant subspaces that the following dual of Proposition 5 holds:

Proposition 6: {(ADDEP) *is solvable*} $\Leftrightarrow$ { (L^*) *is solvable over* $\mathbb{R}(s)$} $\Leftrightarrow$ $\{S^*_{b,\mathrm{im}\,G} \subset \ker H\}$.

THE MAIN RESULT

22. Combining Propositions 4, 5 and 6 immediately yields the result

which we are after:

Theorem : $\{(\text{ADDPM})$ *is solvable*$\} \Leftrightarrow \{\operatorname{im} G \subset V^*_{b,\ker H}$ *and* $S^*_{b,\operatorname{im} G} \subset \ker H\}$

Perhaps it is useful to outline once more the facts which led to this result:

(i) Using a linearization lemma it was possible to set up a *linear* equation which was to be solved approximately;

(ii) Using some results from the theory of almost controlled invariant subspaces it was possible to reduce this approximate solvability to the *exact* solvability of equation (L^*_*) over the field of rational functions;

(iii) Using a lemma on the solvability of matrix equations over a field it was possible to reduce the solvability of (L^*_*) over the field of rational functions to the solvability of *both* (L) and (L^*) over the field of rational functions;

(iv) From the theory of almost invariant subspaces we then obtain necessary and sufficient conditions for the solvability of (L) and (L^*).

23. An interesting consequence of the above theorem is the following *separation principle:*

> *Approximate disturbance decoupling by measurement feedback is possible if and only if (i) approximate disturbance decoupling by state feedback is possible, and (ii) approximate disturbance decoupled estimation of the controlled output by means of the measured output is possible.*

It is logical that (i) is a condition which should enter the picture but why (ii) is needed is a bit more difficult to argue heuristically, even though it is rather straightforward to see where it comes in mathematically.

24. It is interesting the compare the theorem in 22 with the result on exact disturbance decoupling by measurement feedback (DDPM). The necessary and sufficient conditions for this problem are [1, 2]: $S^*_{\mathrm{im}\,G} \subset V^*_{\ker H}$. If we add stability (DDPMS) this becomes $S^*_{g,\mathrm{im}G} \subset V^*_{g,\ker H}$ while the problem with pole placement leads to the condition $N^*_{\mathrm{im}\,G} \subset R^*_{\ker H}$. Here $S^*_{g,\mathrm{im}\,G}$ denotes the infimal detectability subspace (relative to $\mathbb{C}_g$) containing $\mathrm{im}\,G$, $V^*_{g,\ker H}$ denotes the supremal stabilizability subspace contained in $\ker H$, $N^*_{\mathrm{im}G}$ denotes the infimal conplementary observability subspace containing $\mathrm{im}\,G$, while $R^*_{\ker H}$, finally, denotes the supremal controllability subspace contained in $\ker H$. Note in particular that the separation principle explained in 23 holds only for the approximate and not for the exact version of the disturbance decoupling problem by measurement feedback.

25. It is well-known and easy to prove that in the family of all plants Σ_p with n,m,p,q, and ℓ fixed and satisfying $HG = 0$, (DDPM) is generically solvable if and only if

m = # *of controls*	$\geq \ell$ = # *of controlled outputs*
p = # *of observations*	$\geq$ q = # *of disturbances*

We call this the *law of requisite variety in control action and measurement capability*. However, because of the fact that (ADDPM) is concerned with the set of almost controlled and almost conditionally invariant subspaces which form the closure of the set of controlled and conditionally invariant subspaces, respectively, we see that often what is generically true for the exact problem (DDPM) will be exactly true for the approximate problem (ADDPM). Thus it may be shown that (ADDPM) is solvable if either:

1. $\dim \mathrm{im}\, B \geq \operatorname{codim} \mathrm{im} \ker H$, $\dim \mathrm{im}\, G \leq \operatorname{codim} \mathrm{im} \ker C$, $\mathrm{im}\, G \subset \ker H + \mathrm{im}\, B$, and $\mathrm{im}\, G \cap \ker C \subset \ker H$
2. $\dim \mathrm{im}\, B \geq \operatorname{codim} \mathrm{im} \ker H$, $\dim \mathrm{im}\, G \leq \operatorname{codim} \mathrm{im} \ker C$, $R^*_{\ker H} = \{0\}$, and $S^*_{\mathrm{im}\,G} = \mathbb{R}^n$

or 3. $\mathrm{im}\, G \subset \mathrm{im}\, B$ and $\ker H \supset \ker C$. Note that (ADDPM) is hence solvable for the system $\dot{x} = Ax+B(u+d)$, $y=z=Cx$ which is a situation often considered in theoretical applications.

REFERENCES

[1] H. Akashi and H. Imai, "Disturbance localization and output deadbeat control through an observer in discrete-time linear multivariable systems, *IEEE Trans. Automat. Contr.*, vol. AC-24, pp. 621-627, 1979

[2] J.M. Schumacher, "Compensator synthesis using (C,A,B)-pairs", *IEEE Trans. Automat. Contr.*, vol. AC-25, pp. 1133-1138, 1980

[3] J.C. Willems and C. Commault, "Disturbance decoupling by measurement feedback with stability or pole placement" *SIAM J. on Contr. and Optimiz.*, vol. 19, No. 4, pp. 490-504, 1981

[4] H. Imai and H. Akashi, "Disturbance localization and pole shifting by dynamic compensation", *IEEE Trans. Automat. Control*, vol. AC-26, pp. 226-235, 1981

[5] J.C. Willems, "Almost invariant subspaces: an approach to high gain feedback design - Part I: Almost controlled invariant subspaces", *IEEE Trans. Automat. Contr.*, vol. AC-26, pp. 235-252, 1981

[6] J.C. Willems, "Almost invariant subspaces: an approach to high gain feedback design - Part II: Almost conditionally invariant subspaces", *IEEE Trans. Automat. Contr.*- to appear.

Lecture Notes in Control and Information Sciences

Edited by A. V. Balakrishnan and M. Thoma